PORTABLE CONSUMER ELECTRONICS

PACKAGING, MATERIALS, AND RELIABILITY

PORTABLE CONSUMER ELECTRONICS

PACKAGING, MATERIALS, AND RELIABILITY

Sridhar Canumalla, Ph.D.
Principal Engineer
Entertainment & Devices Division
Microsoft Inc.
Redmond, WA, USA

Puligandla Viswanadham, Ph.D.
Retired Principal Scientist
Nokia Research Centre
Irving, TX, USA

PennWell®

Disclaimer: The recommendations, advice, descriptions, and the methods in this book are presented solely for educational purposes. The author and publisher assume no liability whatsoever for any loss or damage that results from the use of any of the material in this book. Use of the material in this book is solely at the risk of the user.

PennWell Corporation
1421 South Sheridan Road
Tulsa, Oklahoma 74112-6600 USA

800.752.9764
+1.918.831.9421
sales@pennwell.com
www.pennwellbooks.com
www.pennwell.com

Marketing: Jane Green
National Account Executive: Barbara McGee
Director: Mary McGee
Managing Editor: Stephen Hill
Production Manager: Sheila Brock
Production Editor: Tony Quinn
BookDesigner: Susan E. Ormston
Cover Designer: Kelly Cook

Library of Congress Cataloging-in-Publication Data

Canumalla, Sridhar.
Portable consumer electronics : packaging, materials, and reliability / Sridhar Canumalla, Puligandla Viswanadham.
p. cm.
Includes bibliographical references and index.
ISBN 978-1-59370-125-3
1. Microelectronic packaging. 2. Miniature electronic equipment--Design and construction. I. Viswanadham, Puligandla. II. Title.
TK7870.15.C36 2009
621.381'046--dc22

2009049063

Printed in the United States of America
1 2 3 4 5 14 13 12 11 10

Contents

Foreword

Anyone who has walked the showroom floor at a recent Consumer Electronics Show (CES) event and taken in the sheer volume and scale of consumer electronic devices available today cannot deny that a revolution has taken place. Faster, cheaper, smaller, sleeker, and more powerful have become buzzwords, and the demand for high-function portable electronic devices continues to grow at a dizzying pace. Where portable devices in the past provided singularly vertical experience, they now provide a dizzying array of integrated scenarios such as reading, browsing, photographing, communicating, viewing, listening, and productivity. The quality and the expectation for high fidelity on all of these device experiences is dramatically different today than five years ago, and will continue to develop exponentially in the coming years. Adding to the complexity of these product categories is the rapidly shrinking product development cycles and product life cycles of most portable consumer electronic devices.

The evolution of electronic packaging has been one of the fundamental driving forces of the explosive growth in portable consumer electronics. The continued evolution to 3-D packaging (die on die and package on package) are examples of the key role packaging evolution has played in the ability to drive smaller and more integrated devices that provide the powerful experiences once only found on desktop and tabletop products. In addition to the technological advance in packaging specific design technology, the corresponding advances in areas such as packaging reliability, materials, assembly technology, and manufacturing test have all come together to make today's portable experiences possible. In the future, new technology areas such as nanotechnology and advances in display technologies will spur the need for further evolutions in packaging and related sciences.

Despite the fact that portable consumer electronic devices constitute one of the fastest growing market segments in the industry, very few treatises and reference books have been written specifically on this market category. There is a definite need for a source which covers the unique aspects of materials, design, and reliability comprehensively for both the novice who is interested or entering portable electronics field, as well as the practicing engineer who looks to continue to develop in this area. This timely book by Sridhar Canumalla and Puligandla Viswanadham fulfills this growing need. Those who are just now embarking on a career in consumer and portable electronic packaging will benefit immensely from this book. Those of us who are longtime practitioners will find it a relevant reference and compelling read while giving us a glimpse into the future of portable electronic devices.

This book will be a valuable addition to the growing electronic packaging literature, especially as related to portable electronics. Sridhar Canumalla and Puligandla Viswanadham bring to bear more than 50 collective years of academic and industrial experience to this effort. Their combined experience covers a broad spectrum of electronic packaging development including mechanics, materials, failure analysis, chemistry, and reliability at companies such as Nokia, Texas Instruments, Raytheon, Sonoscan, and IBM. Viswanadham is currently an adjunct professor of mechanical engineering at the University of Texas in Arlington and Sridhar is a principal engineer in Microsoft's hardware business. The passion that both of the authors have for this field will be readily apparent to all that read this book.

It is my sincere hope that the electronic packaging industry will benefit from the fruits of their collective experience, expertise, and effort, and I wish the book the success it truly deserves. Oh and one last thing: if you have a sincere interest in portable consumer electronics, in addition to reading this book, take a trip down to CES in Las Vegas; if you have shiny object syndrome, it doesn't get any better.

Brian Tobey, Corporate Vice President
Manufacturing, Supply Chain, Information & Services
Entertainment & Devices Division
Microsoft

Preface

Portable consumer electronic devices and appliances have been experiencing an explosive growth with ever-increasing consumer acceptance, market penetration, and growth in recent years. The demand for cheaper, faster, and better products continues unabated, accompanied by inexorable trends in miniaturization and integration. In several parts of the world, they are already an integral part of our daily lives. Yet, for this class of consumer electronics, it is safe to say that the field is still in its infancy with immense possibilities for new applications, and the future looks very bright. Electronic packaging for these products often has to break current barriers and enter new frontiers on several fronts and is driven by market pressures to manufacture these devices cheaper, smaller, lighter, and with increasing functionality. This book encompasses aspects of material, design, and reliability engineering, high-volume assembly technology, and failure analysis for this exciting and nearly ubiquitous class of consumer electronics.

We write this book to provide a single, comprehensive account of the key aspects of packaging for portable consumer electronic devices including first- and second-level packaging, printed wiring board technology, assembly technology, reliability statistics and engineering, and failure analysis. Aspects of industrial design are considered outside the scope of this book partly because it is more art than science.

We want this book to be useful to practicing engineers, technologists, and designers, and have included several real-life examples of failures and also some discussions of current industry practices. Methods to prevent failures and enhance product reliability are also discussed to reflect our passion for making high quality, reliable, and robust products.

Organization and Scope of Book

This book is written at a level that assumes only a basic knowledge in the fundamentals of physics, chemistry, and engineering at an undergraduate level. The book consists of 10 chapters addressing the various aspects of portable electronics. Adequate cross references are provided at the end of each chapter to enable the uninitiated reader to acquire more detailed information on any given aspect or topic. Since, the focus of the book is portable electronics, a prior understanding of general electronic packaging aspects can be beneficial.

The first chapter introduces and defines the landscape of portable electronic products. Some of the unique aspects of portable electronics are introduced in this chapter and both user and original equipment manufacturer (OEM) perspectives are discussed. In the second chapter, different packaging challenges for achieving higher integration and miniaturization are discussed with particular emphasis on the design, materials, and manufacturing aspects.

Printed wiring boards (PWBs) are the backbone of most current portable electronic products, and different aspects of PWB technology are described next. Both conventional PWB and high-density interconnection technologies are discussed with additional discussion on flexible PWBs. Also, embedded integrated modules are discussed, as they occupy a prominent place on the horizon of PWB landscape.

First-level packaging, namely, packaging the semiconductor device, is common across both portable and non-portable consumer electronic products and is relevant to portable electronic products; hence, it is reviewed in chapter 4. In addition to leaded and leadless packages, area array packages including ball grid array package (BGA), chip scale package (CSP), as well as flip chip and land grid array packages are discussed. Techniques to achieve higher levels of integration using stacked Si and stacked packages, as well as emerging trends are discussed in detail.

Interconnection technologies, the methodologies of attaching two different parts of an electronic assembly, are described in chapter 5. It provides a general overview of the various interconnection schemes that include mechanical, alloy, and adhesive interconnections. While insertion mount technology is discussed only contextually, surface mount technology with different lead and termination configurations is described in some detail with reliability implications. These include leadless chip carriers, quad flat non-leads (QFNs), J-, gull wing, thin small-outline packages (TSOPs), ball grid array (BGA), and chip scale package (CSP) area array interconnections. Owing to the ROHS and WEEE initiative, tin–lead alloys are being replaced with multicomponent lead (Pb)-free alloys. Tin–silver–copper alloys have been discussed in some detail with their merits and demerits. A discussion of surface finishes and their importance is also included. Adhesive interconnections include isotropic, anisotropic, and intrinsic adhesives and their current status. The chapter concludes with a brief description on optoelectronic connections.

Because the focus of the book is portable electronics where surface mount assembly is the norm, insertion mount or through-hole assembly technology has not been discussed. Even though through-hole assembly is a more robust technology, it is very inefficient in terms of packaging density and hence is not practiced in portable electronics manufacturing. Different aspects of surface mount assembly such as paste printing, component placement, reflow, etc. are described. Rework and repair operations as well as reliability enhancement with underfills are included. A discussion of flexible electronic assemblies is also included in this chapter. This completes the building of the product, and subsequent chapters focus on the reliability and failure analysis of portable electronic products.

The background mathematical concepts for reliability evaluation and engineering are described and discussed in chapter 7. The emphasis is on practical working knowledge, such as techniques to ascertain if there is an improvement in reliability after design changes. General recommendations on interpretation of probability plots for reliability data analysis are also made using several real-life examples. In chapter 8 we discuss the state-of-the-art in reliability engineering of portable consumer electronic products and the environments that are unique to them. A review of the latest practices and trends in assessment of thermo-mechanical, mechanical, and environmental reliability is provided, along with a section devoted to practical considerations in accelerated reliability testing.

Typical failure mechanisms found in portable electronic products and a brief review of analytical techniques commonly employed to understand these failures are presented in the next chapter. Also included are sections on best practices to avoid failures and on a framework for product development to prevent failures.

In the last chapter, we discuss several near-term emerging trends and extrapolate future trends in IC packaging, PWB technology, display technology, energy sources, nanomaterials, wearable electronics, sensor technology, compliant mechanics, self-powered electronics, and biodegradable electronics. As the technology evolution is extremely rapid, it is thought reasonable to provide a glimpse in to the future, fully recognizing that in very few years these technologies will be the order of the day.

We are grateful to our many of friends and colleagues across the electronic industry in general and our esteemed colleagues at International Business Machines, Motorola, Texas Instruments, Raytheon, Nokia, Sonoscan, and Microsoft in particular. Our association and interactions with them enriched not only our personal lives but also our scientific and technological background. The valuable discussions we had with many of them contributed to a better understanding of the complexities and nuances in the emerging technologies.

We thank our families for the constant and enduring support and encouragement throughout the preparation of this book.

We are also grateful to Stephen Hill of PennWell Publishing for constant guidance and support from the inception to the completion of this book.

Acknowledgments

This book is the culmination of a labor of love and was three years in the making, but it is no understatement that it would not have been possible without the support and help of several friends, colleagues, and family. We thank the editorial staff at Pennwell: Steve Hill, Tony Quinn, and Susan Ormston for their professionalism, diligence, and patience. It has been a delight to work with them. We also thank the reviewers for their review and helpful suggestions.

We have learned much from our colleagues at every step, and a significant portion of that learning has made its way into this book on portable electronics in the form of pictures, discussions, philosophy, and approach. We are indebted to them for their contributions, support, and encouragement. We especially thank Seppo Pienimaa, Kari Kulojarvi, Ramin Vatanparast, Dewey Brooks, David Corkum, Tim Fitzgerald, Colin Martin, Srinivas Rallapalli, Seong Cheol Kim of UNT, and Robert Champaign, Don Cullen, Sesil Mathew, Murali Hanabe, Laura Foss, Jason Wu, Jianjun Wang, Santosh Shetty, Nael Hannan, Steven Dunford, Outi Rusanen, Tuula Stenberg, Mike Wellborn, Mukul Saran, Mark Trahan, Joel Dobson, David Buraczyk, and Darvin Edwards.

In particular, Sridhar would like to thank Raj Master and Brian Tobey at Microsoft for ensuring an environment that enabled the preparation of this manuscript. He is especially grateful to Mike Lane for the constant encouragement and support, many suggestions for improvement while writing this book and for diligently reading the drafts cover to cover.

We would like to thank our families for their encouragement, patience, and love during the preparation of this manuscript. Sridhar is thankful to his wife, Anu, for her constant support throughout his working career and for the occasional call to action that was needed to complete this book, and their two children, Anirudh and Vishal, for bringing joy every single day just by being. It is his immense pleasure to see their never-ending curiosity and passion for understanding why things are the way they are. Thank you. Viswanadham is greatly indebted to Santha his wife for her enduring moral support, consideration, patience, and for allowing him to pursue this endeavor. He is also immensely thankful to Usha his daughter, Sayi his son, Joe Robillard his esteemed son-in-law, and the sweet and adorable grandchildren Jaya, Elizabeth, and Jackson, as they have been a source of constant support and encouragement.

1

Introduction

Few products in the history of mankind with the technological complexity of portable electronics can compare with the mass market penetration achieved by these nearly ubiquitous devices in the last score or so years. The world has gone digital first and subsequently mobile. Indeed, it is no understatement that the field of electronics, especially portable electronics, has revolutionized several aspects of the way we live, work, entertain ourselves, and communicate to such an extent that almost any prediction of the future state is fraught with large uncertainties. For example, in several developing countries phone penetration rates have exploded primarily as a result of people rejecting the traditional land-line phones in favor of an en masse adoption of mobile phones with concomitant precipitous drops in the cost of connecting individuals. The market disruptions due to portable electronics have been unprecedented and have occurred faster than anticipated. For example, within approximately 10 years, the business model of pay phones at public places has been all but abandoned in direct homage to the exploding mobile phone usage. The manufacturing of consumer electronics is by far the largest and fastest growing industry in the world. It is not surprising that it surpasses many of the traditionally large industries, such as automotive, within a very short time since inception.

Indeed, it is fair to say that mankind has been having an increasingly torrid love affair with portable electronic products, and we carry a multitude of these devices on our person. A businessman might carry a personal digital assistant (PDA), a mobile phone, a laptop computer, an electronic translator, a portable projector, digital music player, portable gaming device, a Bluetooth technology headset, etc. We are also increasingly using electronics in cars, inside homes, in the classroom, and in work places. It is instructive to examine the context of the term "portable electronics."

By definition, "portable" means "capable of being carried or moved about" [Webster's New Collegiate Dictionary]. But, in the context of electronic products, the word also signifies that the device can be used normally while the user is moving about and without being tethered to a base. Power for operation is achieved using a battery or other power sources. In that context, cordless telephones are not completely portable electronics because they are operational only within a radius of a few meters from the base set.

So what makes a consumer electronic product a portable consumer electronic product? Since there is no existing definition for this class of products, let us define the term by examining the characteristics of current, commercially available consumer electronic products. Although, several makes and models for each product are available, representative products were selected without regard to the particular brand of the product. The weight of the product in grams and the usage life in hours were obtained from the specification sheet. Further, we collect information whether the product can be used outdoors or only indoors. It would be useful to construct a model of "portability" of these devices to take into account the weight, continuous use time, and whether the product can be used indoors or outdoors. The portability index is calculated using the following formula:

$$\mathrm{P} = \frac{T \times U^I}{w} \tag{1–1}$$

where T is the usage time on a single charge, U^I represents the ubiquity of usage (1 for only indoors, 5 for outdoors or indoors), and w is the weight of the product in grams with battery and without any cable attachments.

Using this non-dimensional measure, at one end of the spectrum the portability index of a Bluetooth earphone is 4.6, while at the other end the portability index of a 17-in. laptop is approximately 7,500 times lower at 0.0006045. In between the two extremes is the portability index of a smart phone at 0.375, while that for an ultraportable laptop is about 0.006. The portability index serves as a relative measure of how "portable" a particular product is considering three parameters important to the user: weight, battery life, and whether it can be used everywhere. From figure 1–1, it appears that the term "portable electronics" can be more easily justified for products with a portability index greater than approximately 0.1. Since, it includes three different customer requirements, the portability index can also serve as a convenient goal for developers of portable electronic products.

On the basis of a purely mass perspective, electronic products with weight less than approximately 500 g can be considered truly portable in that they are often carried on-person, whereas products heavier than about 2,000 g are better classified as mobile computers than portable electronics. The products in the range of 500–2,000 g, such as portable DVD players, are normally carried conveniently in a bag but not typically used while the user is walking or otherwise moving around (table 1–1).

The exceptions are pocket and digital single lens reflex (SLR) cameras, which, by virtue of their unique and narrow role, are used outdoors and while the users are moving about, even though their portability index is lower than 0.1.

Table 1–1. Survey of current commercially available portable electronics with their weight and continuous battery usage time

Product	Weight in Grams (w)	Usage Time in Hours (T)	Ubiquity Index (UI) (Indoors=1; Outdoors=5)	Portability Index	Legend in Figure 1–1
1 17″ laptop PC	3,311	2	1	0.00060405	17″PC
2 15.4″ ruggedized laptop PC	3,084	6	1	0.00194553	15″PC
3 Ultraportable laptop PC	1,497	9	1	0.00601202	12″PC
4 Portable DVD player	1,225	6	1	0.00489796	Portable DVD Player
5 Netbook	1,134	5.25	1	0.00462963	Netbook
6 Digital SLR	690	650 pics	5	0.04347826	SLR
7 Digital reader	625	10	1	0.016	e-Reader
8 4.5″ UMPC	499	2	1	0.00400802	UMPC
9 Large mobile phone	221	6	5	0.13574661	Large Mobile Phone
10 Handheld MEMS projector	220	1.5	1	0.00681818	Projector
11 Portable GPS	210	2	5	0.04761905	GPS
12 Internet tablet	193	6	1	0.03108808	I-Tablet
13 MP3 player	139	36	5	1.29496403	MP3
14 Digital camera	135	200 pics	5	0.03703704	Camera
15 Smart phone	133	10	5	0.37593985	Smart Phone-2
16 Music phone	104	9	5	0.43269231	Music Phone
17 BT earphone	12	11	5	4.58333333	BT Ear

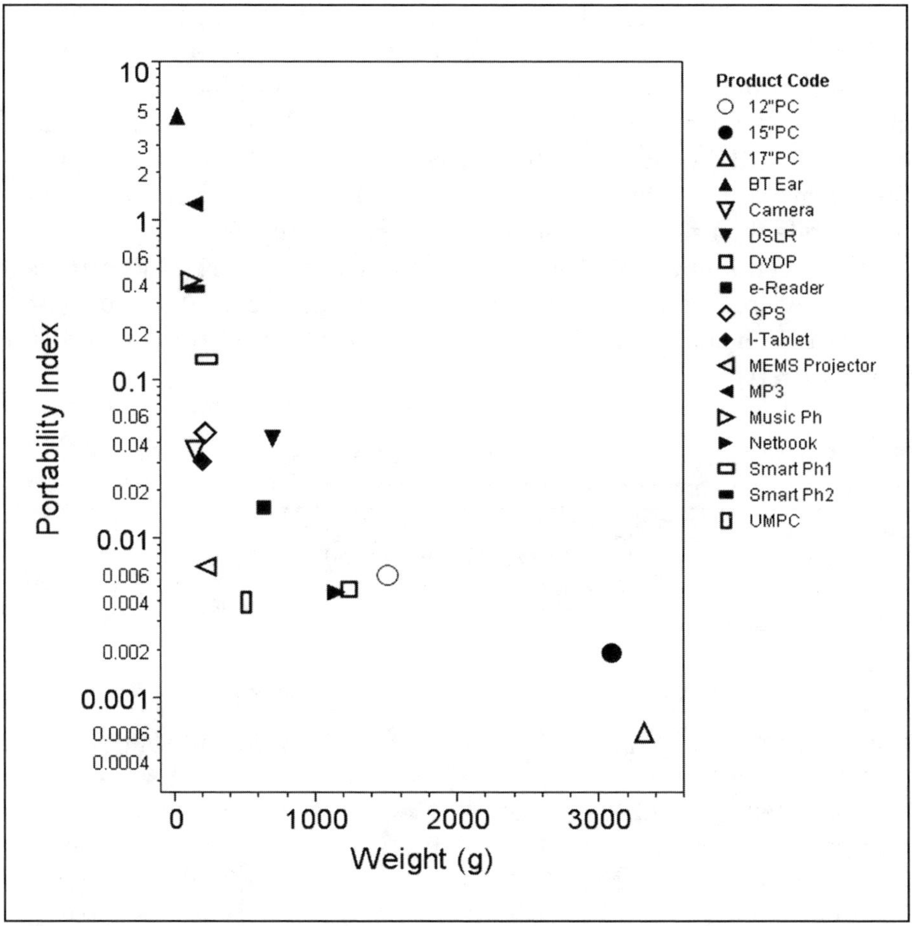

Fig. 1–1. Plot of the weight of several consumer electronic products versus the portability index.

The focus of and emphasis in this book is on the packaging, design, and reliability of portable electronics typically lighter than about 500 g. It can be seen that different kinds of mobile phones fall under this definition of portable electronics, and will serve as the proxy to discuss the attributes of all portable electronic products.

Basic Parts of a Mobile Phone

An introduction to packaging for portable electronic products has to begin with the parts of a typical product of current vintage. Although, there are several hundred types of portable electronic products, an entry-level mobile phone serves as a good example (figs. 1–2 and 1–3). The Nokia 2610 is described by the manufacturer as a classic design with speakerphone and without camera, weighs about 91 g and is about 10.4 × 4.3 × 1.78 cm in dimension. It is a dual-band Global System for Mobile (GSM) phone that can operate in E900/1800 or E850/1900 frequency bands. The main sections of the phone are the baseband, radio frequency (RF), and user interface sections. The baseband portion primarily consists of (a) a power processor that integrates a microcontroller, a digital signal processor, and other digital control functions; (b) the universal energy manager (UEM) that integrates the power supply functions, voltage regulation, and battery charging circuitry; and (c) flash memory to store data and settings. The RF portion of the mobile phone consists of a transceiver, a power amplifier module and associated antenna switch, etc. The user interface of the mobile phone is comprised of the keypad, display, speaker, microphone, and ports for charging, etc. The basic functions described above are summarized in table 1–2 (Source: Prismark).

Table 1–2. Basic functions of a mobile phone

	Transmit	Receive	Components
Baseband	Sampling and digitization of voice Channel coding, encryption and signal modulation	Conversion to analog signal to drive speaker Demodulation and decoding of incoming data	· Microprocessor · DSP · Memory · AD, DA converter
Radio frequency (RF)	Upconversion of carrier frequency to final RF frequency Amplification of RF signal and connection to antenna	Down conversion of the RF signal to the baseband frequency Reception and amplification of RF signal	· Mixer, LNA, PLL · VCO,TCXO, LC filter · SAW etc., filter · Duplexor, power amplifier, antenna, antenna switch
User interface			Display, keypad, audio components, charging and data connectors, volume control, and power switch

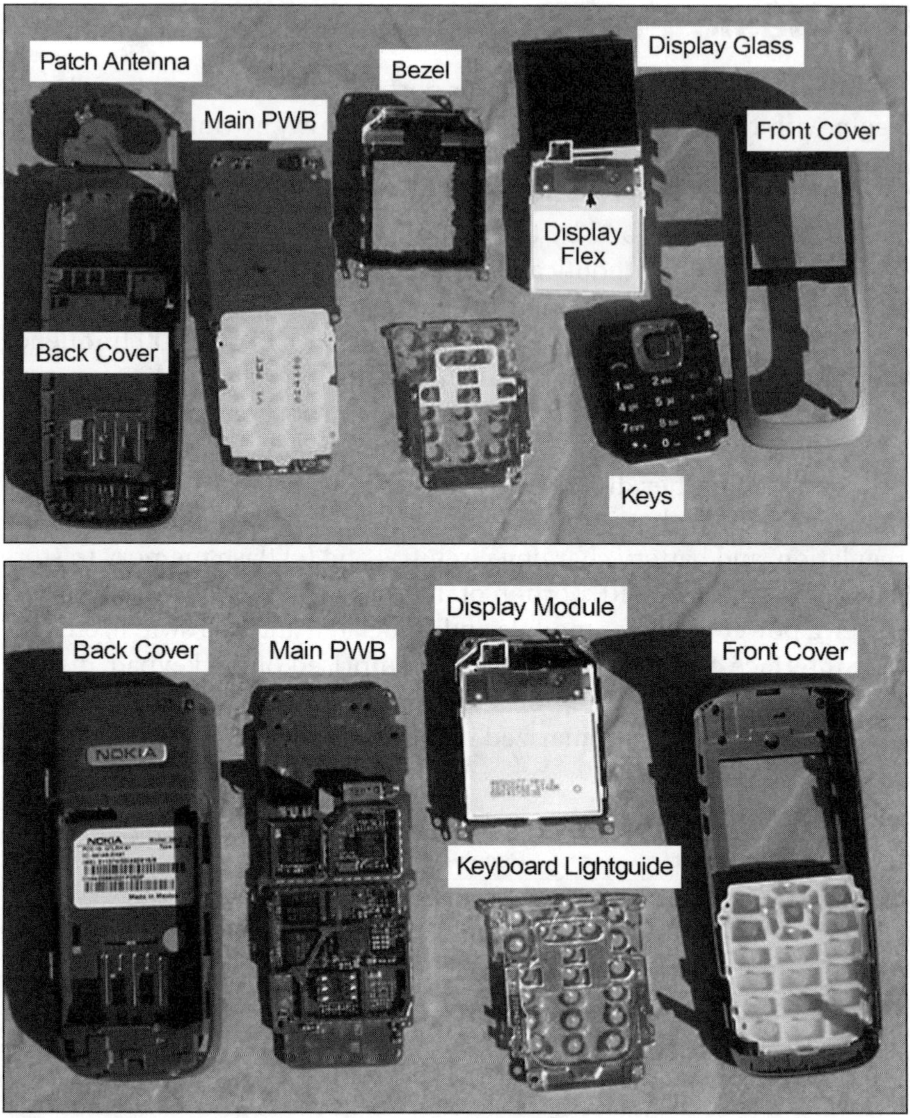

Fig. 1–2. A disassembled entry-level mobile phone with most of the key components except the battery

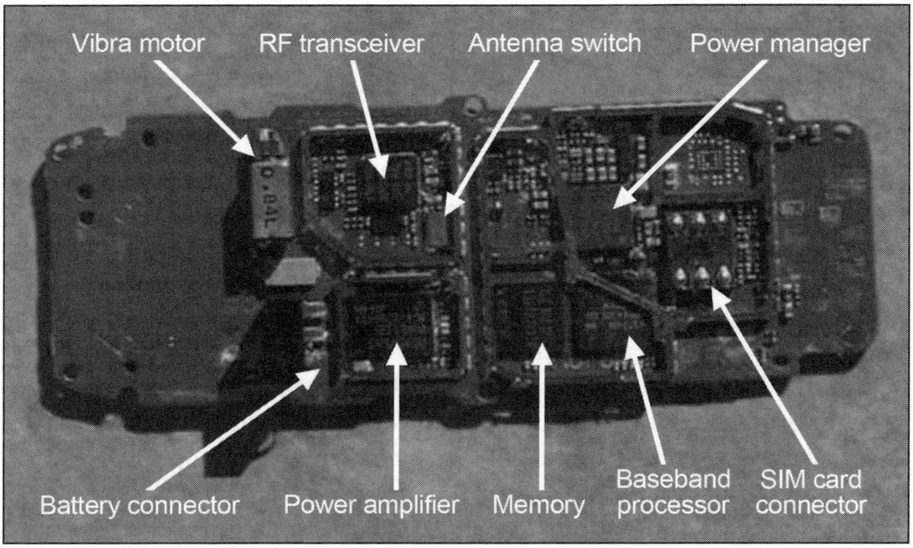

Fig. 1–3. Main printed wiring board (PWB) of an entry-level phone with major components identified

Key Attributes of Portable Electronics—A User's Perspective

Some attributes of portable electronics from a user's perspective can be listed as follows (not in the order of importance):

1. **User friendly.** Key attributes of portable electronics include the size and weight because bulky or heavy devices are not convenient for use by people on the move. In addition, most users gravitate to an ergonomic user interface, and this includes both hardware and software aspects of the product. For example, the scroll wheel commonly found in most e-mail devices or versions of the click-wheel found on most portable music players are user-interface variants that have survived natural selection in the evolutionary battle of the marketplace where only the fittest survive. Long battery life or usage times are also highly prized by users who have to operate far away from recharging stations, sometimes for several days at a time. Ubiquitous usage capability also sets apart the best of the breed from the rest of the pack, and this includes the ability to see the display under a variety of lighting conditions as well as surviving humid conditions,

dust, and heat. While the definition of a convenient size for portable electronic devices is something that varies from user to user, the drive to make devices smaller is driven by users' demand for smaller, lighter, and thinner products.

2. **Appeal.** The importance of visual appeal in portable electronic devices cannot be overemphasized because portable electronic products are not only functional devices but also fashion accessories and extensions of the user's persona. Devices with iconic designs and attractive finishes are highly prized by users who want to show their individuality analogous to apparel and jewels. Interchangeable covers with multitudes of designs and color combinations to suit ones personal taste or occasion, or color combinations to match the apparel, have emerged. A given design widely received in a given geographic region may not be as accepted in another region of the globe. Design for visual appeal, in addition to performance and reliability, is a new and emerging concept unique to portable electronic gadgets and is likely to gain prominence in order to gain market acceptance and profitability
4. **Functionality and flexibility.** In addition to user friendliness, portable electronic products have to have multifunctionalities so that the same device can enable the user to perform several unrelated tasks. Examples include buying a train ticket, conducting a financial transaction in a bank, obtaining directions to a place of interest, etc., transferring funds from one account to another, and a host of other chores. These features offer the consumer an immense, enormous flexibility and functionality. While the hardware provides functionality, the software enables flexibility.
5. **Cost effectiveness.** User also places a premium on getting good value for each price point, and prefer devices that are relatively inexpensive to maintain (preferably maintenance free) and stingy on power requirements. Power usage requirements play into meeting these expectations. Most users will want good quality, reliable products that have interoperability without paying a price premium.
6. **Upgradability.** As rapid developments and innovations continue, new functionalities will be incorporated into a given class of portable electronics. Easy product upgradability will be a highly desirable attribute for a product from the consumer point of view. An increasing emphasis on upgradability at the design stage will likely be prevalent.

While meeting each of these attributes is a challenge in itself, designing and manufacturing portable electronic products entails balancing the various aspects while also ensuring that development costs, material costs, and time-to-market targets are achieved. The high shipment volumes of portable electronic products place this segment of the electronics industry in the driver's seat with regard to the development of new packaging technologies, display technologies, manufacturing technologies, etc., which, once established, become available to other segments of the electronics industry as well. For example, chip-scale packages, microvia substrates, low-temperature cofired ceramics (LTCCs), organic light emitting displays (OLEDs), etc., were either developed for portable electronic applications or a by-product of such developments.

What Is Unique to Portable Electronic Products?

The highly personal use profile and mobility for these products imply that consumers will take these products with them wherever they go, and expect the same dependable performance irrespective of the exposure of the product to the elements, for example, rain, snow, and accidental drop. As users move between air-conditioned spaces and spaces where the temperature and humidity are uncontrolled, it is likely that the inside of the portable product is subjected to both condensing and non-condensing humidity. A diurnal temperature variation of 20–30 °F temperature and 50% relative humidity changes are not uncommon. In addition, the operating environment for mobile electronic equipment also differs considerably from that for desktop or business computers, and is often more varied in terms of thermal excursions, exposure to humidity, and corrosive environments. In metropolitan, urban, and industrial areas, corrosive gas pollutants such as sulfur dioxide, hydrogen sulfide, NO_x, chlorides, etc., will be present in varying concentrations. Portable electronic devices are much more prone to exposure to oils, greases, detergents, disinfectants, cleaning agents, etc., than any other electronic product.

Additionally, portability makes the product more likely to experience mechanical loads such as drop, bend, or twist, or mechanical abrasion due to incessant use. Mechanical drops from such heights as a meter and half on hard surfaces are not uncommon. The number of power ON cycles, the operational voltages, and other conditions are also different. Table 1–3 shows a comparison of the typical operating environments for portable hand-held telecommunication devices with conventional desktop and automotive under-the-hood electronics. In coastal and marine environments, portable electronics can also be subject to higher levels of exposure of not only humidity but also salt fog and spray environment.

Table 1–3. Comparison of typical application conditions of desktop, mobile, and automotive hardware

Desktop	Product Life (years)	Power-on Cycles/ day	Power-on-Hours	Relative Humidity %	Environment Temperature Range/°C	Operational Temperature Range/°C	Voltage/ V
Desktop	5	1 to 17	13,000	10 to 80	10 to 30	20 to 30	12
Mobile terminal	5	20	43,800	10 to 100	-40 to 40	32 to 70	1.8 to 3.3
Automotive under the hood	15	5	8200	0 to 100	-40 to 125	-40 to 125	12

*Source JEDEC

Current and Near-term Portable Electronics Landscape from an OEM Perspective

The landscape of portable electronics is extremely dynamic and fluid, with innovation coming from both large and small companies. From an OEM perspective, it is crucial to be able to not only understand the inexorable trends driving the market for portable electronics but also execute almost flawlessly to develop products that people will want to buy amidst very robust competition. The global trends that dominate the landscape in the near term are as follows:

1. **Miniaturization.** Miniaturization refers to the progressive reduction in the volume and weight of portable electronic devices. From the plot of portability index with weight in figure 1–1, it is easy to envision how a lighter device has a greater likelihood of its use becoming ubiquitous. The extreme progression of this concept is a wearable computer. Mann describes wearable computing as constant and always ready, not monopolizing the user's attention, observable and controllable by the user, and useful as a communication tool and an extension of the person [Mann, 1997]. But even in the interim scale, making portable electronics even smaller, lighter, and thinner is a challenging task,

and new materials, technologies, and user interfaces will be needed to manifest the concept. Concomitant reliability, manufacturing, and safety issues will need to be addressed.

2. **Convergence (functionality).** Convergence may be defined as the ability of the device to perform more functions just as well as or better than several individual devices, making it more convenient to carry just one device rather than juggle several separate ones. For example, instead of having a portable music player, a mobile phone, a camera, and a PDA, currently available smart phones can support the convergence in a single, visually attractive, and user-friendly package. Already, location-aware portable electronic products are coming to the market, using either global positioning satellite systems (GPS) or cell tower triangulation. The day is not far off when such a capability will be used to sell services, products, and other location-dependent business transactions. The really interesting aspect of convergence is that it changes user behavior in unexpected ways, and entire ecosystems will spring up to take advantage of new business opportunities.
3. **Integration.** Integration is to packaging what convergence is to user experience. While miniaturization is definitely needed to drive towards smaller and lighter portable electronic products, really big leaps in miniaturization will also require higher levels of integration. For example, the entire baseband section of a mobile phone can be integrated into a module, or the baseband processor can be made powerful enough to eliminate the expense of a dedicated ring tone generator by integrating the audio digital-to-analog converter (DAC) in the analog baseband and executing the software in the baseband processor itself. Integration can save cost in terms of components and also board real estate, and simultaneously improve the reliability by reducing component count. Since passives and discretes make up the highest portion (70–80%) of the component count, integration of passives can save a significant amount of real estate on the printed wiring board (PWB). On a package level, integration solutions such as packaging and die stacking are already in vogue.
4. **Time to market.** The time to develop new products is a critical business advantage if it is better than the competition. The pace of innovation in the field of portable electronics is so rapid that product development times longer than a few months can prove to be a severe handicap to competitiveness. Since, cost and quality typically compete with time to market, particularly creative processes and approaches such as DFx are essential to making good quality products at low

cost and in adequate time. Innovative design concepts, adaptability to emerging technologies, as well as nimble, lean, efficient, and high yield and throughput manufacturing are keys to success. Two important elements in the product development are product reliability and qualification testing. Also, testing consumes the longest time in a product development cycle. Efforts to reduce test time through implementation of innovative test methodologies are essential. Advanced modeling and simulation methods can aid in reducing product development cycle time.

5. **Cost of goods.** The cost of goods is a key component of profitability, and, in addition to good design and manufacturing processes, an agile supply chain is essential to ensuring profitable operations. Manufacturing close to the market place while also balancing labor and transportation costs can ensure that the total cost of goods sold is low enough to ensure good profitability. Good knowledge about one's supply chain also plays a key role in failure analysis activities because companies today are not vertically integrated and the root cause of an issue can sometimes be remedied only at a second- or third-tier supplier.

Reference

Mann, S. 1997. On the bandwagon or beyond wearable computing. *Personal Technologies* 1 (4): 203–07.

2

Packaging Challenges

Introduction

To comprehend the packaging challenges facing portable electronics original equipment manufacturers (OEMs), it is essential to understand the key desires of people buying portable electronics. There are innumerable variations in the features a particular user group prefers in their mobile phone, global positioning system (GPS) device, etc., and it is more instructive to examine the responses from a broad survey, in which the question was asked whether each of the following factors was important in purchasing their next mobile phone: price, battery life, better display, size/weight, ease of use, style/design, brand, data speed, e-mail capability, camera, and music player. The respondents rated their choices from "Not Important At All" (1) to "Very Important" (5). Table 2–1 depicts percentage of respondents at three importance levels in each of the above factors.

Table 2–1. Percentage of respondents in a user survey on how important each of the following factors would be in purchasing their next portable electronic device

Feature	Not Important At All or Not Important (1 or 2)/%	Somewhat Important (3)/%	Important or Very Important (4 or 5)/%
Price	5	15	80
Battery life	4	21	75
Display quality	4	23	73
Size/weight	4	23	68
Ease of use	5	27	71
Style/design	5	24	71
Brand	12	34	54
Data speed	20	35	45
Email	46	27	27
Camera	48	25	27
Music player	59	21	20

(Not important at all =1 to Very Important =5)

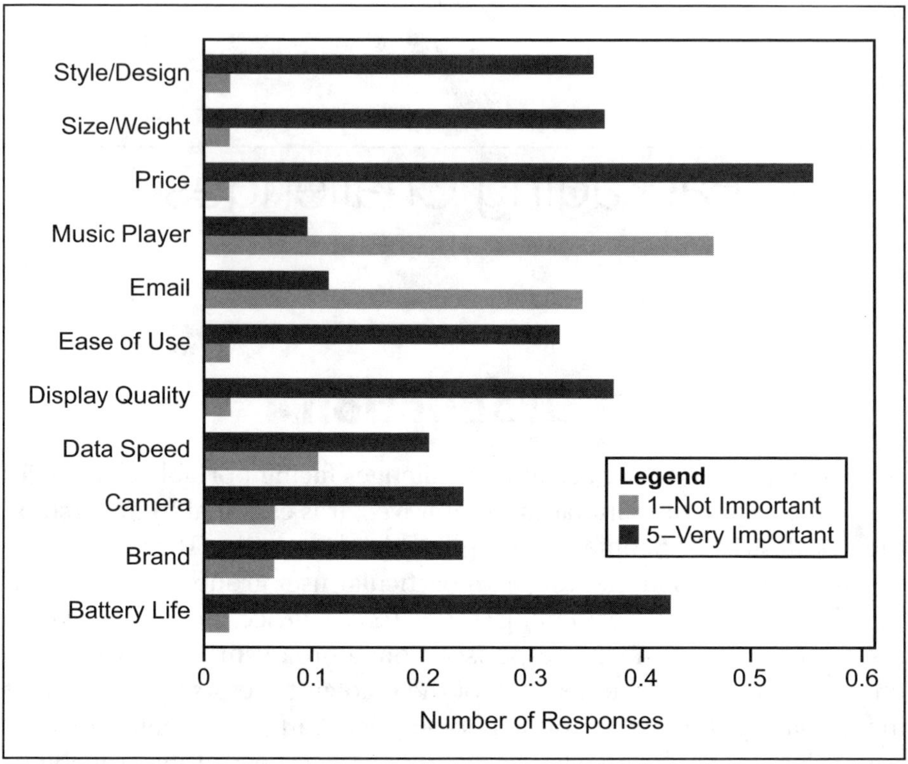

Fig. 2–1. Histogram of purchase priorities for purchasing the next mobile phone.

The different features mentioned in the survey are ranked and plotted in figure 2–1. That price, battery life, display quality, size/weight, ease of use, and style/design were *all* rated as important criteria in purchasing their next phone by more than 70% of the respondents highlights the challenges in designing, packaging, and manufacturing of portable electronic products (PEPs) where numerous competing factors are critical to attract customers and keep them satisfied.

Of course, these priorities are dynamic entities, and as PEPs mature, the relative importance of these features can change.

As portable electronics is one of the most rapidly growing segments of the electronics industry, advances in the semiconductor, sensor, display, battery, and wireless technologies impose innumerable challenges at all levels of packaging. Timely solutions are required to successfully implement the emerging trends with efficiency, but in a cost-effective manner.

Conventional electronic packaging consists of three levels of packaging: namely, first-level packaging consisting of encasing the semiconductor chip in a suitable enclosure to protect it from environment and mechanical damage and also enable attachment to the carrier board; second-level packaging involving assembling the packages on to the board; and third-level packaging consisting of integrating all the subsystems into a functional appliance.

In the scenario of faster, cheaper, and better PEPs, conventional packaging technologies are reaching the limits of their capabilities. Standard first-level packages have already reached their limits of packaging efficiencies. A packaging efficiency of near unity in area packaging is already proving to be inadequate and concepts such as silicon stacking and package in a package are already being implemented at an amazingly fast pace as many of the reliability issues are being worked out. Through-silicon-via technology to reduce the signal path length and to increase performance is vigorously being pursued and is likely to become mainstream high-density packaging with several orders of magnitude higher packaging efficiency. At the rate at which technology is advancing, even this may not be able to meet the industry requirements. New first-level packaging challenges emerge. Innovative concepts such as optoelectronics, micro-nanoelectronics, and nanoelectronics come into play. As nanoelectronic systems and subsystems are introduced, these need to be integrated with microelectronic systems until such time when nanoelectronics becomes the mainstream technology. Such an integration presents several design, assembly, and test challenges. New assembly technologies have to be integrated with the conventional ones without significant increases in cost. Another significant challenge is the integration of opto- and microelectronic systems.

In the second-level packaging, first-level single-chip modules and or multichip modules are assembled on to rigid or flexible printing wiring boards (PWBs). The industry has already witnessed the inadequacy of the conventional multilayer PWB technologies with 4 mil lines and spaces and 8–10 mil via for routing high-performance high I/O devices and has taken recourse to high-density interconnect with smaller nonmechanical, photo-imageable, or laser-drilled vias with 3-mil lines and spaces routing. Industry would soon be encountering the limits of this technology, and newer technologies will need to be explored. Fine-line printable electronics with reel-to-reel processing will be commonplace in the near future. Some of the many challenges include fabrication techniques and reliability evaluations. The challenges encompass discovery, innovation, and development of new laminate materials that meet or exceed the near-term packaging requirements.

Current interconnect technologies are based on alloy interconnections and adhesive interconnects. The most common interconnection alloys are the tin–silver–copper systems. These are used in combination with a host of terminations and laminate surface finishes. Several metallurgical issues still persist with regard to the final interconnection composition, microstructure, morphology, intermetallics, etc., especially with respect to extremely small volume interconnects under different stress loads.

As the package I/O increases and the pitch becomes smaller, placement accuracy and precision affect assembly yields. Rework and repair operations can be very frustrating. The challenges will be in finding cost-effective and efficient assembly techniques.

Elimination of levels of packaging by incorporating or embedding the unpackaged devices into carriers themselves as part of the carrier technology is one way, but the challenges of test and repair still remain. One current approach is to integrate the passive components such as resistors, capacitors, inductances, etc., which is being practiced in a limited way. Large-scale incorporation of active devices is still emerging and in the near future will likely become the preferred practice when the challenges of assembly yields, test, and repair are successfully addressed.

Ergonomic Challenges

Ergonomics is an applied science involving the study of the relationships of people and the associated work environment. Its primary goal is the performance and well-being of the user and the focus is to ensure that the interactions between the user and the machine are safe and effective. Several factors such as the physical and physiological impacts are taken into consideration in the design of the product and also the workplace environment. The safer and more comfortable are the tasks to be performed, the greater is the user satisfaction and the gains in productivity (Haskell 2004).

Ergonomic features are important aspects in the efficient use of an appliance as well as in enhancing productivity and human comfort, as seen from the survey results shown in figure 2–1. In the days of desktop computers and terminals, ergonomic considerations for user comfort, health, and productivity involved the optimization of the furniture design, materials, dimensions, and their adaptability to the user's physical comfort. In other words, the information processing hardware remained stationary on the desk, while the rest of the work environment is fitted to the user's comfort. The scenario with respect to portable electronics is quite different. Portable electronics by their very nature are mobile and of different sizes, volume, weight, and conceptually are of the "use anywhere anytime" type. Hence, ergonomic considerations for these can be complex. Portable products assume a variety of orientations depending on the spatio-temporal activity of the user. For example, a cell phone can be used while the user is jogging in park, working in an office, relaxing in a recliner at home, driving a car, in a flight on an airplane, etc. The same device can experience a variety of climatic conditions during its life. Thus, portable electronic devices can be exposed to much more variable operating conditions in terms of temperature, humidity, gaseous environment, as well as mechanical stresses such as shock, vibration, bend drop etc., than a desktop or mainframe computer that is customarily housed in a controlled temperature and humidity environment.

As regards the products, ergonometric aspects need to be addressed at the design stage rather than later. The following are a few examples of ergonomically significant aspects of a portable electronic device:

- General shape and feel
- Size, weight, and volume
- Keyboard
- Display

The shape of the product is to be such that it does not have sharp corners, should provide a robust feel in the consumer's hands, and should be shaped such that it is capable of being held with comfort and ease for extended periods without undue strain. Size, weight, and volume are generally determined by the functionality of the product, but with increasing emphasis on higher levels of integration and miniaturization the migratory path is towards smaller and smaller form factors. Yet, the distribution of weight within the system can influence the handling of the device and also its propensity for being dropped. The design of the outer case, besides having a visual appeal, should be such as to provide a soft feel with padded features, with comfortable, secure, and optimized outer dimensions.

As the product becomes smaller and smaller, special attention is to be paid to the tactile features of the keyboard, spacing and arrangements of the pads, size, feel of the pads etc. As the functionality increases, a single key with a plurality of functions, multiple keystrokes for specific functions, etc., can be a user's nightmare and a cause for considerable frustration and stress. Ease of use becomes an important design attribute and challenge.

The difficulties associated with the operating of smaller keypads on the keyboards are already being addressed partially by the elimination of keypads on PEPs. The keyboard is being replaced with touch screens.

Among several solutions to address such issues are modular designs with detachable keyboards, screens, pointing devices, wireless interfaces, etc. (Tessler 2006). However, some of these solutions can render these devices less portable or cumbersome to be carried around.

Efforts directed towards generating a set of ergonomic variables through consultations with ergonomic experts, field tests, etc., can yield valuable design input. Also, several tools and methodologies are available in the industry to generate and optimize ergonomic designs. Among these, the analytic hierarchy process (AHP) and multiple decision-making techniques for order preferences by simulating to ideal solutions are popular. One such is called the high-touch ergonomic design process. It consists of (a) identification of functional requirements and operational constraints, (b) arriving at hierarchical structures for ergonomic variables and product functions, (c) development of a relationship matrix, (d) identifying the shortcomings of product weakness, human functions, and possible improvements, (e) development of ideas, and (f) evaluation of ideas and development of a prototype. Examples of independent variables include position of keys, feel of the keys, size of the keys and the spacing between them, font size in displays, screen size, brightness, number of key presses needed to access a given function, etc., The response variables can be the time or number of operations before the onset of discomfort or pain, the time to reach discomforting temperature, time of use, etc. Such a methodology aids in determining an optimal design within the realm of identifiable variables.

Several interesting ergonomic designs have been studied and reported to reduce user strain. A mobile phone with the keypad positioned above and the screen below, a configuration that is opposite to the conventional design, is considered more ergonomic for texting. It was indicated that the phone has an improved grip and the thumb rests in a comfortable position directly over the keypads. In another study, using several brands of mobile phones and several users, the relationship between the thumb size and mobile phone texting satisfaction was examined (Balakrishnan and Yew 2008). Larger thumb circumference decreases the texting satisfaction. With larger thumbs, the possibility of pressing two keys at the same time increases and hence results in improper messaging. Longer thumb length also decreases texting satisfaction because of the difficulty in accessing the 3, 6, 9, and # keys. Also, difficulty in reaching certain keys causes physical discomfort in the first joint of the thumb. Thus, ergonomics plays an important role in the user satisfaction of portable electronics and has to be addressed at the beginning of the product development cycle, namely, the design, and therefore plays a direct role in dictating the packaging solutions necessary in a particular product.

Flexible and Changing Shape

In the current portable electronic landscape, most commercially available devices have a fixed shape. However, there are several ongoing efforts to bend the rules. For example, Nokia has unveiled the concept of a morphing device that would fit easily in a pocket just like a traditional handset in the minimized form. When unfolded or unrolled, the larger size can display more detailed information and incorporate input devices such as keyboards and touch pads. This takes the idea of sliding and folding keyboards much further, and the packaging challenges would not be met by conventional technologies that can tolerate only a limited amount of distortion before failure. Interconnection technologies using nanotechnology, transparent electronics, and flexible displays would be necessary to realize such concepts. Some of these technologies are discussed later in the book.

Display Challenges

In the current crop of portable electronic products, the liquid crystal display (LCD) is the dominant display technology. However, even as PEPs shrink in size and weight, the display size required either remains the same or increases depending on the application. For example, smart phones that are capable of surfing the Internet require a larger display compared to a phone that is used only for voice communications. Already, in some products the keyboard has disappeared in favor of a larger touch screen display. In this particular case, the requirements of a larger display were met by removing the keyboard and integrating the keyboard into the touch screen display. As this product shrinks further in size and power, what else can be integrated into the display? It is only logical to assume that completely new display concepts would be required to maintain ease of use and good visibility of the data. The challenge is not merely in improving existing display technologies but rather in developing entirely new display technologies such as holographic displays, near-eye displays (NEDs), etc.

One way of achieving the increased visual content demands on portable devices is to use virtual displays or to project the image on a surface. Displays smaller than 1 in. diagonal can be defined as microdisplays, and since they are viewed indirectly, optics is usually involved with microdisplays. Microdisplays have a large number of pixels in a small area, and their power consumption is very small. One example of a virtual display is the microelectromechanical system (MEMS) based pocket projector concept based on the digital light processor from Texas Instruments, which can project the content of the web page onto a wall or other flat surfaces. However, this has limitations from a privacy viewpoint and also requires the availability of a flat surface for projection. NEDs overcome this limitation and display the image on eyeglasses or other devices worn on the head (Kimmel et al., 2002). NEDs can be reflective, transmissive, or emissive depending on how the light is modulated. They can also be classified as monocular (only one eye sees the image) or binocular (both eyes see the image). NEDs impose their own packaging and reliability challenges because the devices have to be extremely light and flexible, consume minimal power, dissipate heat without causing discomfort to the user, and be reliable enough under severe environmental stresses in addition to being cost effective.

Sensing Challenges

In the current PEP landscape, phones that are termed "smart" are actually more complex and multifeatured. In the future, as PEPs evolve, "smart" will come to signify devices that are aware of their environment and context (Schmidt, 2001). Context awareness can be defined as knowledge about the user's and device's state, including surroundings, situation, and location, and this context is approximated or estimated by collecting and analyzing sensor data. Although, sensors have been widely used in robotics, machine vision, and manufacturing, issues related to size, power, and cost-effective integration into portable products will be a challenge. Light sensors and audio sensors can be used to determine the ambience and background noise that the user is immersed in. Accelerometer sensors can be used to construct context in terms of whether the user is in a car, walking, running, etc. to provide applications suitable for the occasion.

Location sensing using either cell tower triangulation or GPS signals is already a reality and this will become more sophisticated. If a touch sensor can be directly implemented with conductive planes (e.g., skin conductance, human as capacitor), or indirectly using light sensors or temperature sensors on the PEP, energy consumption can be reduced significantly, especially for devices that only need to be operative in the user's hand. Similarly, motion sensors, gas sensors, air pressure sensors, and biosensors all expand the usefulness and richness of smart phones. Apart from power consumption, cost, unobtrusiveness, robustness and reliability, packaging and manufacturing of these sensors into future smart phones will be a challenge.

In addition to the hardware challenges described above, the system designers will be challenged to determine the context from the sensors and enrich the life of the user based on the context. For example, consider the case of a businesswoman carrying a PEP in a crowded subway train. When she receives a videophone call from home, the phone would sense the context (noisy subway car with privacy issues) and route the video to the microprojector in her eyeglasses and route the audio to the earphone/microphone on her after appropriate noise cancellation so that the call quality is satisfactory. Further, suppose that the caller was requesting the user to pick up some take-out dinner. The portable electronic device would automatically decipher the context of the call and, after analyzing the call transcript, display the menu choices at a few nearby restaurants. The user would then be given the option to place an order of her choice at any of these restaurants via the Internet without actually making a call in the noisy environment. The challenges to achieve this are not only the hardware aspects such as power usage, heat dissipation, reliability, etc., but also software and integration issues.

References

Balakrishnan, V., and P. H. P. Yew. 2008. Effect of thumb sizes on mobile phone texting satisfaction. *J Usability Studies* 3(3): 118–28.

Haskell, B. 2004. *Portable electronics—product design and development.* New York: McGraw Hill Books. ISBN 0-07-141639-0.

Lee, M. W., M. H. Yun, E. S. Jung, and A . Freivalds. 1997. *Int J Industrial Ergonomics* 19 (3).

Schmidt, A., and K. V. Laerhoven. 2001. How to build smart appliances. *IEEE Personal Communications*: 66–71.

Tessler, F. N. January 2006. Laptop ergonomics. *MacWorld* 23 (1).

3

Printed Wiring Board Technology

Introduction

Printed wiring board (PWB) or printed circuit board (PCB) constitutes the basic substructure on which the entire basic product functional elements are assembled. It is a vital and integral component of the total electronic packaging system. The components include active devices such as logic, memory, processors, etc.; passives such as capacitors, resistors, crystal oscillators, inductances, etc.; other components such as power supplies, etc.; and connectors for external and internal interfaces. The essential function of the PWB is interconnectivity that enables interconnection of various elements through circuit traces. As semiconductor technology evolved giving rise to devices with ever-increasing functionality and greater lead counts, the complexity of the PWB increased in terms of materials, design, and fabrication.

As discussed earlier, the key attributes of portable electronic hardware are light weight, speed, inexpensiveness, and better reliability. Increased levels of miniaturization and integration at the various levels of packaging, namely, at the first and second levels of semiconductor device packaging and at the product level, are being implemented at an ever-increasing pace to meet the growing consumer demands. Historically, PWB technology lagged behind semiconductor technology in terms of the capability to accommodate highly complex, fine-pitch devices. Higher wiring densities as well as buried and blind vias were needed. This, in turn, increased the manufacturing complexities and also the number of layers needed to wire the high input/output (I/O) devices. PWBs capable of finer pitch traces and spaces and smaller diameter vias than hitherto possible are needed. Thus evolved a new technology, termed high-density interconnects (HDI). There are a number of variations of this new technology. Some of them utilize the conventional PWB technology as the basis and build the high-density circuitry on the surface layers of the subcomposite structure. Many of the basic PWB process steps are also common to the HDI.

Product performance depends on a number of electrical and mechanical properties of the PWB that play a vital role. These include not only the dielectric constant, coefficient of thermal expansion, and Young's modulus, but also a host of other properties. Table 3–1 lists many of the relevant properties.

Table 3–1. Typical properties of importance for printed wiring boards

1	Dielectric constant	11	Tensile strength (Mpa)
2	Dissipation factor	12	Tensile modulus (Gpa)
3	Electrical strength (volts/microns)	13	Flexural modulus (Gpa)
4	Insulation resistance (mohms)	14	Young's modulus (Gpa)
5	Surface resistance	15	Poisson's ratio
6	Volume resistance (mohms/sq.cm)	16	Resin content (%)
7	Flammability class UL-94	17	Glass transition temperature (Tg)
8	Peel strength (kN/m)	18	Coefficient of thermal expansion (x, y, z) (ppm/K)
9	Water absorption (%)	19	Thermal conductivity (W/mK)
10	Specific heat (J/kgK)	20	Density (g/cc)

The specific value of each of these properties depend on the use environment, the design life, and the reliability requirements of a given product. A product that experiences a harsh temperature and humidity environment may require high-temperature materials with a high glass transition temperature (Tg), low water absorption, etc. A judicious choice of materials with the requisite properties is essential.

The complexities of PWBs are sometimes classified as conventional, advanced, and HDI. PWBs with less than 65 interconnections per square inch are considered conventional; thinner materials for construction down to 0.004" with 65–150 interconnections per square inch that may contain blind and buried vias as advanced; and those that contain blind and/or buried vias of the order of 0.006" diameter produced by techniques other than conventional drilling and having 130 interconnections per square inch or greater are considered HDI boards. According to the Association Connecting Electronics Industries (IPC), any organic substrate with vias of 6 mils or less in diameter and made with any number of techniques is a high-density board.

In addition to the number of interconnections and via sizes, several parameters are used to characterize the density of the PWB and include the following:

1. Layer density—inches of lines/square inches of the board at a given channel width
2. Number of connections per square inch of the board.
3. Pad diameter
4. Circuit linewidth
5. Spacing between adjacent circuit lines
6. Via diameter as drilled
7. Size of the target pad
8. Size of the capture pad for the drilled via
9. Presence/absence of blind vias
10. Presence/absence of buried vias

In addition to line widths and spaces, considerable real estate on the PWB is taken up by the through holes that facilitate interconnection to inner layers as well as between the top and bottom layers. For higher I/O area array package interconnections, the surface routing capability on the PWB is limited by the channel routing, and interconnections are possible only by (a) a dog-bone shaped pad structure where the solder ball sits on one pad of the dog-bone while the via in the other pad facilitates interconnection to the inner layers, or (b) a via-in-pad structure with savings on the card real estate. The higher the I/O count, the more the number of card layers that are required to adequately route the signals. Also, mechanical drilling is limited by the drilling capability and the capture pad diameter, which is generally twice the diameter of the drill via diameter, as well as the thickness of the board. The drilling of fine holes becomes increasingly more difficult with increasing thickness of the board.

The via diameter is limited to 10 mils by the conventional board technology. Thus, the via drill diameter and the associated capture diameter impose severe limitations on the circuit density of the card. Migration to microvias as well as finer line widths and spaces are the key to cost-effective high-density PWBs. Thus, the technology has come to be known as microvia technology. Also, nonmechanical drilling techniques have to be employed, and these techniques are limited to mostly the surface layers.

At the device packaging level, this technology is employed for the redistribution of the device I/Os on the substrate providing the capability to give a higher lead or ball pitch. Thus, the technology has also come to be known as redistribution layer technology, microvia technology, buildup layer technology, etc. Table 3–2 shows the general classification of board technologies in terms of sophistication. It is to be recognized that any classification is subject to getting outdated with time, since what is state-of-the-art technology at one point in time becomes conventional technology as the technology matures.

Table 3–2. General classification of printed wiring board technologies

PWB Feature	Conventional	Advanced	Leading Edge	State-of-the Art
Layer density (in/sq.cm)				
0.050″ channel	23	55	117	200
0.025″ channel	...	...	67	150
Interconnections/sq.in	40	130	330	1000
Contact pad diameter	0.012″	0.010″	0.006″	0.0055″
Line width/line space	0.006″/0.006″	0.004″/0.004″	0.003″/0.003″	0.002″/0.002″
Drill via diameter	0.013″	0.010″	0.009″	0.006″
Capture pad diameter	0.024″	0.020″	0.016″	0.014″
Blind or buried via (drilled)	None	0.013″	0.008″	0.008
μ-via diameter	None	None	0.004″	0.002″
μ-via capture pad	None	None	0.012″	0.008″

In this chapter, we provide an overview of the conventional PWB technology followed by a more detailed description of some of the more commonly practiced HDI structures.

Conventional PWB technology

A conventional PWB is a composite of organic and/or inorganic materials with internal and external circuit patterns, which enables individual functional elements such as electronic and electrical components to be mechanically mounted and electrically interconnected. For most consumer electronic products, it is an organic laminate, while for special high-end, high-reliability applications multilayer ceramic laminates are used. Knowledge of the PWB materials and processes is important for the successful design and fabrication of reliable electronic products.

A surprisingly large amount of chemistry, both inorganic and organic, is involved in organic laminate PWB fabrication. There exist two popular PWB fabrication processes: an additive process and a subtractive process. These are described in some detail in the ensuing sections.

Additive processes

A catalytic substrate core material constitutes the base. A catalytic adhesive is deposited on both sides of this core. A palladium-based catalyst is dispersed into the adhesive resin homogeneously. The particles are fine enough and well separated so as not to cause any surface insulation resistance (SIR) degradation. The resin is usually a thermoset resin, such as bisphenol-A, mixed with rubber and filler. This adhesive application to the core is by dip, transfer, or curtain coating. Through holes are then drilled or punched as required by the design. The adhesive layer is abraded mechanically to enhance the adhesion of the plating resist. A permanent plating resist is then applied either as a dry film or screen-printed in the form of a reverse conductor film. This implies that after the application of this process the cured resist masks areas other than where circutization is needed. The board is then treated with a dilute solution of chromic acid–sulfuric acid mixture. This acid treatment dissolves the rubber in the adhesive and creates a microporous surface structure. This structure increases the contact surface area for better copper adhesion. Electroless copper is then deposited on to the exposed microstructure and the drilled hole-walls. The panels are then baked. The final step is the application of solder mask and any legends before the final test. Figure 3–1 shows a schematic of the additive circuitization process.

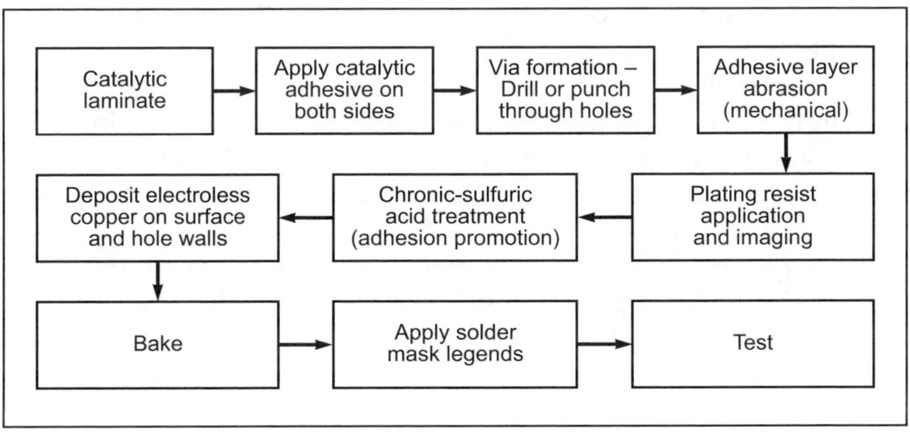

Fig. 3–1. Schematic of an additive printed wiring board process

Subtractive process

The example chosen is the typical multilayer FR-4 board. The designation FR-4 indicates fire retardancy grade in addition to other characteristics of the material. The typical process steps in the PWB fabrication involve the following:

1. PrePreg
2. Lamination
3. Circuitization
4. Chloriting
5. Drilling
6. Desmearing
7. Plating
8. Solder mask application
9. Solderability preservation

Lamination, circuitization, drilling, etc. may be repeated multiple times in the fabrication of complex board structures with several internal layers and also those involving buried vias.

Prepreg. A glass cloth that is impregnated with an appropriate organic resin material and cured is called *prepreg* and constitutes the separating dielectric plane between any two circuitized planes. In one example, the prepreg is made up of glass cloth, brominated epoxy resin, ethylene glycol mono-methyl ether (EGME), dicyandiamide (Dicy), and tetramethyl butane di-amine (TMBDA). The di- or tetra functional brominated epoxy constitutes the resin, whereas the EGME, Dicy, and TMBDA mixture comprises the hardener. A 2:1 ratio of resin and hardener is used.

Several resin systems are in vogue. These include bifunctional and tetra-functional bis-phenol-A epoxy resins, bismaleimide triazine (BT), benzocyclobutene (BCB), polyimides, etc. They differ in their glass transition temperature, chemical resistance, as well as electrical and mechanical properties.

TMBDA, Dicy, and EGME are premixed separately and the mixture is then combined with the epoxy resin to form the resin system. Methylethylketone (MEK) solvent is then added to adjust the viscosity of the mixture. It is sometimes called *varnish*. It is then held in a dip tank.

The glass fabric is a plain weave cloth with warps alternately going above and below each yarn in the fill direction. Glass fibers are available in several styles designated as 104, 106, 108, 116, etc. The designation of the glass depends on its special properties and is indicated as *E* for electrical, S for high strength, C for chemical resistance, D for high modulus, etc. The thickness of the glass varies from 0.03 to 0.089 mm. The weight of the glass cloth is generally in the range 20–109 g/m^2. Glass fibers are generally coated with special silane coupling agents to enhance the adhesion of glass to the epoxy. Excellent adhesion of epoxy to glass is important since any delamination under temperature and humidity stresses can result in interfacial degradation, which will reflect as degradation in the electrical performance.

The prepreg is processed in a treater tower as shown in the schematic in figure 3–2. Glass cloth in reels of 1,000 ft is unwound and is subjected to vacuum-cleaning for removal of any particulate debris and dust. The cloth is then passed over rollers through a tank containing the varnish to pre-wet and thus prepare it for the next step. In this step, the glass cloth is saturated with the epoxy mixture. The sheet then traverses through a set of stainless steel rollers where the cloth is squeezed so that the liquid mixture is squeezed into the cloth and any excess varnish is squeezed out. This is a critical step in the preparation of the prepreg. The spacing between the rollers, speed of the cloth, and pressure are important parameters that determine the amount of resin in the cloth. It is also called the *pick up*, i.e., the amount of resin picked up by the glass cloth. The glass to resin ratio controls several properties of the prepreg, namely, the coefficient of thermal expansion, the dielectric constant, and also some of the mechanical properties. The glass cloth then travels up the tower where in the first zone the solvents are evaporated and in the second zone the epoxy is partially cured. This is called the B-stage prepreg. The cloth is then cooled down to room temperature. Samples of cloth are then cut and tested for flow, resin to glass ratio, and final properties. Too much flow implies insufficient curing and less resin content and hence poorer insulation. The cloth is then rolled and wrapped in plastic for shipment.

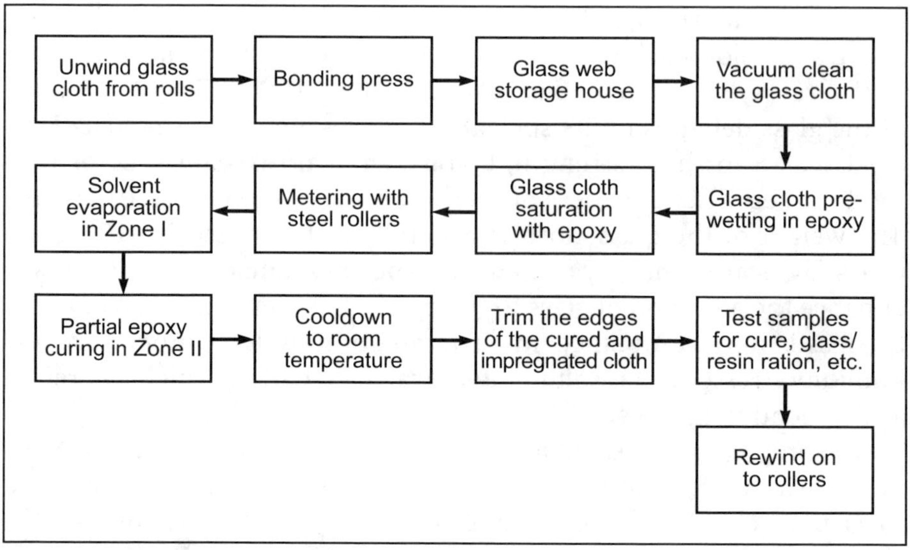

Fig. 3–2. A schematic of the prepreg manufacturing process

PWB structure. The PWB structure depends on the product design and contains several planes of prepreg and copper in a sequence dictated by the product design. Three to four prepreg layers between two conducting planes are not uncommon. The structure may consist of signal splanes, power planes, ground planes, thermal cores, etc. Ground plane is a conductive layer on a substrate or buried within a carrier that connects a number of points to one or more electrodes. Power planes buried within the structure establish voltage levels for the circuits and provide an impedance reference system for the signal lines. They also minimize voltage fluctuations at the device level. The board structures are designated sometimes as xSyP, implying x number of signal and y number of power planes. A 4S2P structure thus will have 4 signal planes and 2 power planes, implying a six-layer board. Figure 3–3 shows the layup of a 4S2P structure.

The copper foil thickness is also dictated by the design. Often, the Cu foil thickness is designated by its weight per square foot. For example a 1-oz. copper implies a sheet thickness of 0.0014" since 1 ft^2 of copper foil of 0.0014" thickness would weigh 1 oz. Thus a 0.5-oz. copper sheet thickness will be about 0.67 mils.

A prepreg sheet or sheets laminated with copper foil on both sides is termed a *core blank*. After being circuitized, the blank core is laminated with additional layers of prepreg, copper, etc. to form the subcomposite. Thus a subcomposite is a product laminated at least two times. On the other hand, a composite is a panel after the final stage of lamination that is ready for drilling and copper plating.

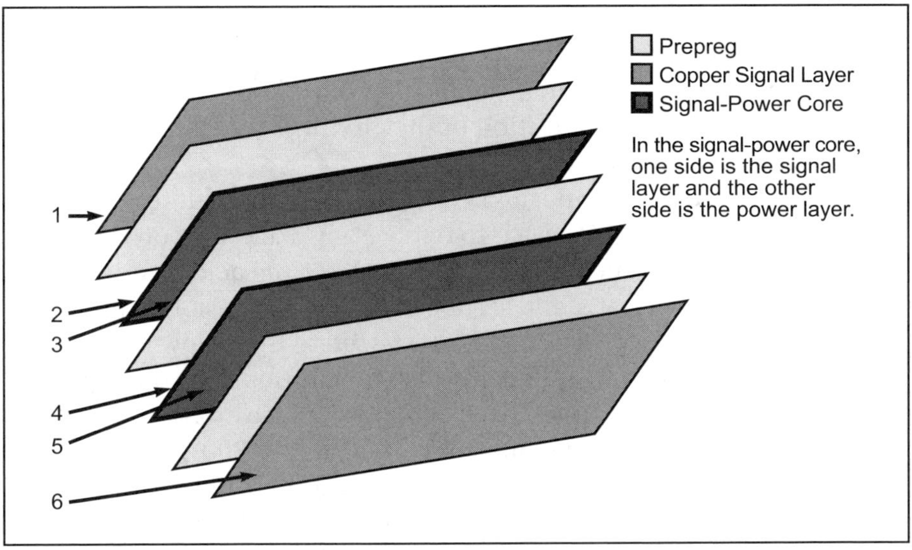

Fig. 3–3. Schematic layup of a 4S2P printed wiring board structure (1 signal, 2 signal side, 3 power side, 4 power side, 5 signal side, and 6 the signal)

In the PWB nomenclature, a page consists of several layers of prepreg and copper or circuitized planes, and prepreg and copper. One page is thus one stack. Also, a book consists of several pages stacked with planishing plates to a desired thickness. Planishing plates are stainless steel sheets that separate the pages.

Lamination. Lamination is carried out using a mechanical press. The lamination temperature and pressure are dependent on the nature of the product being manufactured. For FR-4 products, typical pressure is about 400 lb/in^2, and temperature is in the vicinity of 340°F. Time, temperature, pressure, and humidity are very important process parameters. It is also in the lamination process that final curing of the prepreg takes place. Humidity affects the curing process, and partially cured laminates results in poor drilling. The drill bits get gummed up during high-speed drilling owing to the heat generated in the drilling process. Excess moisture also causes delamination. It is important to ensure that the lamination press, planishing plates, the prepreg, and copper planes are devoid of any particulate contamination. Entrapped contamination can subsequently cause defects or failures depending on the nature of the contamination.

Subsequent to lamination, the laminates, depending on the process stage at which the lamination is performed such as subcomposite stage or composite stage, are subjected to machining to cut into smaller cards, deburring, notching and punching for orientation, stamping, or baking for dimensional stability.

Circuitization. The next step in the fabrication after lamination of core blanks is to circuitize the inner layers of the subcomposite. This process consists of precleaning, application of a photoresist, developing, etching, stripping, and inspection of the circuitry created, as described in the following paragraphs.

The laminated subcomposite is precleaned with a dilute solution of hydrochloric acid. The objective is to remove any oxide layer on the copper surface formed since the lamination step. After the acid treatment, the panels are washed in water and scrubbed with a suspension of pumice to provide a rough surface, dried, and transferred to the yellow room.

The yellow room is so called because it is in this light that a photosensitive resist layer is applied to the copper surface on both sides. This room is maintained at a cleanliness level of Class 10,000. A Class 10,000 environment implies that it contains no more than 10,000 particles of 0.5 mm or larger size per cubic meter of space. The photoresist is a light sensitive organic film of 0.0015–0.0020" thick that is covered with a polyolefin film on the bottom side and a Mylar film on the top. The film is rolled and squeezed on to the product and, as it is rolled, the bottom polyolefin film is peeled off. Heat and pressure are applied as it is laid, and the resist and Mylar are trimmed. The product is then cooled to room temperature and registration and tooling holes of predetermined size and shape are punched. It is important at this stage to ensure that the surface is free of wrinkles, free from any entrapped particles and debris, and properly aligned. The application of film is accomplished on both sides of the laminate subcomposite as required.

At this stage, a diazo film containing the required circuit pattern, with clear sections where circuit features are required and dark sections in the rest of the area, is placed next to the top and bottom of the product surface and aligned. The product is then exposed to ultraviolet light of a specified wavelength for a specific duration. The ultraviolet light passing through the clear areas of the film polymerizes and cures the resist. The top Mylar film is then removed and the product is moved to the next process station.

Develop, etch, and strip (DES). The first step is to remove the non-polymerized resist while leaving the polymerized resist attached to the copper beneath, as this constitutes the required circuitry. A dilute solution of sodium carbonate in water is maintained at a given concentration and temperature. The boards are passed through this solution and then through rinse water tanks. At this stage, the copper underneath the removed, uncured polymer is exposed, while the circuitry is still covered with the cured polymer.

The exposed copper is then etched away. This is accomplished by spraying the boards with a mixture of cupric chloride and hydrochloric acid solution. The following reactions occur between the etchant and the surface copper.

$$Cu + CuCl_2 = Cu_2Cl_2 \tag{3–1}$$

$$HCl = H^+ + Cl^- \tag{3–2}$$

$$CuCl_2 + Cl^- = CuCl_3^- \tag{3–3}$$

Alternatively, a solution of hydrogen peroxide and sulfuric acid can also be used as an etchant. The reaction is given by:

$$Cu + H_2O_2 + H_2SO_4 = CuSO_4 + 2H_2O \tag{3–4}$$

Time, temperature, and the concentrations of solutions are critical in this step. The possible defects that could arise during this step are either overetching or underetching of the copper. The boards are then rinsed and sent to the next station for stripping.

In order to expose the circuitry, the overlying cured photoresist needs to be stripped. The polymerized resist is chemically broken down into flakes, chunks, and pieces. This is accomplished with an aqueous solution of a strong alkali such as sodium hydroxide (NaOH). Thus, the boards are treated with a solution of NaOH at a predetermined temperature for a specific duration. The boards are thoroughly rinsed with water and dried. Inadequate DES processes can lead to a variety of defects as shown below:

1. Excess copper
2. Neck down
3. Dish down
4. Shorts
5. Opens
6. Blooming
7. Spacing violation
8. Too much copper in clearance rings, etc.

Chloriting. The subcomposite inner layer that is circuitized may have to be laminated subsequently along with other circuitized layers and layers of prepreg according to the product design. It is generally difficult to obtain good bonding between the circuitized layer and the adjacent prepreg layer unless the copper on the circuitized layer is specially prepared. One of the processes that prepares the boards for the second pass through lamination is called *chloriting*. It is also called *black* or *brown oxide treatment*. The boards are first cleaned with a hot solution of sodium hydroxide and then subjected to cascade rinses of water. The boards are then treated with a solution of sulfuric acid and sodium persulfate in water. The objective of this treatment is to microetch the copper surface. The boards are then rinsed with water and treated with a solution sodium chlorite and sodium hydroxide at an optimized temperature for a specific duration. The boards are then rinsed and dried with cold and hot air in sequence. The copper surfaces are thus oxidized resulting in either a black or brown surface and hence the name black oxide treatment.

All the different layers of the product are stacked in the appropriate sequence and are laminated together as described earlier to give the final composite.

Drilling. Mechanical drilling of holes through the laminate subcomposite or composite is carried out for a variety of reasons. Some of these are to provide tooling holes, for connecting inner layers, for inserting the through hole or insertion mount components, for locating boards on assembly fixtures, for disconnecting commoning bars, and for other reasons.

The equipment consists of multispindle high-speed drilling machines mounted on heavy granite tables for stability. These machines are capable of drilling several thousand holes simultaneously. Boards are generally stacked and drilled. The drill capability depends on the material, stack height, and drill diameter. Stack thickness depends on the card structure and composition. The drill diameters vary from 0.006 to 0.025". The smaller the diameter of the drill, the thinner the stack that can be successfully drilled. Also, the glass and metal content, the type of the epoxy resin, as well as any thermal planes in the board construction influence the efficacy of drilling operation.

Drilling is a precision operation sensitive to a host of parameters and is carefully controlled. The hole quality depends on the material being drilled, the drill sharpness, the speed, and feed parameters. The boards, after drilling, are optically inspected for missing holes, extra holes, plugged holes, and incomplete drilling. Internal plane registration is generally checked using X-ray imaging techniques.

Desmearing. During the drilling process, the drill bit has to go through the composite material consisting of copper, epoxy, and glass cloth. During high-speed drilling, depending on the drill condition, speed, and feed parameters, considerable amount of heat is generated that can soften the epoxy and smear it along the hole walls. Also, as the drill traverses the inner copper plane, terminations can be dragged up and down along the hole walls. Hence, it is important to desmear the holes in order to clean the holes after drilling and texture the walls in preparation to plating.

The smeared epoxy needs to be softened and dissolved away. The boards are first treated with a mixture of a swelling agent such as butyl carbitol, a surfactant, and sodium hydroxide. They are then water-rinsed and treated with an alkaline solution of potassium permanganate to dissolve the epoxy. The boards are rinsed, treated with a neutralizer to remove the alkali, rinsed again, and steam-dried.

The boards are inspected for hole quality after the desmear operation. Inadequate desmear can lead to poor copper plating and electrical discontinuities at the inner planes and the hole wall.

Copper plating

Electroless copper plating. The next step is to plate the through-holes and the top and bottom surfaces with copper. It is also a very important and complex operation. Copper is plated not only on the top and bottom surfaces but also in the holes, where there is insulating glass with impregnated epoxy, and internal copper plane terminations. This operation establishes through-hole connection to the bottom and top layer as well as the inner plane connections as per the design.

A combination of electroless and electrolytic copper plating processes is utilized. Special seeding and activation steps are included to facilitate electroless copper plating on hard-to-plate glass and epoxy surfaces. The panels are first degreased to remove contaminants, oils, and greases. Then the boards go through a sodium persulfate solution, hydrogen peroxide, and dilute sulfuric acid etch, followed by a treatment of cupric chloride and hydrochloric acid etch and an additional HCl acid treatment to remove the residual chloride. At this point, the boards pass through a seeding step consisting of dipping in a dilute solution of stannous chloride in hydrochloric acid. The boards then pass through a solution of acidified palladium chloride as an activation step. The reactions in these two steps are as follows:

$$SnCl_2 = Sn^{2+} + 2Cl^- \tag{3–5}$$

$$PdCl^2 = Pd^{2+} + 2Cl^- \text{ and} \tag{3–6}$$

$$Sn^{2+} + Pd^{2+} = Sn^{4+} + Pd^0 \tag{3–7}$$

Thus, a thin layer of palladium is deposited on to the surface making the surface electrically conductive.

The boards then pass through an electroless copper plating bath of proprietary formulation. In the plating bath, the source of copper ions is a solution of copper sulfate, and the reducing agent is sometimes formaldehyde, and the pH (hydrogen ion concentration or acidity) of the bath is adjusted by the addition of sodium hydroxide. Complexing agents such as Rochelle salt, ethylenediamine tetraacetic acid (EDTA) are included in the formulation to prevent the precipitation of copper ions as copper hydroxide at high pH. The overall auto-catalytic reaction can be represented as

$$CuSO_4 + 2HCHO + 4NaOH = 2HCOONa + H_2 + 2H_2O + Na_2SO_4 \tag{3–8}$$

In addition, several proprietary additives are added as levelers, grain modifiers, stabilizers, brighteners, etc. These are usually organic compounds containing S, N, CN, etc. and differ from supplier to supplier and formulation to formulation.

After electroless copper plating, the boards go through the electrolytic copper plating step where copper of requisite thickness is plated on the surface and in the through-holes.

Electrolytic copper plating. In this process, the boards are made the cathode and pure copper balls held in a basket are made the anode. Again, there are several plating formulations in vogue in the industry. One common plating composition contains pyrophosphates owing to its long and proven performance. A typical pyrophosphate electroplating may utilize the following materials and processing conditions.

1. Materials
    a) Cu = 2.7–3.5 oz./gal
    b) Pyrophosphate = 19.4–26.3 oz./gal
    c) Orthophosphate = 8 oz./gal
    d) Ammonia = 0.2–0.3 oz./gal
    e) PY61 = 0.25 0 0.75 ml/A.h (PY61 additive is dimercaptothiadiazole)
2. Process condtions:
    a) pH = 8.1–8.5
    b) Temperature = 111–125°F
    c) Cathode current density = 20–35 A/ft^2

The overall plating process is a reduction of Cu^{2+} ions to metallic copper at the cathode as shown in the following reaction:

$$Cu^{2+} + 2e = Cu\ (s) \tag{3–9}$$

Pulse plating

Sometimes, in plating thick PWBs with high-aspect-ratio through-holes and fine line features, some amount of mushrooming occurs in conventional direct current plating. This limits the closeness of the features due to concerns of shorting and spacing violation. This is overcome using pulse plating with higher than conventional current densities.

In conventional plating, the anodes are covered with an unknown layer that is only slightly soluble in the acid medium. This unknown layer blocks the current, thus passivating or polarizing the anode. By reversing the polarity of the current, the PWB is made the anode for a very brief interval. During the forward pulse, copper is deposited on the entire PWB surface. In the case of through-holes, more copper is plated at the entrance and exit of the hole while the hole interiors have relatively thinner plating. The plating assumes a dog-bone shape. When the polarity is reversed, the PWB becomes the anode and the additive is attracted to the edges of the entry and exit areas of the holes, which also assume a dog-bone shape. When the pulse is reversed, the areas not shielded by the additive get plated with copper and the barrier is dissolved in the process during the first part of the forward plating. Copper deposition continues during the second part of the forward plating step. The last half of the forward pulse completes the dog-bone plating, thus eliminating the dog bone and providing uniform plating in the high-aspect-ratio hole. Figure 3–4 depicts the plating sequence in the forward and reverse pulses schematically.

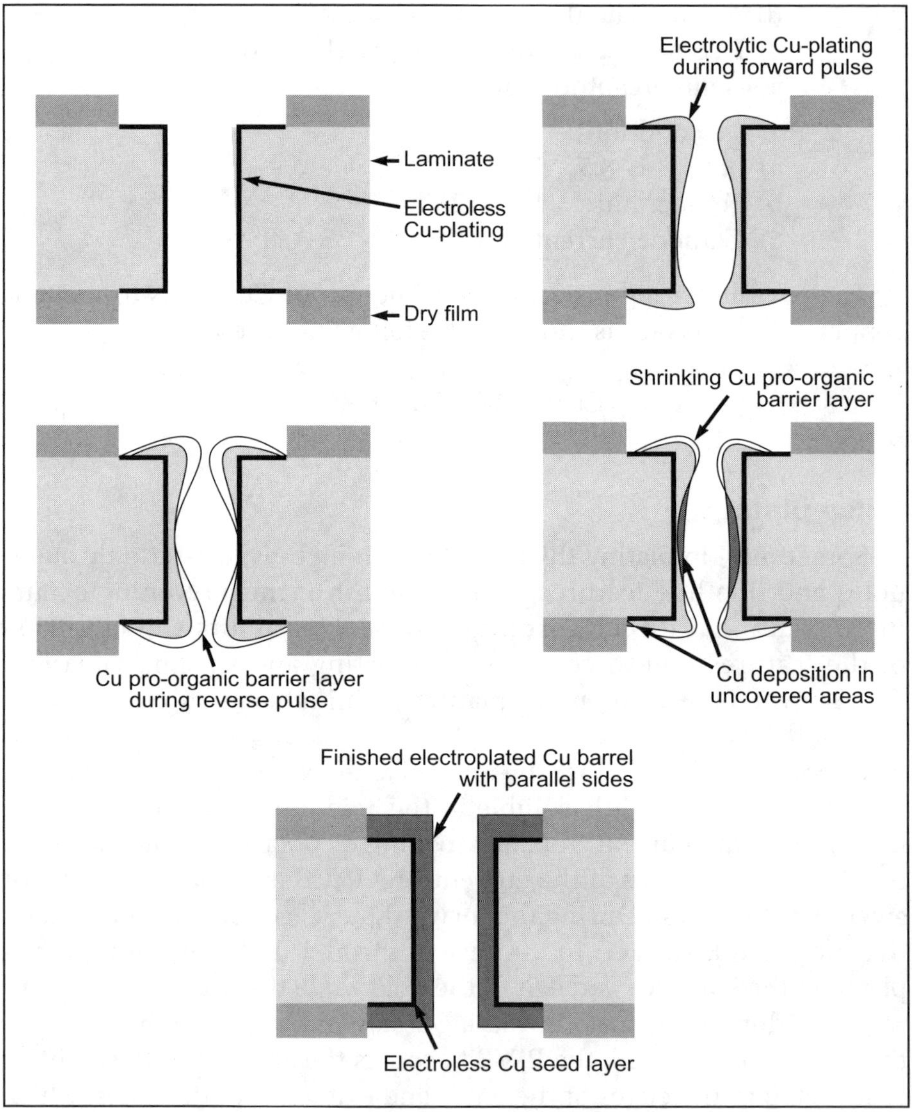

Fig. 3–4. Pulse plating process—forward and reverse plating steps

The plating process essentially eliminates mushrooming of the plating around sharp corners and provides well-defined circuit features, thereby increasing the capability for fine-pitch and high-aspect-ratio copper plating.

After the copper plating, the boards are inspected and checked for any defects. The panels are then sent for final circuitization on both sides. The process is identical to the one described earlier.

A situation that is sometimes encountered is the insufficient adhesion between adjacent copper layers. These may be electroless and electrolytic copper, or foil copper and electroless copper. If the failed interface contains palladium, it implies that the peel has occurred between the electroless copper and the underlying layer. Preplate cleaning steps are essential to avoid copper–copper debonding. The surfaces must be free from minor organic contamination, loose oxides, epoxy desmear residues, developer antifoam, or any other contaminant.

Poor electroless adhesion is sometimes associated with the conditioner chemistry. The possibility that the conditioner chemistry is such as to leave a thin barrier film between the copper surfaces cannot be ruled out (Dietz 2006).

Solder mask

The panels are then treated with a protective coat, alternatively called *solder mask*, to protect the bare copper in areas other than the ones needed for component attachment and test points, grounding areas, etc.

The process consists of a pre-cleaning step with a sodium persulfate solution and application of the epoxy coating by any one of the various techniques such as curtain-coat using a liquid epoxy. A solvent evaporation step is used to expel the solvent, followed by a flashing step. The panels are then flipped over and the liquid epoxy is applied and treated in a similar manner. A diazo film with the requisite pattern is applied, exposed, developed, and cured. The panels are then subjected to final testing.

Solderability preservation/protection

The panels at this stage have exposed copper areas needed for the component assembly and testing. The surface copper is prone to oxidation and the panels have very limited shelf life. In order to preserve solderability of panels until actual second-level assembly, the PWB surface is treated with protective coatings. A number of protective coatings are in use in the industry and these are described later in this chapter.

High-density PWB technologies

There are several HDI technologies that have been developed in recent years. Some of them are briefly described below.

Surface laminar circuitry (SLC)

One of the earliest technologies using a liquid photoimageable dielectric (PID) is by IBM Yasu, Japan (Tsukada, Tsuchida, and Moshimoto 1992; Carpenter and Memis 1996). Subsequently, the liquid dielectric material was replaced by dry film dielectric. The starting point for this technology is a conventional PWB subcomposite, where the requisite inner circuitry layers are laminated. At this stage, one option is to drill through-hole vias, plate with copper, and fill the through-holes with suitable materials. Then the top subcomposite copper layer is masked, exposed, developed, etched, and stripped to generate the surface circuitry on the subcomposite. A PID is then applied in a liquid or dry film format to the bottom and top surfaces. The liquid PID is applied by a curtain coating process, and the dry film is applied by a vacuum lamination process. In both cases, the material is a non-reinforced polymer. The PID is then imaged to form the microvias and cured. The board is then mechanically drilled for insertion-mount component mounting, mounting holes, and inner plane connections, and is inspected. The cured PID is then surface-treated to provide the required surface roughness for good copper adhesion. This is usually a mechanical abrasion process. The surface and the photo-vias are then plated with copper. Photoresist is then applied to the surfaces, imaged for the requisite circuitry, developed, etched, and stripped in the conventional manner. At this stage, there is one microvia circuitized layer on top and one at the bottom. This configuration is termed a 1+1 structure. If additional surface circuit layers are needed, the process is repeated with the application of additional PID layers. After the last PID layer is processed, the top layer is coated with the final solder mask layer, and an organic solderability preservative such as a benzotriazole or polyalkyl benzimidazole is applied. The panels are then singulated into individual product cards, tested, and shipped.

Figure 3–5 shows a schematic block diagram of the SLC process and figure 3–6 illustrates the 1+1 SLC process flow. As liquid PID utilizes a curtain/slot coater and a leveling tool to planarize the surface, an after-cure is required. The photo-vias made with liquid PID will have tapered walls (bath tub) and hence provide good plating uniformity. Photo-vias obtained with dry film will have vertical walls and hence allow vertical walls and also smaller via openings. However, dry film PID process produces good planarization, but requires a vacuum lamination tool.

PID permits all vias to be formed simultaneously and there is no associated incremental per-via cost and is considered well-suited for extremely high density via applications

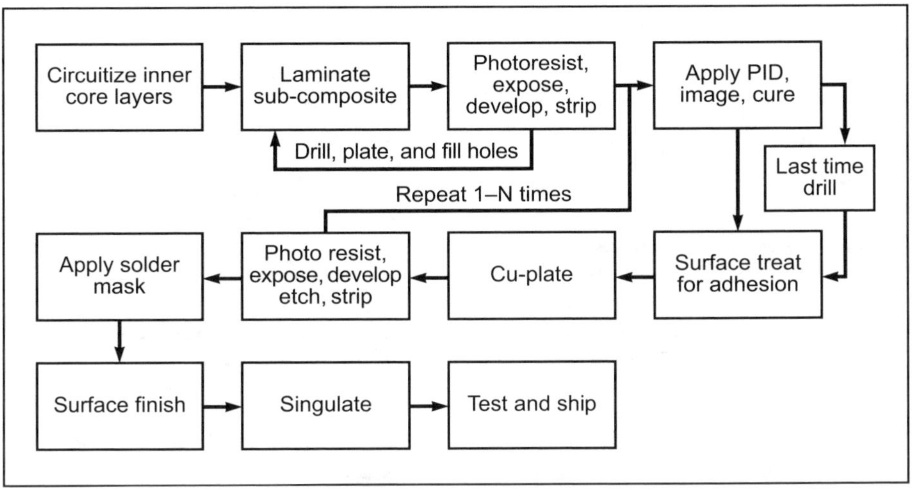

Fig. 3–5. Schematic process flow of surface laminar circuitry technology

Laser micro-via technology

Another most widely practiced HDI technology utilizes laser ablation to produce the micro-vias. Laser micro-via generation is a photochemical and/or photothermal processes. In the photochemical process, the organic polymer molecules undergo photodecomposition due to high-energy (2 eV) irradiation of <400 nm wavelength. In the photothermal process, the thermal energy due to irradiation with infrared energy in the range of 500–10,600 nm melts, decomposes, and vaporizes the organic polymer. The dielectric materials that are amenable to the laser ablation are resin-coated copper (RCC), FR-4, aramid-impregnated epoxy (Thermount), Polyclad, Speedboard, etc.

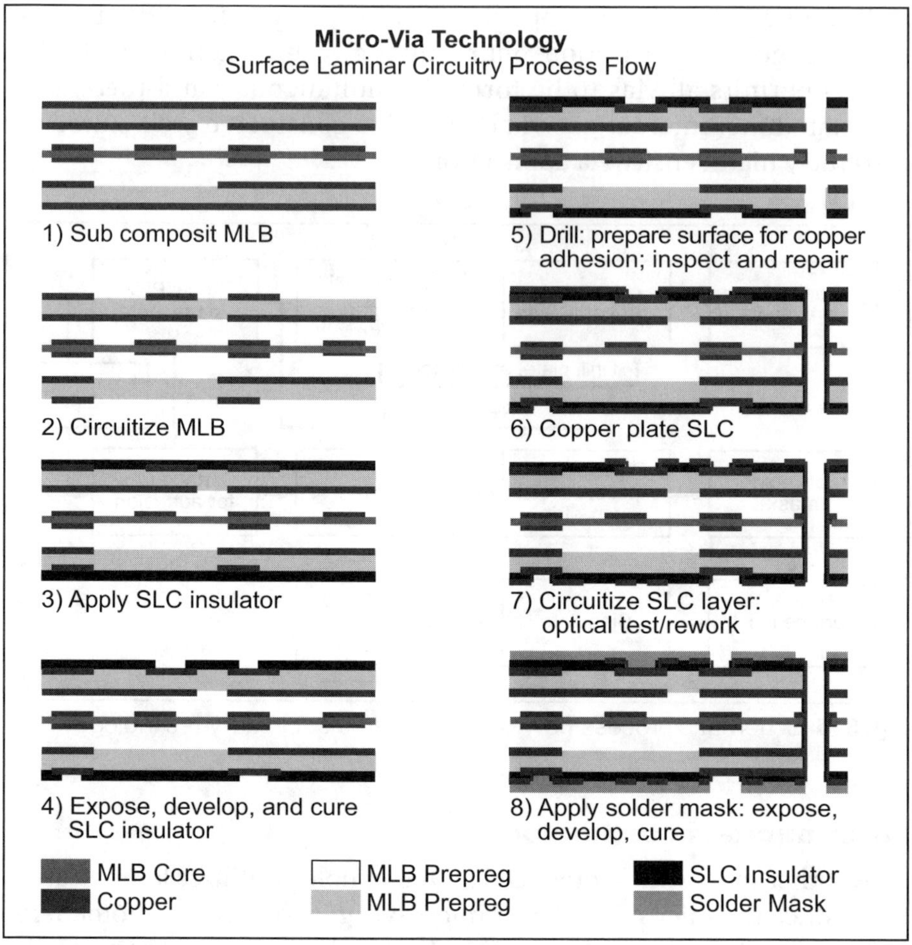

Fig. 3–6. Process flow of 1+1 surface laminar circuitry

Typical lasers used in the laser vias processes include the Nd:YAG, CO_2, and excimer lasers. The high power lasers, namely UV lasers, are capable of ablating reinforced laminates as well as copper, whereas the CO_2 laser is better suited for non-reinforced and aramid-reinforced laminates. The Nd:YAG lasers operate at 256, 355, 532, and 1064 nm, while the CO_2 lasers operate at 10,600 nm. Transversely excited atmospheric CO_2 lasers operate at 1–150 Hz repetition rates and 9–11 μm wavelength with a beam size 0.5 mm × 0.5 mm and can produce 200–600 vias/s. Almost 90% of the energy is absorbed by the PWB materials. Copper penetration is difficult without charring the organics. Pulsed CO_2 lasers are used at 1–400 Hz at 10.6 μm wavelength with 0.1 mm beam size and can produce 200–1000 vias/s. UV YAG excimer lasers operate at 193, 248, 308, and 351 nm at the pulse rate range of 1–200 Hz.

HDI boards are lower in cost due to the reduced number of layers and reduced size. They can also be brought to market faster due to design efficiencies, improved component escapes, and vias-in-pad designs. They have better performance through higher wiring densities, smaller geometries, as well as thinner, smaller, and lighter boards. In addition, high-density laminates have lower electrical parasites, lower distributed capacitance, and closer ground planes.

Until recently, implementation of HDIs in high-volume manufacturing had to overcome several impediments related to infrastructure and technological readiness. High capital investments, manufacturing readiness for high-volume production, yield issues, and difficulties with regard to electrical testing had to be overcome. Also, most materials are relatively new and have not been well characterized for high-temperature and high frequency applications. Lack of standards and simulation tools also played a role. Reliability data was also slow in coming.

ALIVH technology

ALIVH is an acronym for any layer inner via hole HDI laminate technology (Boggio 2000). As the name implies, via holes can be incorporated into any desired layer to interconnect adjacent circuit planes. The technology was first developed by Matsushita, and several modifications to the original concepts have been developed recently to meet the product needs. The technology comes under the category of conductive ink or paste-filled micro-via technology. The prepreg is an aramid-filled high-temperature epoxy (Tg ~ 160–180°C). The starting point is the thin aramid-impregnated epoxy prepreg sheet. Microvias are formed by laser drilling as per the product design, and the holes are then filled with a conductive paste or ink containing silver particles and cured. Other filler materials include copper, solder, Cu/Pb, Sn-coated Cu, etc. The particle loading is such as to provide good electrical conductivity as well as processability, and is generally in the 40–50% by volume. The prepreg is then laminated with copper foil on either side, and is then circuitized using standard develop techstrip process to generate the desired circuitry. The layers are tested for defects and electrical integrity. Other layers are similarly fabricated and tested. The tested individual layers are then laminated together. If interconnection of the top and bottom layers is desired, the composite is drilled, Cu-plated, and circuitized. Solder mask coating and solderbility preservative application follow to obtain the final product. Figure 3–7 shows a process flow schematic of the ALIVH fabrication process, and figure 3–8 depicts the sequential layup of ALIVH.

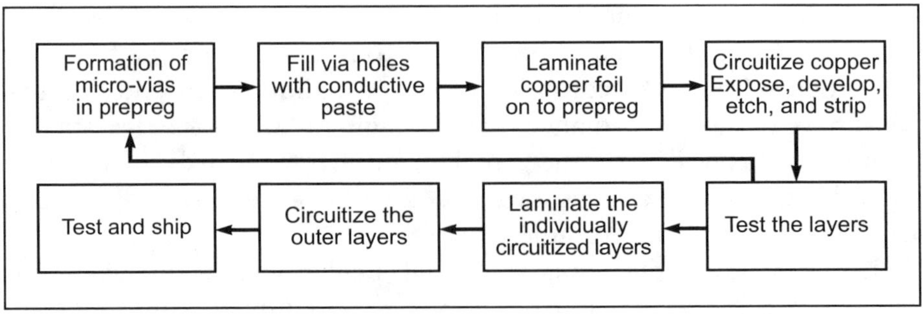

Fig. 3–7. Schematic of a typical any layer inner via hole (ALIVH) PWB process schematic

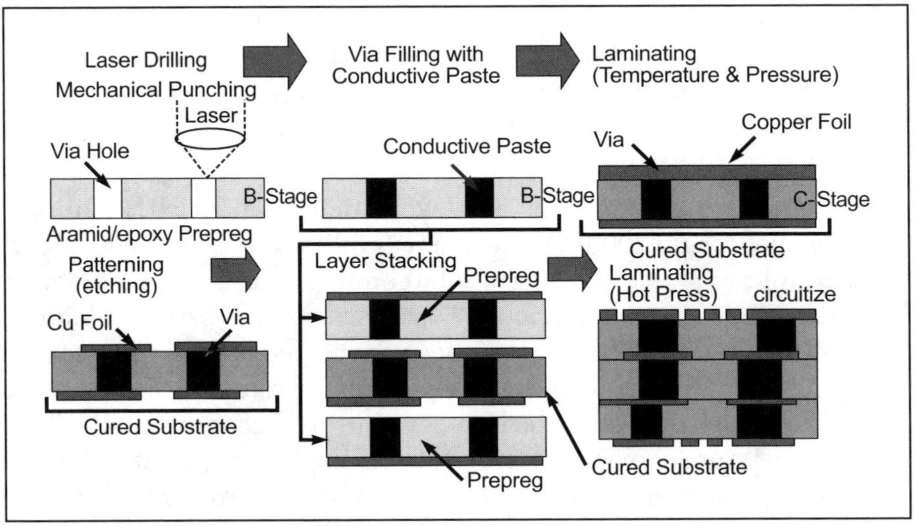

Fig. 3–8. Example of fabrication of ALIVH with buried and blind vias

The choice of the conductive paste is critical. The material should be screen-printable, have low shrinkage on curing, have good adhesion to the laminate, contain low volatiles, and have low viscosity, high strength, and high electrical conductivity.

Thus ALIVH technology offers considerable flexibility for fabricating blind and buried vias compared to the conventional process (Ishamura et al. 1999). Pad densities in excess of 100 pads/cm^2 are possible. The process is, to a greater degree, a drier and more mechanical process. The technology is well suited for radio frequency (RF) applications and has been shown to have high reliability. Two other versions, namely ALIVH-B and ALIVH-FB, were developed to accommodate the needs of high-density packaging of original equipment manufacturers (OEMs).

Flexible Laminate Circuits

Electronic circuits on thin laminates amenable to bending around sharp corners and kinematics are called *flexible circuits*. An example of such an application is the head-arm assembly in storage devices such as hard disc drives. With the proliferation of hand-held portable information processing and communication devices, such as digital cameras, camcorders, personal digital assistants, and camera cell phones, the demand for flexible PWBs has increased significantly in recent years. Three-dimensional flex circuits often provide volumetric efficiencies that are essential in portable electronics. Low dielectric constant, low coefficient of thermal expansion, excellent adhesion characteristics to conductor metallurgies, low moisture uptake, good thermal stability at the processing temperatures, low chemical resistance, high fracture toughness, and good mechanical properties are some of the desirable attributes for the polymer films. While many of the process steps for their fabrication are generically similar, the material sets and process details are different. The processes are significantly modified in terms of equipment and parameters to enable handling thin foils and films. In this section, a brief overview of design, materials, and process aspects of flexible PWB fabrication are provided.

Design

Two concepts that are important in flexible circuits design are (a) the three-dimensional nature and, (b) the dynamic operational environment.

Conventional rigid PWBs operate in a static mode. On the other hand, many flexible PWBs operate both in static and dynamic modes. In the static mode, the PWB is bent into the desired form factor and is held in place in the enclosure, thus forming a three-dimensional architecture. In the dynamic mode, the circuit card assemblies may undergo translational, rotational, flexural, and cyclic motions during the product use. Cyclic fatigue due to flexure is not uncommon. The traditional design rules applicable in the conventional PWB design have to be modified appropriately in the design of flexible circuits taking into account not only the unique properties of these materials but also the manufacturing aspects.

Flexible laminate materials are dimensionally unstable and undergo shrinkage during the various process conditions. It is important to take into account, while designing a flexible PWB, the likely changes in dimensions and the dimensional tolerances. Shrinkage can be asymmetric, i.e., the extent of dimensional change in the machine direction can be greater than in the cross-machine direction. Design for processing is important.

If packages are to be mounted on flex circuits, localized stiffening using compatible adhesive materials prior to drilling and punching may be necessary. Flexible members mounted or fastened to three-dimensional rigid structures need to be provided with appropriate strain relief. Traces that cross a bend should be designed so that they cross orthogonal to the bend. Rigid PWB product designs cannot, in general, be retrofitted for flexible products. It is preferable to start afresh in design. Design concepts need to embrace a three-dimensional product perspective. It is important to optimize the amount of copper on the laminate after circuitization, along the borders and unused areas, as it will enhance the stability of the material.

Special design rules are applicable in the case of flex circuits in terms of bend radii, copper thickness, copper patterning, pad shapes etc., and the reader is referred to the guidelines of IPC, indicated in appendix A.

Materials—laminates

There is a choice of materials that depends on the application, use conditions, and cost. These include:

1. Polyimides
2. Polyester
3. Aramid
4. Polyetherimide
5. Polyester–Epoxy blends
6. Teflon

Of the above, polyimide and polyester are perhaps the most prevalent laminate materials in the flex industry. Other materials are selected on the basis of specific requirements such as frequency, dielectric constant, moisture resistance, and temperature of use. An important aspect in regard to flex materials is the coefficient of thermal expansion (CTE) along the z-axis, which can be different from the CTE in the x and y directions. Stresses in the vertical axis can impact the through-hole integrity, and steps should be taken to minimize barrel cracking and delamination under mechanical and thermal stresses.

Polyimides. Polyimides are an interesting group of polymers that are extremely strong and astoundingly heat and chemical resistant. Their strength as well as heat and chemical resistance is such that they often replace glass and metals, such as steel, in many demanding applications.

Polyimide is a thermoplastic material. Among the important attributes of this material are its ability to withstand soldering temperatures and resist solvents without degradation. The material is available in different versions with a range of electrical and thermal properties. The material is available in a range of thickness from 0.0005" to 0.005". Some forms of polyimides are loaded with fillers such as alumina powder to enhance their thermal conductivity. In others, they are loaded with materials to modify the electrical conductivity in specialty applications. Table 3–3 depicts the typical properties of polyimides.

Table 3–3. Typical properties of polyimide films for electronic applications

Property	Values
Tensile strength	25,000 psi
Tensile modulus	4.3 x 10^6 psi
Coefficient of thermal expansion (CTE)	20 ppm/°C
Dielectric constant	3.4
Dissipation factor (1MHz)	0.03
Dielectric strength	600 v/mil
Volume resistivity	1 x 10^{16} ohms-cm
Water absorption (saturation)	1.2%
Melting temperature	388°C (crystalline)
Density	1.33–1.43 g/cc

Polyesters. Generically, polyesters are condensation products of alcohols and acids, in this case that of terephthalic acid and ethylene glycol. Devoid of any plasticizers, polyester films do not lose their pliability over time, and possess high tensile strength, good solvent resistance, and good dielectric strength. It has, however, a relatively low service temperature range of 70–150°C. Thinner films are likely to be sensitive to soldering temperature, especially to lead-free interconnection processes, but they can be successfully soldered. It has a CTE of 17 ppm/°C and moisture absorption of less than 0.8%. Its tensile modulus is 550,000 psi and a tensile strength of 25,000 psi. Polyester films have a dielectric constant of 3.0–3.7 with a strength of 7,500 v/mil (25°C/60 Hz). A dissipation factor of 0.005 and a volume resisitivity of 10^{16} Ω-cm are typical for his material.

Polyesters are relatively inexpensive in comparison to polyimides and find use in several large-volume, low-cost automotive and communication product applications.

Aramid. The term aramid is an abbreviated form of aromatic polyimide. Aramid was developed by Stephanie Kwolek in 1961. It is “a manufactured fiber in which the fiber-forming substance is a long-chain synthetic polyamide in which at least 85% of the amide linkages are attached directly to two aromatic rings." It is used in aerospace and military applications as a body armor fabric and also as an asbestos replacement.

An especially fire-proof variant of aramid is Nomex, which is a Dupont registered product. It can be considered as an aromatic "nylon". It is sold in both fiber and sheet forms and is used as a fabric wherever resistance from heat and flame is required. It is actually a calendered paper and made in a similar fashion. Nomex type 410 paper is the original grade type made, mostly for electrical insulation purposes. In sheet form, it is supplied as a nonwoven fabric in varying thicknesses in the range 2–30 mils.

It has a negative coefficient of thermal expansion (–5 ppm/°C), and a 5-mil thick sheet has a dielectric constant of 2.4 with dissipation factor of 0.005 and a dielectric strength 730 V/mil. The material has a tensile strength of 83 and 39 lb/in. in the machine and cross direction, respectively. It softens above 300°C.

Nomex is a high-temperature product that withstands second-level circuit card assembly processes. An undesirable property of aramid cloth is its propensity for water absorption, and it is generally suggested that the product be properly baked after each wet processing step in PWB fabrication. Moisture absorption during the final product life can still be a concern in terms of propensity for corrosion, electromigration, delamination, etc. Recent developments indicate significant improvements in lowering the moisture absorption potential of aramid.

Aramid-impregnated PWBs are preferred in the fabrication of communication products owing to their better performance in the RF range.

Adhesives

The common adhesive systems are epoxies, acrylics, and polyesters. Epoxy adhesives are generally stable not only at processing temperatures but also in high-temperature operating environment. Phenolic-butyrals and nitrile phenolics are some of the modified epoxies used in the industry. Polyester adhesives are especially suited for use with polyester flex laminates. They have a lower heat resistance and are generally used where extensive soldering processes are not encountered. Acrylic adhesives resist short high-temperature exposures such as soldering operations. Table 3–4 shows the comparative physical properties of the adhesive materials.

Most of the materials have surface resistance in the range of 100 MΩ. However, PTFE adhesives have even higher surface resistance of 1000 MΩ.

Table 3–4. Comparison of select properties of some common adhesive materials

Adhesive Type	Peel Strength After Heat exposure (N/mm)	Mositure Absorption (%)	Dielectric Constant Max at 1 MHZ	Adhesive Flow mil/mil x 10^{-2}
Polyimide	1.0	3.0	4.0	1
Polyester	....	2.0	4.0	2
Acrylic	1.6	6.0	3.5	2
Epoxy	1.4	4.0	4.0	6
PTFE	>1	0.01	2.2	0.07
butyral-phenolic	1.0	2.0	3.0	2.5

Metal foil

Two types of copper foils, namely, electrolytic copper and cold-rolled copper, are in vogue in flex circuit fabrication. The electrodeposited copper is fabricated by a drum plating process where a rotating stainless steel drum constitutes the cathode. The copper coated on the drum is separated in a coil form, where the surface facing the drum is smooth while the outer surface has a rougher surface. The rougher surface promotes adhesion of copper to the flexible organic laminate. The grain structure of the copper is vertical owing to the atom-by-atom electrodeposition. It should, however, be noted that electrodeposited copper is less ductile than rolled copper.

Rolled-annealed copper is made from copper ingots. The ingots themselves are made by fusing and chill-casting the metal. These ingots are hot-rolled to an intermediate thickness and milled to remove surface defects. The sheets are then cold-rolled to the required thickness and annealed, before the final roll-milling. Rolled copper is more ductile than electrodeposited copper. The grain orientation is horizontal and is more tolerant to flexure. On the other hand, the horizontal grain structure and smoother surface can have poor bondability, and special surface treatments are employed to enhance bondability.

Process

Flex circuit PWB fabrication, owing to the design, materials, and construction, requires special considerations in the fabrication process.

Fabricating a mockup is an important flex circuit implementation methodology to ensure a greater degree of success in flex designs. These can be done by computer aided design (CAD) tools. The designs are biased towards copper as they would provide better dimensional stability. However, this may be possible only with single-sided flexes. Nesting circuits in a panel affords better panelization efficiency.

Flex circuit design considerations can be quite different from those of the rigid PWBs. Flex circuits can be either static or dynamic in nature. As indicated earlier, the magnetic head-arm assembly in a disk drive and the print-head assembly in a printer are examples of dynamic flex circuits. The flex is subject to thousands of cyclic bend operations. During the bend operation, the inside layer is subjected to compression while the outer layer to tension. One has to take into consideration the fatigue life requirements in the design of the flex assembly. Use of staggered length designs is recommended where bending and flexing is involved. Generally, 1.5 times the thickness of flex is added to each successive layer. Also, circuit lines are routed perpendicular to the fold or bend. Acute or right angles are avoided in a bend. Trace entry to the pads in a flex needs to be tear-drop-shaped to avoid high corner stress. As the flex is routed, it is important to avoid sharp right angles and provide rounded corners. On double-sided flex, it is important to stagger traces to avoid the I-beam effect. Cross-hatched ground planes are preferred to solid ground planes as they provide better flexure.

As flex circuits are thin and prone to tear, tear-resistance features such as providing metal corners, radiusing internal corners, use of radiused slots, laminating with glass fabrics in the corners, etc., are employed. All flex connectors are generally provided with strain-relief stiffeners.

With regard to second-level assembly, flex assemblies need to be paid special attention. These materials are moisture sensitive and absorb a few percent of moisture, and it is recommended that they be baked prior to assembly to avoid undesirable outgassing of occluded moisture at assembly temperatures. Product-specific board holders or plates are used for assembly. In case of wave soldering, special cut-outs are incorporated to expose the component leads to the solder wave. Surface-mount assemblies also require board holders. Also, vacuum fixtures are used to hold the flex circuits planar for solder paste screening and component placement.

As flex circuits are more likely to predominate in portable and mobile electronic systems, it is important to be cognizant of the implications of this technology in order to use them efficiently and cost effectively.

Surface Finishes

The primary objective of a PWB surface finish or component termination finish is protection and connectivity. In electronic packaging, the leads or terminations of the components are connected to the pads on the footprints of the PWB. The surfaces of the PWB, as well as the surfaces of the component terminations, have to be protected and their solderability preserved until they are assembled together.

After the PWB fabrication, the exposed copper areas such as component mounting pads, plated-through holes, fiducials, test points, etc., need to be protected from oxidative degradation until the boards are finally assembled. Oxidation of copper degrades its solderability and hence results in unacceptable interconnections and reduced solder joint reliability. In the case of components, lead metallurgies such as alloy-42, Kovar, Cu alloys, phosphor bronze, Be-Cu alloys, etc. are also prone to atmospheric oxidation. The termination metallurgies are coated with a suitable surface finish to preserve solderability.

Until the advent of lead-free solder technology, tin or tin–lead based alloys have been used as surface finish. The recommended lead-free interconnection alloys have a 35°C higher melting temperature than the eutectic Sn–Pb solder and electronic assemblies are processed at a higher temperature than in the past. Also, the proliferation of portable electronics and newer product designs with fine-pitch high-density packages imposed constraints and limitations on the choice of surface and termination finishes. In this section, the important surface finishes are discussed.

The integrity and reliability of interconnection between any two surfaces with an alloy depend not only the nature and properties of the alloy but also on the nature of the surfaces to be joined. Any interconnection between the alloy and the surfaces to be joined involves a metallurgical bond through the formation of an intermetallic layer at the interfaces. To accomplish this, the PWB and the component terminations are provided with appropriate surface finishes. Surface finishes can be divided into two groups: those that are used on the PWBs, and those that are used with components. It is preferable to have the same surface finish on the component terminations and the PWB surface. However, this is always not possible. Different surface finishes are currently in use in the industry. There are a number of considerations that need to be taken into account in the choice of a suitable surface finish. The following is a partial list of considerations for the choice of PWB finishes.

1. Nature of the PWB surface
2. Nature of the component leads and terminations
3. Shelf-life requirement of the PWB
4. Number of assembly reflows requirements
5. Surface planarity requirements for component assembly
6. Corrosion and chemical resistance
7. Contact resistance
8. Hardness and wear resistance
9. Cost of the surface finish application

For components or packages, some additional requirements such as the compatibility with lead metallurgies and wire bondability need to be considered.

These coatings fall into two categories, namely, organic and inorganic. Organic coatings are follows:

1. Benzotriazole
2. Imidazoles
3. Polyalkyl benzimidazoles, etc.

Inorganic coatings are follows:

1. Hot air solder leveling—vertical or horizontal (HASL)
2. Immersion tin
3. Immersion silver
4. Electroless nickel/immersion gold (ENIG)

Each of these coatings has its merits and demerits. Hot air solder leveling has been in use for a long time and has been the oldest solder preservative coating with long shelf life. However, the surface tends to be convex and is not well suited for fine-pitch surface-mount assembly. Organic solder preservatives, such as benzotriazole, provide extremely thin coatings with a very flat surface but have a rather shorter shelf life. ENIG has excellent solderability and gives a very planar surface. It is relatively more expensive and is sometimes afflicted by sporadic joint embrittlement due to a phenomenon called *black pad*. Immersion tin and immersion silver have also been used with success in recent years owing to their compatibility with lead-free interconnection alloys. Some of the more common surface finishes are given in table 3–5. Each of the above surface finishes is described and discussed in the following paragraphs.

Table 3–5. Some common printed wiring board finishes

Metallic Surface Finishes	Organic Surface Finishes
Vertical hot air solder leveling	Benzotriazoles
Horizontal hot air solder leveling	Imidazoles
Electroless nickel/immersion gold (ENIG)	Polyalkybenzimidazoles
Electrolytic nickel/gold	Carbon inks
Immersion silver	
Immersion tin	
Electroless Ni/Pd/immersion gold	

Hot air solder level (HASL)

HASL has been one of the oldest of surface finishes to preserve the solderability of PWBs and has served the industry for well over a half a century. The process of applying this finish generally consists of three basic steps. The PWB is first dipped into a warm bath of flux followed by dipping the panel into molten near-eutectic solder for a few seconds, and as the board is being withdrawn, the excess solder on the surface features and in the plated-through holes is swept off by two hot air knives positioned on either side of the board. The shape of the solder on each pad assumes a nearly convex surface. The interface of the solder and copper surface consists of a thin metallurgical bond made up of Cu_6Sn_5 intermetallics of approximately 40 μin. (Jackson 1988)

HASL finish has excellent shelf life in excess of one year and excellent wettability with very short wetting times.

In order to improve the uniformity of solder leveling, another method was developed known as horizontal hot air solder leveling. In this technique, the PWBs pass continuously through the solder melt at 250°C, entering on one side, passing between pinch rollers, and exiting on the opposite side. The hot air knives set at 235–245°C sweep the excess solder off. The copper surfaces attain a temperature of 240°C while the dielectric remains below 180°C. Such an arrangement ensures that each section of the board is treated more uniformly, with smaller solder thickness variations (Cardenes 1983, Goodell 1988). The average solder thickness was claimed to be 0.003" with a standard deviation of 5×10^{-5}

in. Both vertical and horizontal HASL have a thermal impact on the board and can affect the flatness. While HASL may have better thickness control, the intermetallics can still play a significant role in the success of fine-pitch surface mount technology (SMT) assembly (Viswanadham, Evans, and O'Hara 1990a,b)

As the packaging density increased with higher I/O and finer pitch, the HASL surface posed several assembly challenges. The surface of solder leveled surface, as indicated earlier, is convex due to the surface tension. In addition, as the hot air sweeps across the board surface, solder thickness on any given pad is higher on the leading edge than on the trailing edge. On rectangular quad flat pack footprint pads, pads aligned in the direction of the air leveling will have a slightly different solder volume compared to solder on pads aligned in the perpendicular direction. If excessive solder is swept off, the pads are enriched with the Cu–Sn intermetallics. Thus for fine-pitch surface-mount assembly, two problems are encountered. First is the nonplanarity of the surface and the associated difficulty of placing a component on a curved surface. The second is the difficulty to solder to an intermetallic surface if much of the solder is depleted (fig. 3–9).

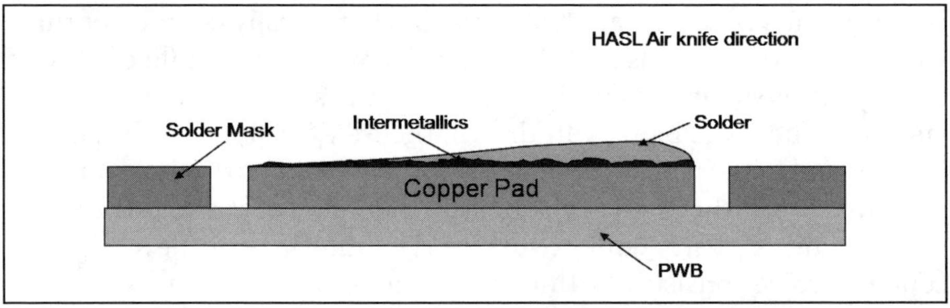

Fig. 3–9. Solder depletion on passive component pad due to HASL

Organic solderability preservatives (OSPs)

Much of the chemistry related to the OSPs remains proprietary. The exact chemical composition, the chemical reactions of the preservatives with the substrate metallurgy, and the flux and activators remain a trade secret. As a class of compounds, these are heterocyclic organic compounds such as alkyl or aryl azoles, imidazoles, etc. (Cotton 1963). Benzotriazole (BTA) is one of the first compounds to be used a surface solder preservative (Cotton and Scholes 1967). Later developments included polyalkyl benzimidazoles and arylphenyl imidazoles. BTA coatings can be as thin as 50–100 A, whereas alkyl benzimidazoles are much thicker. Also, conventional long-chain alkylimidazoles and substituted benzimidazoles can leave thin film residues on surfaces such as Au, Sn, Sn/Pb etc. and can impact subsequent processes, wire bonding for example. A compound that selectively reacts with Cu is preferable.

Arylphenyl imidazoles represent a later development in the pursuit of OSPs. These films are about 1,000 A thick. The decomposition temperature is around 350°C and is well adapted to lead-free soldering. It has been shown to have superior properties compared to traditional OSP finishes (Carano and Saeki 2005).

These materials are designed to have at least 6 months to 1 year shelf life and also to withstand at least two to three reflow cycles in the assembly process. During the soak stage of the reflow process, the preservative thermally degrades, exposing pure copper surface to the active solder paste flux. The metallurgical bond between the solder and the copper is made up of copper–tin intermetallics.

OSPs films are extremely thin and provide the flattest surfaces for fine-pitch SMT devices and package assembly. The OSP application process also does not involve any thermal shock to boards. Short shelf life, thermal stability for multiple reflows, abrasion resistance, etc. can be the limitations.

Immersion silver

Immersion silver coating appears to be a viable surface finish and is becoming increasingly popular. Silver is one of the constituents of the popular lead-free alloy, namely, the tin–copper–silver and hence it is compatible with the interconnection alloy. Also, it provides a planar surface suitable for fine-pitch leaded and area array package assembly (Brunner, et al. 1998).

However, several failure modes such as due to galvanic attack, atmospheric corrosion (tarnish), ionic contamination, less than optimum solderability, and microvoids have been reported in the literature although not with the same severity or consistency. Of these, galvanic attack (32%) , microvoids (27%), tarnishing (16%), and exposed copper (16%) seem to account for most of the failures.

Immersion silver plating, as with most other immersion plating processes, is a self-limiting displacement process where the silver ions are reduced to metallic silver on reaction with copper. The reaction slows once the surface is covered. Plating solution formulations contain proprietary organic additives that render the silver surface corrosion resistant under normal storage conditions (Reed 1998).

The thickness of immersion silver coating is generally in the range of 8–12 μin. as measured on a 60 × 60 mil pad, and a minimum of 5 μin is recommended (IPC 4553).

Entrapment of a plating solution in a crevice caused by undercut, overdevelopment, poor Cu–solder mask adhesion, etc. with a limited supply of Ag^+ ions etches copper in the crevice [$Cu(s) = Cu^{2+} + 2e$] to provide electrons for the reduction of silver on to copper outside the crevice by the reaction $Ag^+ + e = Ag(s)$. The less noble copper dissolves at the expense of the nobler Ag.

Microvoids by definition are smaller than 25 μm in diameter and are sometimes observed at the pad–silver interface. The microvoids that are associated with immersion Ag are called *champagne voids*. Figure 3–10 shows the type of microvoids observed with immersion silver. These voids can have a significant impact on the interconnection reliability both under thermal and mechanical loading. Although not all thick platings consistently exhibit voids, they are reported to be observed when the silver thickness is greater than 15 μin. It is also hypothesized that the voids are the result of decomposition of organic additives during the soldering process. The root cause for the formation of champagne voids is still unclear. Process control in terms of organic additive concentration, the topography of the copper surface due to microetch processes, controlled silver deposition, pH, drag-out rinses all seem to play a role in minimizing the potential for the microvoids. Keeping the process well in control generally precludes the formation of champagne voids (Cullen 2006). A tightly packed immersion silver plating of 6–12 μin. is considered to be resistant to tarnishing and also very conductive.

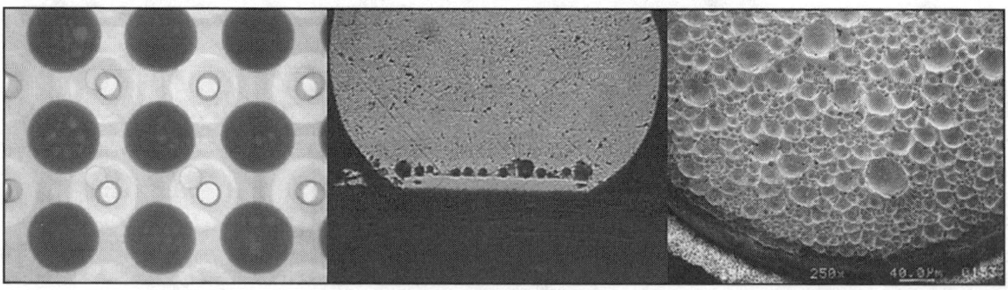

Fig. 3–10. Examples of champaign voids on immersion silver surface finish joint (Courtesy of D.P. Cullen, MacDermid, Inc.)

Sulfur contamination arising out of sulfur-containing packaging materials and atmospheric hydrogen sulfide transform the Ag to yellow or black silver sulfide, rendering the surface non-solderable. Exposed copper can result from residual contamination such as solder mask on the copper surface inhibiting the silver deposition (Ormerod and Yau 2006).

Immersion tin

Immersion tin is sometimes considered as an ideal finish owing to its lubricity and thickness uniformity, especially for compliant pin and press-fit connector applications. A minimum thickness of 40 μin. is recommended for a six-month shelf life without significant formation of Cu–Sn intermetallic (Ormerod 2000).

Immersion tin coatings, as in the case of immersion Ag, may contain some organic inhibitors and the process is autocatalytic. A concern in the utilization of tin as a surface finish is the propensity for whisker formation, which can lead to intermittent electrical shorts between adjacent conductors. Another concern is the formation of Cu–Sn intermetallics prior to the interconnection process (Roberts et al. 2007). It has been shown to be an acceptable surface in many applications. The affinity between copper and tin can result in reactions between them at ambient temperatures and in intermetallic formations, and hence plating thickness can play a significant role.

Gold-based finishes

While ENIG has been one of the more prevalent of gold-based finishes, a variety of other gold-based finishes have been introduced in recent years. These include electroless nickel/electroless palladium immersion gold (ENEPIG) and direct immersion gold (DIG). Of these ENIG is the most popular and widely used finish in the PWB industry.

ENIG finish

Electroless nickel-immersion gold consists of plating electroless nickel on the copper features of the PWB and is followed by a plating of a very thin film of gold by an immersion process. Electroless nickel plating is autocatalytic and does not require an electric voltage and current to effect the deposition. Two formulations of electroless nickel exist. In one, the reducing agent is amino borane and in the other sodium hypophosphite ($NaHPO_2$). For most electronics applications, phosphorous-based electroless nickel is employed. The purpose of nickel is twofold. First, it prevents the oxidation of the underlying copper. Second, it provides a hard nickel surface that is resistant to abrasion. Also, a nickel layer serves as a diffusion barrier for the underlying copper. Owing to the nature of the plating formulations, the plating contains a certain amount of included phosphorous. It normally contains about 7% phosphorous. However, the phosphorous content can vary in the range of 6–12% depending on the formulation. The nickel plating is about 5 μm thick. The gold plating on nickel is by immersion plating. As the job is immersed into the gold plating bath, gold deposition starts and proceeds by a displacement reaction. Gold atoms replace the nickel atoms. Once all the surface nickel atoms are replaced by gold, the reaction essentially ceases. Thus, the plating is rather self-limiting as opposed to electroless plating where the plating continues with time, and is about 0.05 μm in thickness. The purpose of gold is to preserve the solderability of nickel until soldering. However, it should be recognized that the gold layer can be porous. During the second-level assembly, the gold rapidly dissolves in solder and the tin in the solder reacts with nickel forming a metallurgical bond consisting of the Ni–Sn intermetallic phase.

Occasional solderability problems with the ENIG surface have been reported in the literature in recent times. Solder interconnection failures on the circuit card assemblies are detected either after the assembly, during reliability testing, or on an assembly returned from the field, where the fractured surface shows areas that are black in appearance and hence the name "Black Pad." The defects cannot be repaired or reworked by standard surface-mount procedures. The problem is sporadic, and becomes obvious only after the board assembly and subsequent testing and/or in the use environment. Efforts to repair the interconnection with aggressive flux and heating only seem to aggravate the problem.

Figure 3–11(a) shows a defective solder joint that has separated at the base of the joint. An aerial view of the fractured surface, shown in figure 3–11(b), indicates only the soldered areas, which are white. The non-soldered areas appear black and gray. A magnified image indicates that the surface is characterized by crevices and cracks similar to mud cracks on a parched land and is depicted in figure 3–12. A vertical cross-section micrograph shows the crevices and channels in the nickel layer created by the hyper-corrosivity of the Au plating solution. In extreme cases, it is possible that the channels reach the copper surface. With the immersion Au layer on the top of the nickel layer, the tendency of copper to migrate to the gold layer cannot be ignored.

The nickel layer can contain 8–12% P, and any heating process tends to precipitate Ni_3P forming noncoherent boundaries in the nickel matrix. And extended heating promotes Ni_3P enrichment and consumes increasing amounts of nickel. The surface can become enriched in Ni_3P. In fact, Ni_3P and other oxides of nickel are black in color and are difficult to solder. In addition, immersion gold plating formulations are generally corrosive. These plating formulations induce galvanic corrosion. Other aspects that can contribute to the severity of the problem include: (1) poor process parameter controls

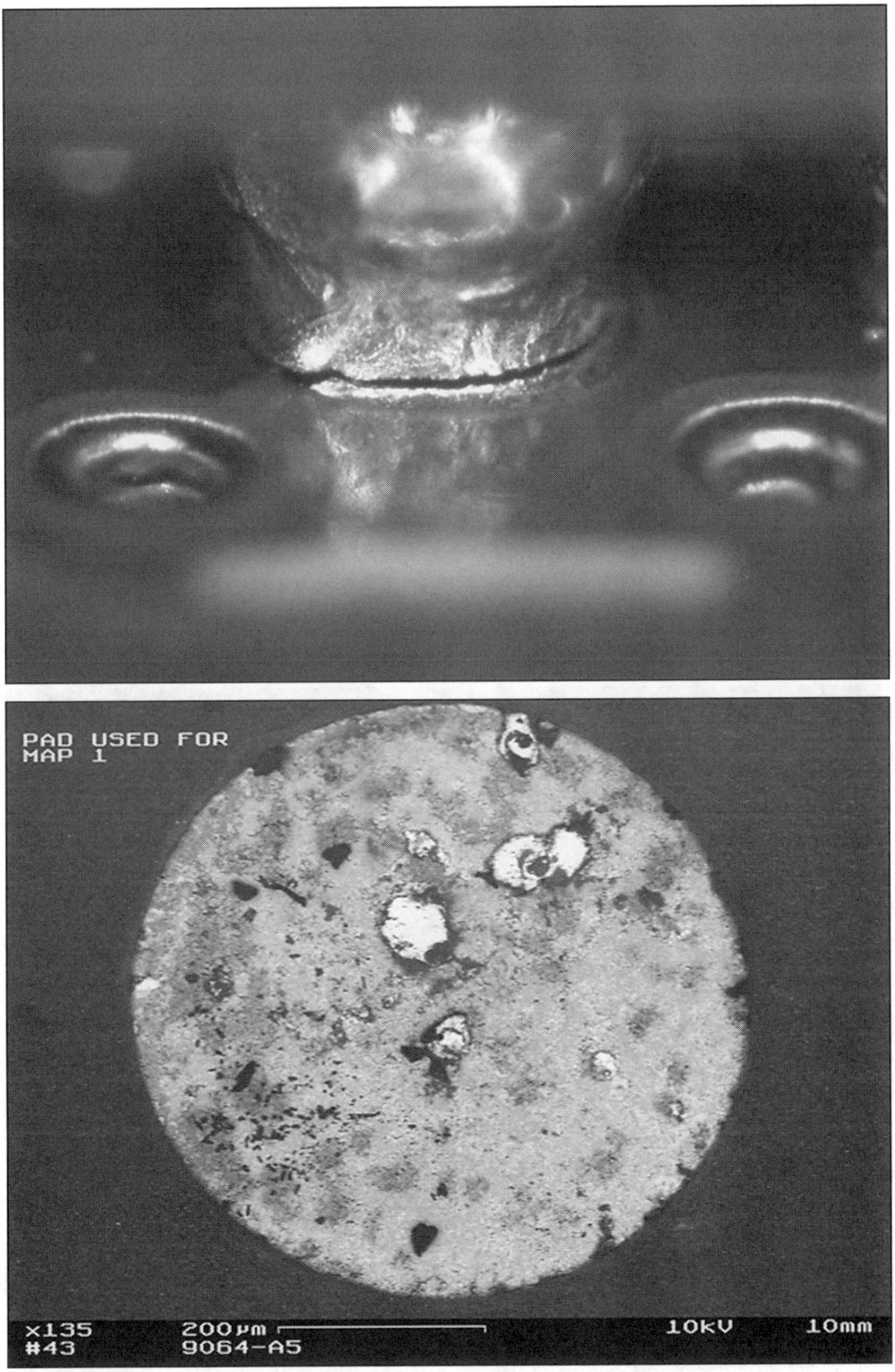

Fig. 3–11. (a) Defective BGA solder joint separated at the base. (b) Fractured surface of the joint indicating black pad features (Courtesy of Steve Dunford, Nokia, Inc.)

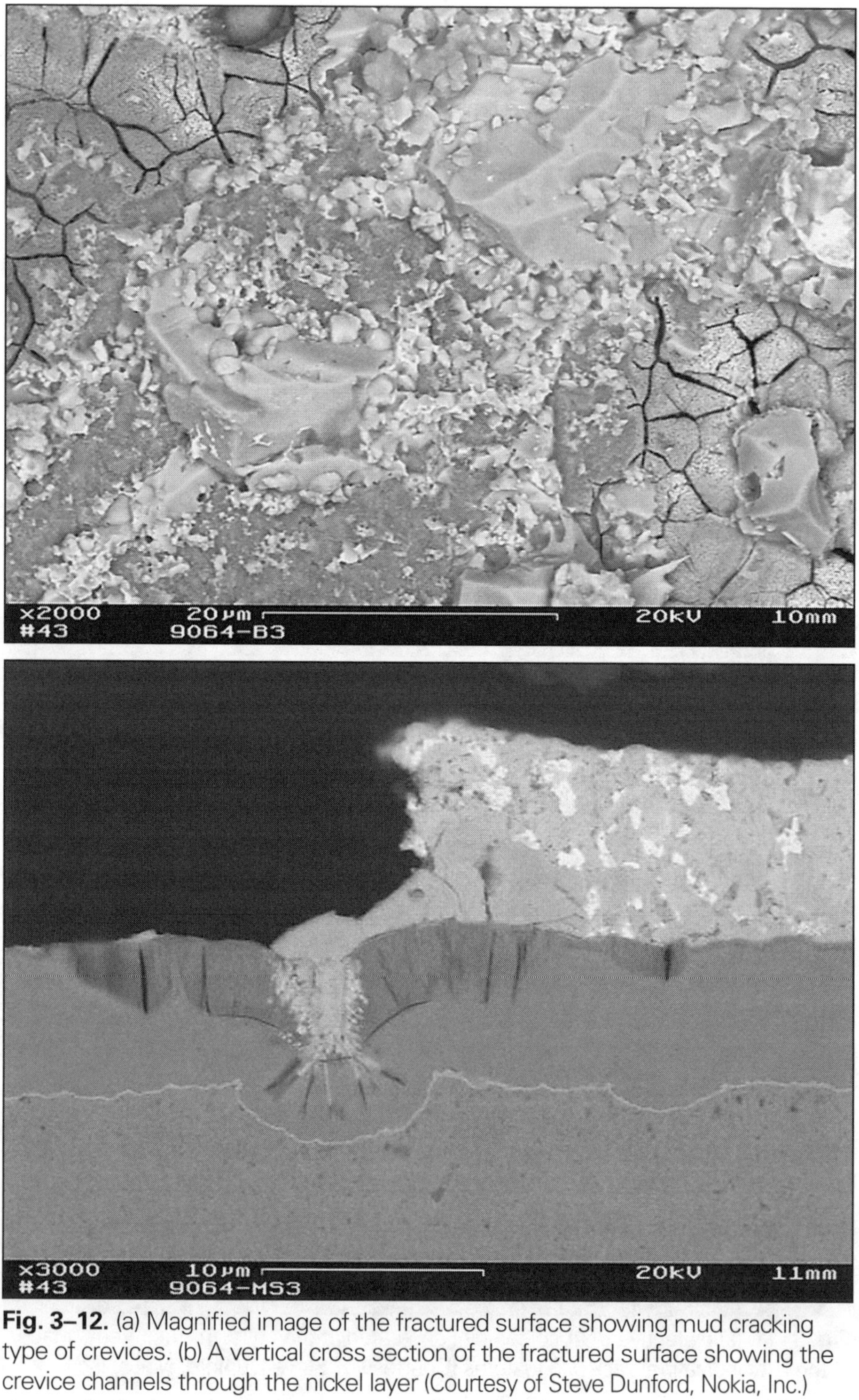

Fig. 3–12. (a) Magnified image of the fractured surface showing mud cracking type of crevices. (b) A vertical cross section of the fractured surface showing the crevice channels through the nickel layer (Courtesy of Steve Dunford, Nokia, Inc.)

such as pH, temperature, time, etc.; (2) inadequate rinsing that can leave residues of grain modifiers, brightners, and other additives on the plated surface; (3) heavy-element contamination; (4) accumulation of decomposition products of the bath when the plating baths are idling at the operating temperature; and (5) variations in bath loading.

In the past, the interconnection pad sizes were relatively large, as the industry was not practicing fine-pitch high-density package assembly. Thicker and less porous gold surface was acceptable. More aggressive organic solvents or water cleanable fluxes were in use. Pad to defect ratio was large and hence there was still enough good pad area to provide an acceptable interconnection. On the other hand, with the advent of fine-pitch surface-mount technology and the introduction of 0.5 and 0.4 mm pitch chip-scale packages, the feature sizes and pad sizes have become smaller and smaller. Use of thicker and less porous gold is a cause of concern in joint embrittlement. Migration to no-clean, less aggressive fluxes made the problem even more obvious. In addition, the pad to defect ratio is much smaller and there is not enough good pad area to give good interconnection. Cost competitiveness may lead to processing more boards through the line and less rigorous process control and monitoring etc. Since the first reporting of this solderability problem, industry has taken several steps to minimize and alleviate the problem through better process control, monitoring, and changes in plating formulations. Still, an occasional appearance of black pad defect cannot be ruled out. Other solutions include migration to electroplating of nickel, two-stage Au plating consisting of immersion and electroless Au, etc.

Typical thicknesses of the common PWB surface finishes are shown in table 3–6.

Table 3–6. Typical thicknesses of the various PWB finishes

Surface Finish	Typical Thickness μ/(μ-inches)
Electroless Ni/electroless Au	3–6 /(120–240) Ni and 0.25–1.3/(10–50) Au
Electroless Ni/immersion Au	2.5–5 /(100–200) and 0.05–0.23 (2–9) Au
Ni//Pd/Au	25 (100–200)Ni/0.2–0.6(8-24)Pd/0.02–0.05(1–2) Au
Immersion Ag	0.15–0.45 (6–8)
Immersion Tin	0.6–1.6 (25–60)
Immersion Pd	0.1–10 (4–400)
Hot air solder level	0.65–50/(25–2000)
Organic solder preservative	0.2–0.6 (8–24)

Embedded integrated module

This technology aims to eliminate a level of packaging and embeds the active devices inside a conventional FR-4 laminate utilizing the chip-on-board technique. The chip interconnection and the associated wiring on the layer are accomplished by an additive plating process. The fabrication process starts with an FR-4 subcomposite. A hole is drilled into the laminate to accommodate the chip. A polyimide tape is attached over the hole, and a glass mask is placed under the substrate. With a chip placement machine, the device is positioned aligning the active component to the contact pads on the glass mask. The glass mask is then removed and the chip is embedded with an epoxy molding compound. At this point, the active surface of the chip is at the same level as the substrate. The epoxy is cured and the adhesive tape is removed. The generation of desired circuit pattern, passive component circuitry, and interconnection are accomplished by surface activation, photoresist application, and imaging and generating the pattern, followed by electroless copper plating. Colloidal Sn–Pd solution is used to activate the surface to receive the electroless copper. The above process is repeated as many times as required for the product design, and the surface is finally laminated. The final step involves the circuitization of the surface layers and application of the solder mask. The surface layers still have the facility to mount additional components.

This concept was demonstrated with a board containing two high-density active devices and 58 passive components (Tuominen and Kivilahti 2000).

Emerging Trends

With ever-increasing demand for portable personal electronics of high functionality, the emerging trends are likely to be in the areas of further miniaturization and integration with greater emphasis on reducing manufacturing costs.

Embedding passive devices such as resistors, inductances, and capacitors in multilayer PWBs for mobile phone applications has already been demonstrated and is gaining acceptance. Embedding resistors and inductances has been shown to be far easier than embedding capacitors. Understanding the effects of electrostatic discharge and addressing and elimination of related risk are important issues.

The concept of the integrated module demonstrates the potential for integration and convergence of first- and second-level packaging. While implementation of this concept may not gain universal acceptance, niche areas will be developed for specific low-cost high-volume applications.

As silicon devices become thinner and thinner, reaching thickness limits of 25 μm and below, embedding active devices in flexible laminates such as polyimide and polyester laminates is a distinct possibility.

Roll-to-roll manufacturing of multilayer flexible laminates and circuitization techniques involving printing techniques such as ink-jet printing, screen printing, etc., are likely to emerge in the near future.

The proliferation of portable electronics continues at huge volumes with shorter and shorter product development cycles. And, newer products will be emerging at a more rapid pace replacing the earlier ones. Disposal, reclamation, and recycling will be important. Paradigm shifts in PWB technologies are imminent, leading to biodegradable laminates and electronics.

Going green is one of the concepts that will be of increasing relevance to electronics industry in terms of materials, and processing in all areas of packaging and PWB technology is no exception. It was inferred that, under certain combustion scenarios, PWB with halogenated fire-retarding materials can give off highly toxic dioxins and furans. "Halogen-free" in PWB industry generally means free of brominated fire retardants and polyvinyl chlorides. Alternative phosphorous-based, nitrogen-containing, aluminum- and magnesium-based hydrates, and even materials such as zinc and borate compounds, are being investigated for possible use (Bedner 2008). While no specific material is finally arrived at, investigations continue to assess the impact of these additives on the ultimate physicochemical properties of the laminate in various applications.

Suggested Reading

1. Coombs, C. F., Jr. 2001. *Printed Circuit Handbook*. 4th ed. New York: McGraw Hill Book Co.
2. Lau, J. H., and S. W. Ricky Lee. 2001. *Microvias ForLow Cost, High Density Interconnects*. New York: McGraw Hill.
3. Tummala, R. R., E. J. Rymaszewski, and A. G. Klopfenstein. 1997. *Microelectronic Packaging Handbook*. 2nd ed.. New York: Chapman and Hall.
4. Harper, C. A. 2001. *Electronic Packaging and Interconnection Handbook*. 2nd ed. New York: McGraw Hill.
5. Puttlitz, K. J., and P. A.Totta . 2001. *Area Array Interconnection Handbook*. Kluwer Academic Publishers.
6. Fjeldtad, J. 2006. *Flexible Circuit Technology*. 3rd ed. Seaside, OR: B. R. Publishing Co.

References

Bedner, D. 2008. New halogen-free materials: their time has finally arrived. *CircuiTree* 21 (9): 26–29.

Boggio, B. 2000. The any layer interstitial via hole process. *The Board Authority* 2 (1): 91–91.

Bora, Y., and P. Viswanadham. 1988. Comparative study of copper benzotriazole and hot air solder level surfaces using SMT hybrid assembly process with ECAT/PMC test vehicle, Paper presented at the proceeding of the SMT-ITL Conference: 99–104.

Carano, M., and K. Saeki. 2005. Next generation organic solderable preservatives (OSP) for lead-free soldering and mixed metal finish PWB and BGA substrates. *The Board Authority*: 48–52.

Cardenes, O. 1983. Hot air solder leveling and the product. Paper presented at theNEPCON, p. 76.

Carpenter, R., and I. Memis. 1966. SLC: An organic packaging solution for 2000. Paper presented at the proceeding of the NEPCON West Conference,

Cotton, J. B (1963). Control of surface reaction of copper with organic reagents. Paper presented at the Proceeding of the 2nd International Congress on Metallic Corrosion, New York: 190.

Cotton, J. B., and I. R. Scholles. 1967. Benzotriazole and related compounds as corrosion inhibitors for copper. *British Corrosion Journal* 2: 1.

Cullen, D. (2006). Eliminating micro-void risk—An optimized Imm silver process. Paper presented at the proceeding of the International Conference on Pb-free soldering, Toronto, Canada.

Dietz, K. 2006. Fine lines in high yield, copper-copper peelers. *CircuiTree* 19 (11): 26–27.

Goodell, S. 1988. What is new in solder leveling, IPC Paper # 875.

Ishimura, Y. et al. 1999. Advanced ALIVH substrate with fine design rules for high density packaging. Paper presented at the proceeding of the printed circuits world convention 8, Tokyo, Japan.

Jackson, M. W., Hayden, T. F., Bakos, P., and Viswanadham P. (1988). Solder application techniques for fine pitch SMT interconnections. Paper presented at the IBM SMT-ITL conference proceedings: 72–78.

Ormerod, D., Y. H. Yau, and J. Wynschenk, 2006. Elimination of immersion silver plating void defects. *CircuiTree*. 19(11): 10–15.

Reed, J. (1988). Immersion Ag as a replacement for solder finish. Paper presented at the IPC/SMTA electronics EXPO. Proceeding.

Roberts, H., S. Lamprecht, E. Bevan, J. Coates, and S. Prosser. 2007. Imm Sn as a cost effective and reliable surface finish for Pb-free automotive electronic applications Phase II investigations. Paper presented at the proceeding of the technical conference of SMTA-Int.

Tsukada, Y., Tsuchida, S., and Moshimoto. Y. 1992. Surface laminar circuitry packaging. Paper presented at the proceeding of the ECTC Conference: 22–27.

Tuominen, R. and Kivilkahti J. K. 2000. A novel IMB technology for integrating active and passive components, Paper presented at the proceeding of the 4th International Conference on adhesives and coating technology in electronic manufacturing, Helsinki University of Technology, Finland: 269–273.

Viswanadham, P., Evans, H. E., and O'Hara, J. P. 1990a. Comparison of solderable surfaces in second level electronic packaging. Paper presented at the proceedings of the International Society of Hybrid Microelectronics Symposium, Chicago.

Viswanadham, P., Evans, H. E., and. O'Hara, J. P. 1990b. Importance of solderable surfaces in second level electronic packaging. Paper presented at the proceedings of SMT-CON, Atlantic City, NJ: 149.

4

Component Technologies—First Level Packaging

Introduction—Anatomy of a package

A monolithic microelectronic circuitry comprised of multiple interconnected transistors, resistors, capacitors, inductances, and other functionalities fabricated in-situ on a single substrate such as silicon is called an *integrated circuit* (IC). Several ICs are fabricated in an array format on a substrate called a *wafer*. The shape of the wafer is circular and its size varies from 6.7 cm to 30 cm in diameter depending on the manufacturing facility. The individual ICs are separated from the wafer by dicing or singulation with a diamond saw and are variously called *chips, bare dices, devices, ICs,* etc.

These singulated devices are subsequently packaged to facilitate their assembly on to a printed wiring board (PWB). Chips when packaged into single chip modules or multichip modules are considered *first-level packaging*. Assembly of single chip or multichip modules on to a printed circuit card is called *second-level packaging*.

Packaging is an enabling technology and constitutes an essential bridge between the semiconductor device and the PWBs. Semiconductor devices, also loosely called *chips*, when packaged into single chip modules or multichip modules are considered first-level packaging and the packages are variously and interchangeably called *components, modules, chip carriers,* etc.

A package, essentially, is an enclosure that provides a platform for component mounting to a PWB protecting the components from moisture, contaminants, and mishandling. It also provides electrical function, a path for heat removal and thermal management, mechanical support, and protection from environmental and physical damage. Packaging prevents harmful radiation escaping or entering the enclosure. In addition, a package supports the system's organizational requirements, facilitates repair and rework operations, and enables electrical and functional testing.

Components can be categorized as passive or active components. Active components are those that can operate on an imposed electrical signal so as to change its basic characteristics such as amplification, switching, rectification, etc., Examples of this are transistors, diodes, etc. Passive components are resistors, capacitors, inductors, etc. These do not change their basic characteristics when an electrical signal is applied.

The materials and design aspects of a package depends on the device, the number of inputs and outputs, its function, the PWB or carrier on to which it is assembled and the application environment. Other important considerations are electrical, thermal, and mechanical properties, corrosion resistance, phase stability, and manufacturability. In principle, the silicon device is attached to a metallized substrate, and the active surface of the device and the interconnections to the carrier are protected by encasing it in an enclosure or encapsulated with a molding compound. Several interconnection schemes of the package to the PWB have evolved over the years.

Historically, the first semiconductor carriers were hermetically sealed ceramic packages for use in high-reliability military, government, and business applications. With the introduction of personal computers and proliferation of consumer electronics and ever-increasing demand for faster, cheaper, and better products, plastic encapsulated microcircuits (PEM) have evolved. These have advantages of cost, performance, high-volume manufacturing, and availability.

Two chip mounting configurations, namely, cavity up or cavity down, are practiced in component packaging. The active side of the IC device faces up in the cavity-up package, while in the cavity-down configuration the active side faces down as shown in figure 4–1(a) and (b). Each configuration has specific attributes and advantages as described elsewhere in this chapter.

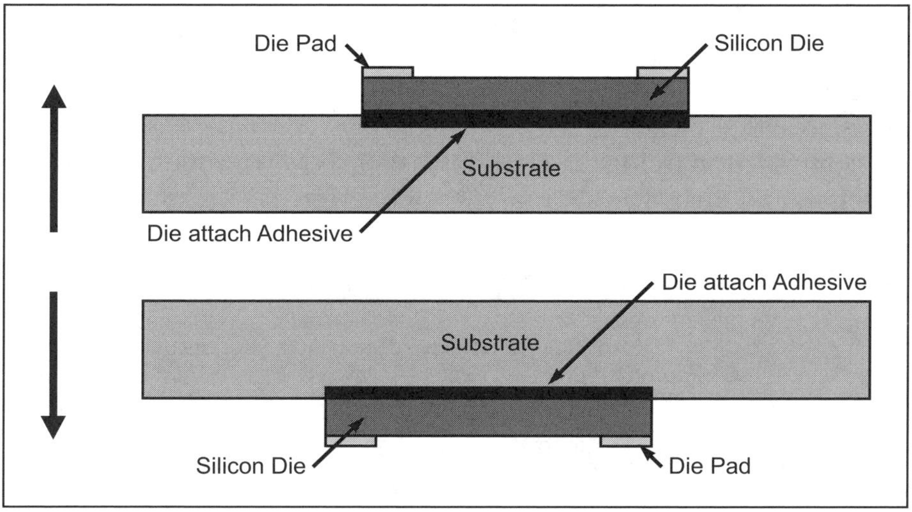

Fig. 4–1. Schematic of (a) cavity-up package, (b) cavity-down package

A generic fabrication process of a plastic packaging of an IC device consists of the following steps:

1. Lead frame fabrication
2. Device attach to the lead frame
3. First-level interconnect—device I/O to lead frame
4. Lead frame load into mold cavity
5. Molding
6. De-flash
7. Lead plating
8. Dam bar removal
9. Trim and form leads
10. Test

Some of these steps are described in some detail in the following paragraphs.

All the components utilize two common technologies for the internal chip-to- carrier/lead frame interconnection, namely, either wire-bonding or the controlled collapse chip connection, also known as C4. Subsequent to this, the assembly is either hermetically sealed with a metal or ceramic cap or encapsulated with an epoxy molding compound (EMC), before the lead forming or application of termination metallurgy. These processes are described below.

Lead Frame Fabrication

A variety of lead frame materials have been in use in the industry. These are either nickel- or copper-based alloys; Alloy 42 (42%Ni, 58%Fe) is a common lead frame material. Table 4–1 shows common lead frame materials and their typical properties of interest.

Table 4–1. Typical lead frame materials and their properties

Alloy	CTE/ppm/C	Melting point/C	Density g/cc	Thermal Conductivity W/m/C	Electrical Conductivity/ %IACS	Elastic Modulus/ GPa
Fe58-Ni42	4.3–4.7	1425	8.15	12–16	2.5–3.0	144.83
Cu0.1Zr	17.7	1000	8.94	359.8–380	90–95	120.37
Cu-0.1P Fe2.3, Zn0.12	17.4–16.3	1009	8.8	260	60–65	120.37
Cu,Ag.034P-.058, Mg0.11	17.7	1002	8.91	344–347	86	117.43

CTE = Coefficient of Thermal Expansion; IACS = International Annealed Copper Standard

The lead frame material and geometry depends on the package design. It comprises an area called *die paddle* or *flag* to mount the silicon device and several lead fingers. One end of each lead finger is used for the first-level interconnect with the device and the other end constitutes the lead for the second-level interconnect to the PWB. Several attributes and properties of lead frame material are important and include the following:

- Coefficient of thermal expansion (CTE)
- Electrical conductivity
- Thermal conductivity
- High strength
- Formability
- Adhesion to the molding compound
- Adhesion to the die attach material

It is not always possible to achieve a good match of CTE among all the material sets involved. When a lead frame material is chosen to match the CTE of the molding compound, it leads to a mismatch between the lead frame and the silicon device. The choice of materials is always a compromise and involves an optimization process.

The fabrication of lead frame for a given functional device involves a series of metal stamping with die-punch machines, or photochemical etching steps starting with reels or sheets. Photochemical etching is preferred for high I/O fine-pitch leaded packages. After stamping, the sheets are deburred to remove excess metal and nonuniformities. In the photochemical etching process, first a photoresist is applied to both sides of the lead frame metal. The desired lead pattern is imaged, exposed, developed, and stripped to generate the final pattern. These process steps are very similar to the circuit generation steps on a PWB, except that the chemicals used for developing, etching, and stripping can be different. The inner lead fingers are then plated with such metallurgies as electroless nickel–gold, palladium, or silver to facilitate the first-level interconnect. Figure 4–2 shows an illustration of a typical lead frame.

Fig. 4–2. Illustration of a typical lead frame before trim and forming operations

Device/die attach to lead frame

The device, the silicon die, is bonded to the die paddle of the lead frame with a variety of die-bonding materials. The choice of the material depends on, among others, the thermal management design aspects of the product. Considerations such as whether to use convective cooling or conductive path through the PWB are pertinent. The two common materials of choice are binary or ternary solder alloys, or filled conductive adhesives.

Alloys. Binary solder alloys such as Au–Sn, Au–Ge, Au–Si, etc. with high flow stresses offer excellent fatigue and creep resistance. It should, however, be recognized that lack of plastic flow can induce stresses in the silicon device. Also, soft solders such as those made up of Sn/Pb (5/95), Sn/Ag/Sb (65/25/10), and Sn/Sb (92/8) are also used. Ternary solders such as Pb/Ag/Sn are not uncommon. In solder attachment, the backside of the die is metalized and can consist of barrier layers of Cr, Ti, or V, followed by Ni and Ag. During the attachment process, silver dissolves into the solder and a nickel–tin (Ni_3Sn_4) metallurgical bond is formed. Thickness of the individual metal layers is to be maintained at the optimum values. Silver-filled specialty glass materials are also used for die attach in specific applications. These require high processing temperature, some times as high as 400°C, and their use is limited to high-performance, high-reliability applications.

Organic adhesives. Metal filled epoxies, polyimides, cyanate esters, etc., are extensively used in die bonding to the lead frames. The most common filler material is silver, which meets the necessary thermal and electrical conductivity requirements, although gold is occasionally used. The filler metal is usually in the form of flakes and the mixture can have 70–80% metal loading. The actual cure temperature and time routine depends on the material set used. The temperature is generally in the 150°C–180°C range and cure times are as short as a few minutes in the case of improved advanced materials. The materials are available as two-component or single-component formulations.

A number of attributes and properties are important in the choice of the adhesive and include the following:

- Shear strength
- Cure temperature
- Cure time
- Thermal stability
- Outgassing propensity
- Viscosity
- Ionic content
- Shelf life
- Glass transition temperature

Each of these properties, if not selected appropriately, can have an impact on the manufacturability, quality, and/or reliability of the product. Table 4–2 shows some of the properties and process attributes.

Table 4–2. Some typical die attach adhesive properties and application parameters

Property/Parameter	Typical Values
Viscosity	20, 000–30 000 cps
Specific gravity	~1.85
Percent solids	75–87
Glass transition temperature	60–100°C
Coefficient of thermal expansion	Below Tg 29–30 ppm Above Tg >125 ppm
Tensile modulus	6000–900 MPa
Shear strength	1500–1800 psi
Weight loss	< 3.00%
Thermal stability	> 300°C
Dielectric constant	4–5
Thermal conductivity	5–7 W/mK
On-screen life	8–12 hrs
Ionic contamination	< 1 ppm
Cure time	< 20 min
Cure temperature	150°C
Shelf life at -10 C	6–8 months

The typical range of die shear strength is 7–35 MPa and the thermal conductivity is in the range of 7–70 W/m °C. The ionic content is kept below 10 ppm level for most applications. It is important to avoid any voiding in the die bond, as this will affect shear strength, thermal conductivity, electrical conductivity, and hence the overall quality and reliability of the package.

The choice of the die attach adhesive is influenced by the nature of the final product, die size, the application conditions, thermal requirements, etc. and should be selected with care.

First-Level Interconnect

The first-level interconnect is the process of enabling the connection between the semiconductor I/O pads to the lead frame fingers or the pads on the substrate. Two techniques are in vogue. In one method, the device I/O pads are connected to the lead frame or substrate I/O pads using gold or other metal wires. The active side of the devices faces up. In the second method, the silicon device with solder bumps is flipped and attached to the circuitized substrate face down. Both these techniques are described in some detail in the ensuing paragraphs.

Wire bonding

Interconnection of the device bond pads to the lead frame or substrate pads is performed using metal wires. A variety of metals are used. These include gold, aluminum, silver, copper, etc., gold being the most common. The interconnection is effected by one of the three common techniques, namely, ultrasonic (US), thermosonic (T/S), or thermocompression (T/C). Two types of bonds are in vogue. On the die pad the bonds are generally ball bonds, while on the lead frame they are wedge or stitch. While T/S and T/C bonding produces either ball or wedge bonds, the U.S. method produces wedge bonds.

In US bond, the bonding wire is fed through a nozzle on to the bond site and pressed on to the pad and, as the bond interface is under compressive stress, bursts of ultrasonic energy in 20–60 kHz range are applied to the wedge. Thus, the application of pressure and ultrasonic energy results in a cold weld. The bonding time is generally about 20 ms. This type of bonding is used mostly with aluminum wires, although gold and copper can also be bonded. After bonding is effected on the die pad, the bonding tool is moved to the lead frame pad and a similar bond is made on the lead frame. As the nature of the bonding suggests, only wedge bonds are possible with this technique. Figure 4–3 shows a typical wedge bond made by this method.

In T/C bonding, as the name suggests, the wire is fed through a refractory capillary of alumina or tungsten carbide and the tip is heated with a hydrogen flame or by a capacitive discharge technique to form the ball. The tip is lowered on to the bond pad and pressure is applied (Spencer 1982). The bonding process involves plastic deformation and interatomic diffusion. The bonding pad itself is heated to about 300°C–400°C either by the bonding tip or by an external device. The thermal energy and the mechanical force result in the breakdown of the interface layers and also an increase in the bonding area. After the ball bond is made, the capillary bond head is moved on to the lead frame and a wedge is bond is made and the wire is clipped. The bonding head then moves on to the next bond pad on the chip.

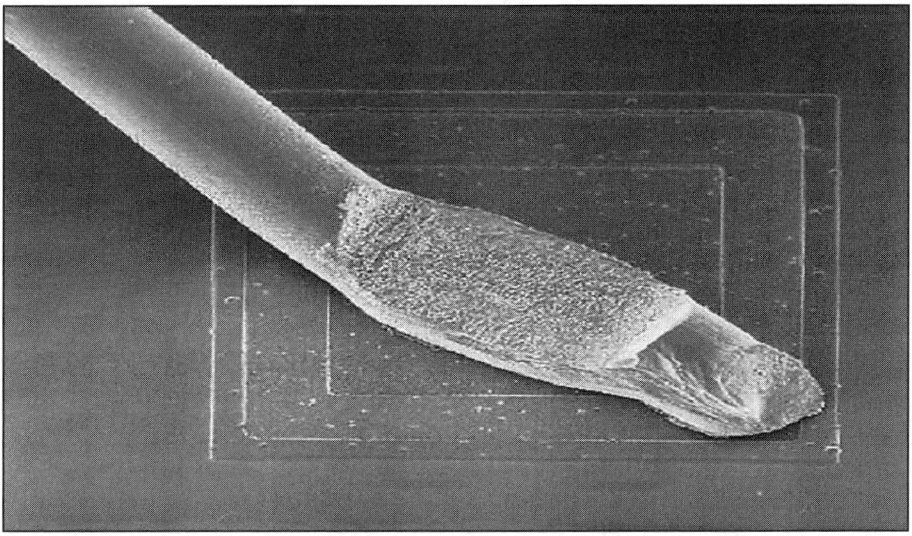

Fig. 4–3. Typical wedge bond on die pad

In T/S bonding, the features of US and T/C bonding techniques are combined. In this technique, the bonding capillary carrying the wire is subjected to bursts of ultrasonic energy. In this method, the bonding surface is heated to about 100°C–150°C. Typical bonding parameters for a 25-μm gold wire bonded on to a 1.3-μm-thick Al–1%Si metallization on a chip that is 0.8-μm-thick oxide layer under the pad have been reported to be 30–80 gF, 10–25 mW power, 300°C temperature, and 10 ms ultrasonic times (Singer 1984). A shield gas consisting of argon + 10% hydrogen at 1 l/minute for ball bonding and nitrogen +10% hydrogen at 6 l/min for wedge bonding is used. A loop height of 80 μm is typical.

Use of ultrasonic energy increases the dislocation density of the wire and pad metalizations. Microscopic slip planes form and shear across each other, and as they slide over each other, fresh clean oxide-free and high energetic material rises to the surface. As thermal energy is applied, solid-state diffusion is facilitated and a diffusion weld is formed. Continued supply of ultrasonic energy increases bond area, and thermal energy promotes solid-state diffusion.

The critical parameters for bonding are the ultrasonic power, temperature, force applied, and bonding time. A possible failure mechanism in ultrasonic and thermosonic bonding is the *cratering* of the bond pads. High ultrasonic power and low bonding force can, in fact, induce cracks in the underlying silicon. Contamination of the bond pad, if not detected, can lead to improper bond parameter settings. Thin metalizations are more prone to cratering than thicker ones. Too high or too low bonding force can result in cratering in wedge bonds. The propensity of cratering also depends on the bond pad hardness. Softer pads tend to distribute the energy more uniformly, while harder ones tend to transmit the energy to the underlying substrates. It is desirable to match the hardness of the wire with the bond pads for better bonding. Figure 4–4 shows a defective wedge bond due to improper bond settings.

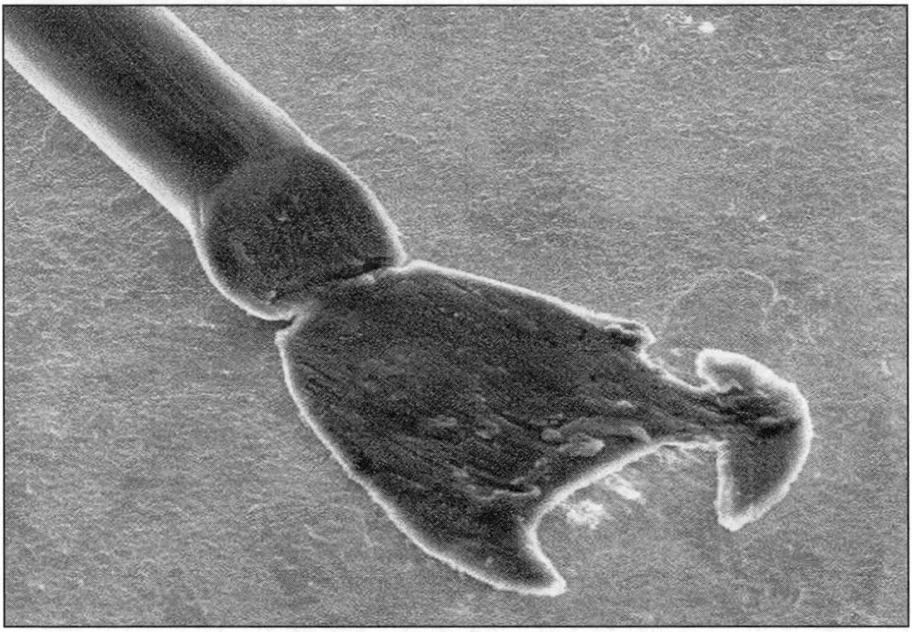

Fig. 4–4. Poor-quality wedge bond due to improper bond settings

Several wire materials are in use, but gold and aluminum are the most common. In the case of gold wire, it is alloyed with 5–10 ppm Be and or 30–100 ppm Cu for better drawability and workability. Beryllium alloying renders the wire 10–20% stronger than Cu doping. The Au wire diameter is in the range of 25–75 μm. As-drawn Au wire begins to weaken in the first two weeks and breaking load decreases by as much as 15%–20%. Therefore, Au wire is annealed prior to use in order to prevent unwanted wire breakoffs. Repeated thermal cycling flexes the wires. Undispersed Si in Al grows as stress rises. Sixty micrometer pitch bond pads can be successfully bonded with good yields in a manufacturing environment, while 50 μm pitch for leading edge applications and even 35 μm pitch bondability for ultrafine pitch application have been demonstrated. Gold wire also has excellent loop formation capability. Figure 4–5 shows a typical ball bond formed by T/C bonding.

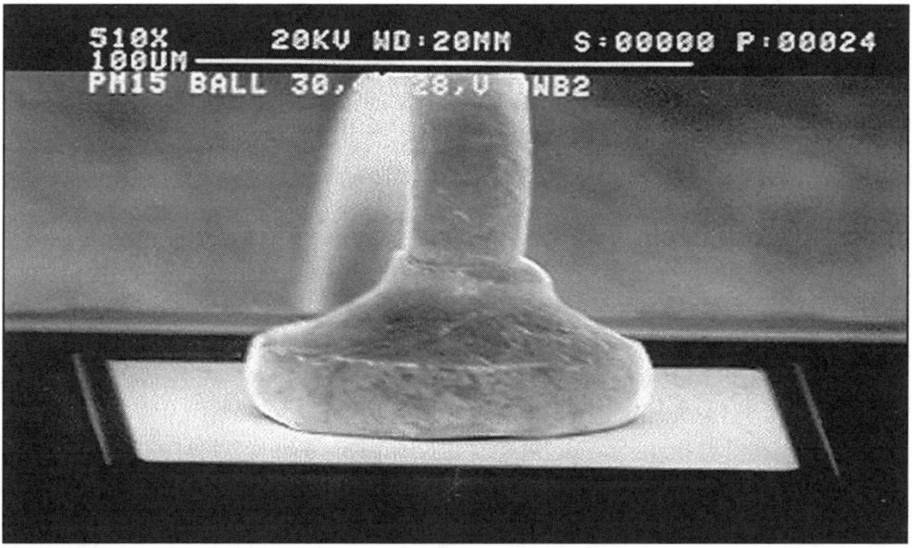

Fig. 4–5. Typical ball bond formed by thermosonic bonding

Wire bonding with gold wire to aluminum metalization gives rise to several potential intermetallics, namely, $AuAl_2$, Au_5Al_2, AuAl, Au_4Al, and Au_2Al. The formation of intermetallics is generally accelerated at temperatures above 175°C. Many Au–Al intermetallics are susceptible to flexure damage.

Both $AuAl_2$ that is purple in color and Au_5Al_2 that is white form under suitable conditions when gold wire is bonded to aluminum bond pad metalization. Since, the intermetallics are brittle, the bond integrity is compromised, leading to metal fatigue and stress cracking especially due to wire flexing under vibration and thermal cycling stresses. Early wire bond failures due to the $AuAl_2$ intermetallics are commonly referred to as *purple plague*. At higher operating temperatures, rapid diffusion of Al into $AlAu_2$ phase leaves voids along the Au/$AuAl_2$ phase boundary. Similarly voids are formed at the Au and Au-rich Au_5Al_2 phase boundary due to rapid diffusion of Au to the phase boundary. When excessive intermetallics are formed, these voids coalesce leading to an electrical open. Thus differential diffusion fluxes lead to voiding. By minimizing the dwell time at elevated temperatures and working at the lowest temperature that is practical, risk due to this type of voiding can be contained. This type of voiding is called *Kirkendall voiding* and is described below.

Au-to-Au bonding is superior since no intermetallics are formed. Gold wire bonding to silver metallization is also considered acceptable since no intermetallics are formed. However, the possibility of sulfide corrosion cannot be ignored.

Au wire bonding to Cu bond pads gives rise to CuAu, Au_3Cu, and Cu_3Au intermetallics and decreases bond strength at temperatures above 200°C.

Void formation such as Kirkendall voiding and cleanliness are also important considerations in wire bonding.

Kirkendall voiding. Differences in the diffusion fluxes of metals A and B result in a vacancy flux. This vacancy flux then moves in the direction of the slower moving atom. Where there exists enough sources and sinks for the vacancies to maintain an equilibrium value, this vacancy flux represents a shifting of the layer of atoms from one side of the reference plane to the other. Thus, the reference plane migrates relative to the ends of the sample. This phenomenon, discovered in 1947, is termed the Kirkendall effect (Smigelkas and Kirkendall 1947; Darken 1948).

Voids formed in the solid state are due to differential inter-diffusion coefficients in dissimilar metal couples. Al diffuses rather rapidly into the Al-rich phase $AuAl_2$ and leaves behind voids at the Al–$AuAl_2$ interface. This type of void is also formed at the Au and Au-rich Au_5Al_2 phase boundary.

When intermetallics formation is predominant or excessive, voids tend to coalesce resulting in a contiguous void, with the appearance of a crack. Consequently, bond lifting, and hence an electrical open, could result.

Impurities such as Fe, Co, Ni, and B have also been known to be deleterious. Au–Cu couples have also been reported to exhibit Kirkendall voiding at elevated temperatures.

Aluminum wire bonding. Aluminum besides being a cheaper metal is also very soft. It is generally alloyed with 1% Si or Mg. Silicon-doped aluminum is industry standard for Al wire bonding owing to its good fatigue resistance and reliability (Bischoff 1984). It is added to meet the breaking load and elongation requirements. The equilibrium solubility of silicon in Al is 0.02% at room temperature and reaches 1% at about 500°C. Therefore, silicon precipitates forming a second phase, resulting in hardening and constituting a source of fatigue crack initiation. Small diameter Al wires are annealed to make the silicon dispersion more uniform and stable before bonding. Sometimes more than one treatment may be necessary. Also, Al–Al bonds do not form intermetallics. However, ball formation is rather difficult with Al, and wedge bonding is time consuming with US. T/S and T/C bonds are considered better. Wires of 50–75 μm diameter are employed. It is important to recognize the potential for aluminum corrosion in the presence of chlorine. Aluminum wire bonding on Au can give rise to Kirkendall voiding. Al–Ni bonds are considered more reliable than Al–Au or Al–Ag bonds and are less prone to Kirkendall voiding.

The equilibrium solubility of magnesium in silicon is about 5%, and hence Mg doping is sometimes considered better than Si for better fatigue resistance.

Silver wire bonding. Silver wires are sometimes used in specific applications. Silver/aluminum intermetallic growth rates are about 2% lower than Au/Al intermetallics at a given temperature. Silver balls are harder than gold balls and are prone to cratering. The shear modulus is 25% greater than that of gold balls. The mean shear strength of silver ball bonds is 10% less than gold wire bonds after 100 h exposure at 85°C and 85% relative humidity. Silver–aluminum bonds generally require annealing at 300°C for 1 h for adequate diffusion and intermetallic formation. Also, molten silver dissolves about 2% oxygen at 1 atm pressure. Silver wire bonding is generally considered for high-speed devices (Kamijo and Igarashi 1985).

Copper wire bonding. Copper, like aluminum, is an inexpensive metal and has been considered for wire bonding infrequently (Mori et al. 1988; Onuk et al. 1987; Singh et al. 2005). It is harder than gold, the bonding operation is more difficult, and Cu–Cu bonding is sensitive to surface contamination and can degrade by as much as 35% and is prone to cratering. The bonding forces can also cause metal splash if the forces are too high as shown in figure 4–6. As the bonding parameters are increased, the ball bond becomes thinner and wider. The fatigue properties are dependent on the dissolved oxygen content in copper. $CuAl_2$ intermetallics are brittle. Copper wire bonding interconnections are also prone to corrosion in the presence of moisture, chloride ions, etc. (Nguyen et al. 1995; Toyozawa et al. 1990). However, copper wire bonds are more resistant to wire sweep. A comparison of Cu and Au bonding is reported by Khoury et al. (2007)

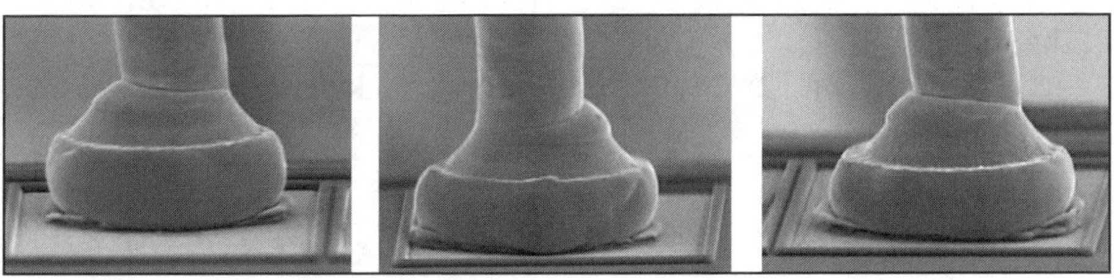

Fig. 4–6. Cu wire bonding at (a) low (b) intermediate, and (c) high bonding parameters showing increasing amount of Al splash at the higher settings.

For all wire bonding applications, the purity and/or composition of the metal, wire roundness to have good capillary feed, cleanliness of operation, and good handling practices are important.

There are several concerns and issues that one needs to be cognizant of in plastic packages. Failures can occur due to a variety of factors. These include the following:

- Delamination of epoxy molding compounds
- Delamination of die attach adhesive
- Microcracks generated in the silicon die due to wafer dicing, back grinding, lapping, etc.
- Passivation breakdown (passivation materials include silicon dioxide, silicon nitride, polyimide, phosphosilicates, benzocyclobutene, etc.)
- CTE mismatches
- Wire fatigue
- Any residual stresses in the system

Wire sweep is a phenomenon that sometimes occurs in the plastic wire-bonded packages (Nguyen 1992). It is a visible lateral deformation of the bond wires that occurs in the direction of flow of the epoxy molding compound. It can cause wire bonds, both ball and stitch bonds, to develop kinks at the bonding locations near the ball and the wedge. The exposures due to this phenomenon include defects such as current leakage, shorting of the adjacent wires, and even wire breakage.

Improper choice of material and process parameters can lead to wire sweep. High viscosity, high flow velocity, filler particle collision, void transport, and unbalanced flow are some of the causes for wire sweep. Unbalanced flow fronts in the mold cavity below and above the die paddle can lead to what is generally known as the *race track effect,* where one flow front overtakes the other and blocks the vents before completion of the filling.

In addition, bond fracture, lift off, and package cracking are also potential exposures.

Flip chip bonding

The second most important first-level interconnect scheme is the flip chip bonding. As the name implies, in this scheme the active side along with the bond pads is face down, and faces the substrate. In the case of a high I/O device, the bond pads are arranged as a matrix of rows and columns in an area array. As the number of I/O per row increases, the total number of possible interconnection in an area array increases significantly compared to the perimeter connections that are possible. Table 4–3 shows a comparison of the total number of interconnections possible for a perimeter configuration versus an area array configuration for different I/Os per row. As can be seen, while the number of interconnections in both cases is the same for a device with two I/Os per row, the area array interconnections are almost tenfold considering a 40 I/Os per row configuration.

Table 4–3. Comparison of possible total interconnections in perimeter and area array configurations

I/Os per Row	Total Perimeter Connections	Total Area Array Connections
2	4	4
6	20	36
10	36	100
20	76	400
50	196	2500

The silicon wafer containing multiple devices is first subjected to a 1-MHz screen testing and then subjected to the wafer bumping process (see below) where each bond pad is suitably prepared for receiving the solder balls of appropriate metallurgy. The wafer is then diced or singulated into individual devices. The singulated devices are then vacuum-picked, fluxed with suitable solder flux, and then precisely placed on the respective footprints on the circuitized package substrate. The substrate with the devices is then placed in a reflow oven set to approximately 350°C to effect the chip-to-substrate interconnection. In cases where a solvent-cleanable flux is employed, the assembly is cleaned with appropriate aqueous or nonaqueous solvents.

Subsequent to the cleaning operation, the assemblies are baked at 125°C for several hours to expel any absorbed moisture and are then underfilled with a suitable underfill material. The devices are then marked, and system tested for performance. A schematic of the flip chip packaging process is shown in figure 4–7.

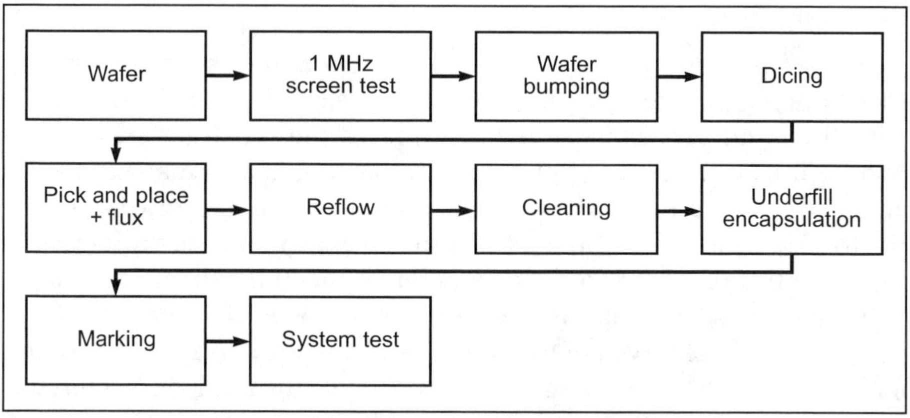

Fig. 4–7. Schematic flow of the flip chip packaging process

Flip chip bonding has several advantages over wire bonding. It enables extremely high density packaging within a minimum area and thus constitutes an efficient packaging method. Owing to the shorter interconnection length, it provides superior electrical performance. As the package to substrate interconnection is through short solder balls, it also provides efficient thermal management through the substrate/carrier in addition to the ability to attach a heat sink if needed. The interconnection materials are integral to the package. Also, the interconnection scheme is well suited for multichip packaging. All the chip to substrate interconnections are effected simultaneously in contrast to the sequential nature of the wire bonding process (Broffman et al. 2001). Further, the assembly yields with the flip chip attach process are very high, closely approaching the six sigma owing to the self-centering nature of the bonding process which comprises controlled collapse chip connection often referred to as the C4 process (Miller 1969; Koopman and Totta 1988; Fried et al. 1982).

Under bump metallurgy (UBM). One of the prerequisites for flip chip bonding is the preparation of the aluminum die pads appropriately. Since solder does not wet aluminum this involves the application of a solder wettable termination metallurgy which would also determine the area of the solder connection on the chip side. It generally consists of multiple metal layers. This is variously called *under bump metallurgy* (UBM), *pad limiting metallurgy* (PLM), *ball limiting metallurgy* (BLM), etc. The ball limiting metallurgy has to fulfill several requirements. The relevant metallurgy should have good adhesion to the chip passivation layer, be it silicon dioxide, silicon nitride, or polyimide, as the case may be. It should have low ohmic resistance to the final interconnection metallurgy, act as good diffusion barrier to the underlying metal layers, and impose little or no stress on the silicon. In addition, the under bump metallurgy should protect the chip metallurgy from the environment and should be capable of being applied on already probed wafers. Under bump metallurgy also provides a slightly larger bonding pad for solder ball attachment than the die pad.

Several under bump metallurgies are in vogue. These include aluminum/nickel/copper (Al/Ni/Cu), electroless nickel/gold (Ni/Au), Ni/Cu/Au, Cr/Cu/Au, Ti/W/Cu, etc. Metalizations like Cr, Ti, etc. are employed for good adhesion. As can be seen, on the chip side the termination metallurgy such as copper or nickel is sandwiched between the adhesion layer such as Cr, Ti, W, etc. and a passivation layer such as a thin film of gold. The bump metallurgies are applied by well-controlled sputter deposition techniques.

The thickness of copper termination metallurgy has to be carefully controlled. Too thin a copper may result in the consumption of all the copper in the formation of copper–tin intermetallics resulting in ball fall off, while too thick a copper layer may result in the need for a slightly higher reflow temperature and formation of excessive Cu–Sn intermetallics causing joint embrittlement. Figure 4–8 shows the schematic of an under bump metallization process.

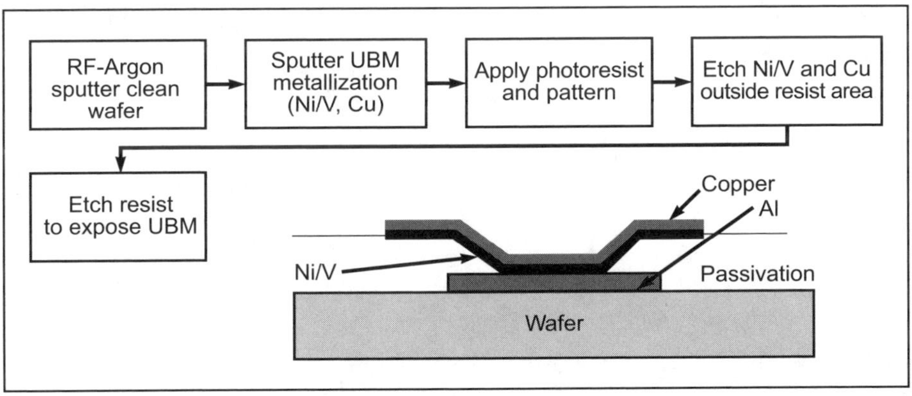

Fig. 4–8. Schematic of the under bump metallization process flow

Solder bumping. Once the under bump metallization is performed, the next step is to apply the interconnection material to the solderable UBM. The interconnection materials are either a high-melt solder such as 90Pb/10Sn, 95Pb/5Sn, or 97Pb/3Sn or eutectic Sn/Pb, or Sn/Ag/Cu. A variety of processes are used.

In one method, lead (Pb) and tin (Sn) are sequentially sputter-deposited in the appropriate ratio on a suitably masked UBM and then reflowed to form the solder bump. The reflow is performed in a reducing atmosphere at 350°C for high-melt solder and at 220°C–240°C for eutectic solder. In another method, a T/S ball bonder is used to deposit and thermosonically attach the solder, just as is done in a wire bonding operation. The wafer or the single chip is subsequently reflowed to alloy the bump to the pad and attain proper shape as shown in figure 4–9. The interconnection metallurgies are also deposited by electroplating the individual metals in the appropriate ratio on the suitably masked UBM and subsequently reflowed to form the solder bumps as shown in figure 4–10) (Kawanobe et al. 1981). Another method of bumping is by an ink-jet process where molten and pressurized solder is dispensed on to the UBM vias as droplets using a piezoelectric pump dispense head and subsequently reflowing the solder as shown in figure 4–11. This technique is currently only suitable for dispensing eutectic tin–lead solder due to temperature limitations (Hayes and Wallace 1996; Wallace 2005).

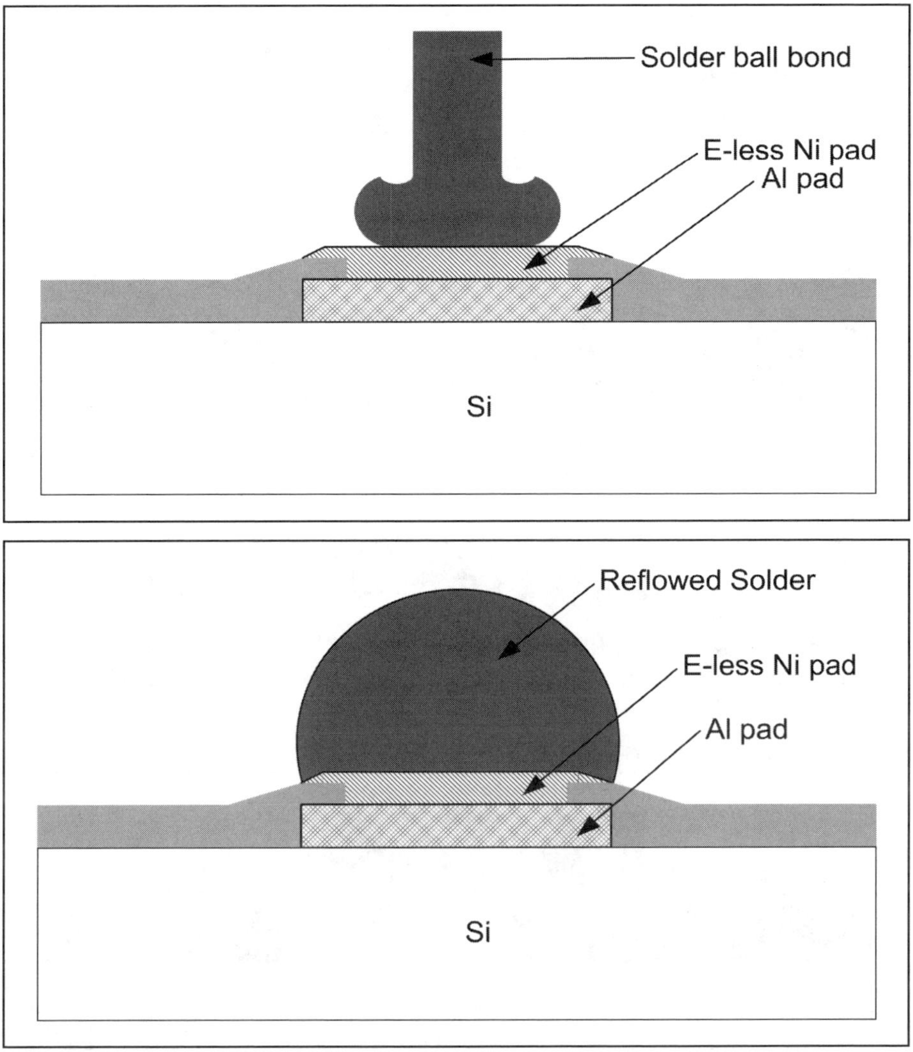

Fig. 4–9. Solder bumping by thermoscopic bonding

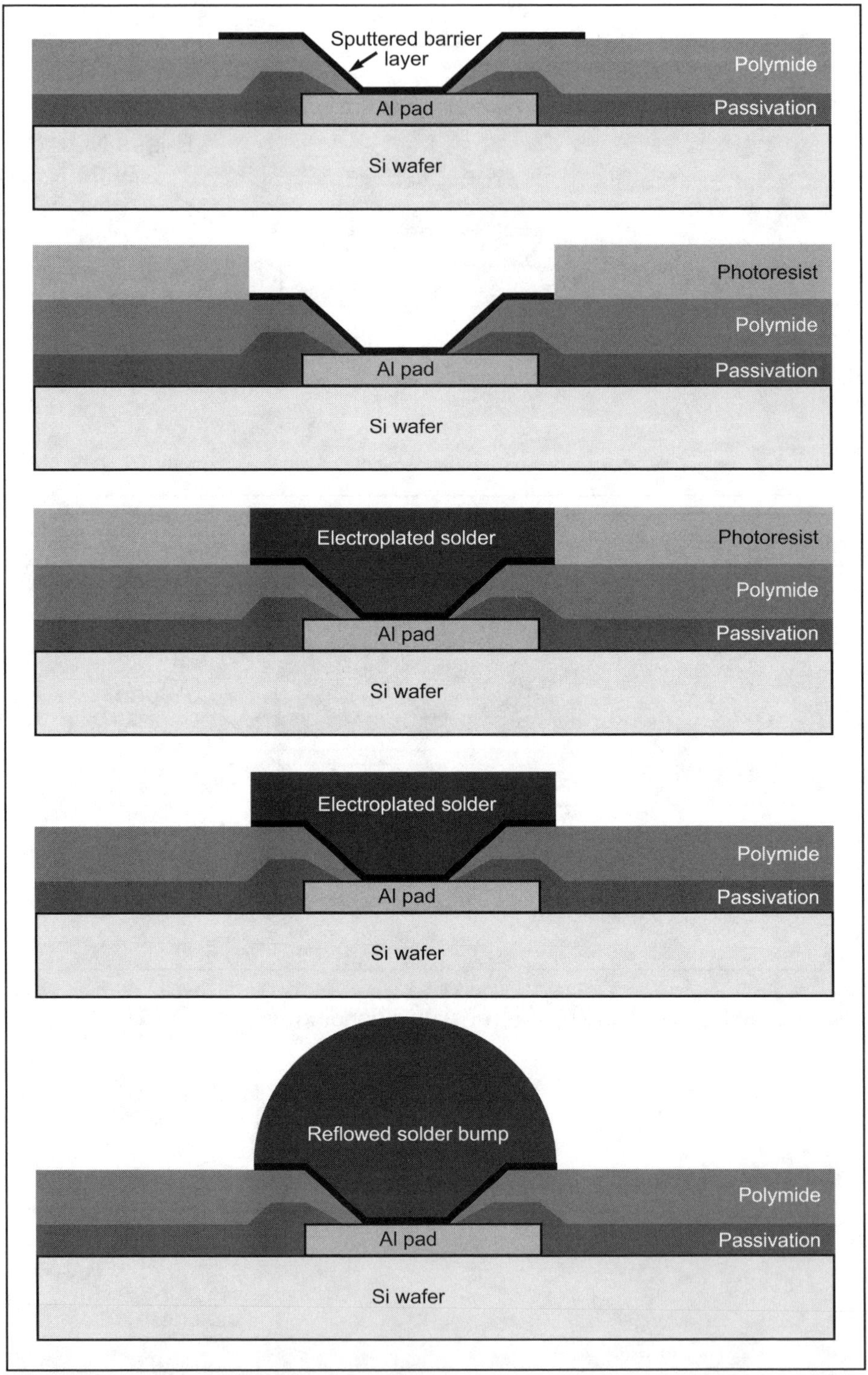

Fig. 4–10. Solder bumping by electroplating

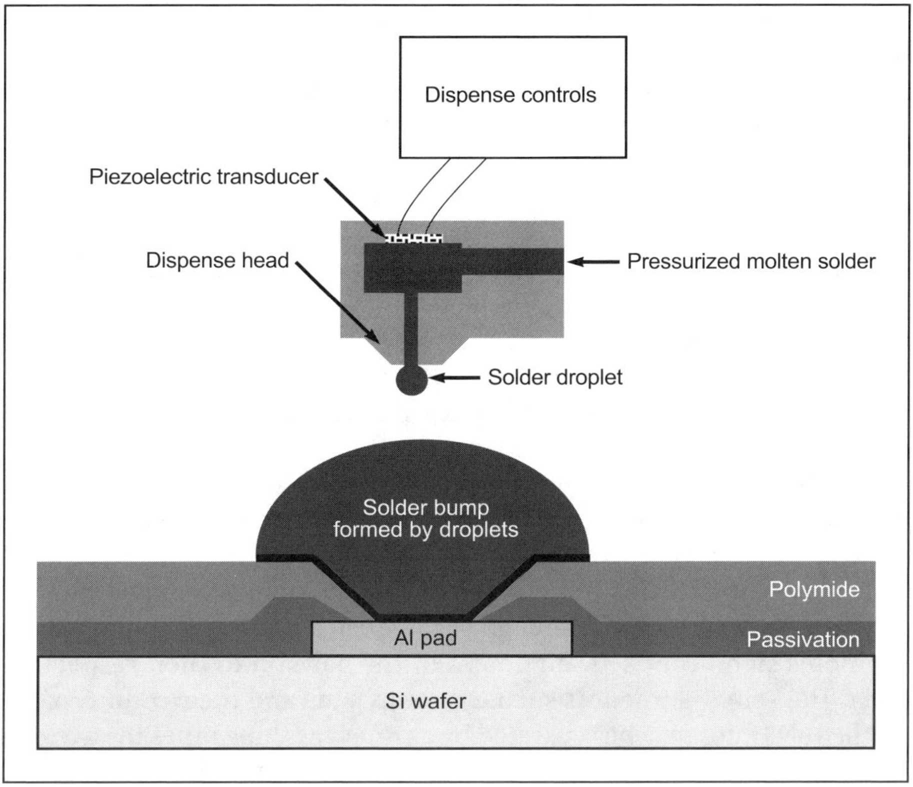

Fig. 4–11. Solder bumping for "ink-jet" dispensing

C4NP. C4NP refers to Controlled Collapse Chip Connection New Process. In this process, the entire wafer is replicated as a mirror image of cavities in a glass mold. It is made up of borosilicate glass to match the CTE of the silicon wafer. The patterning and etching of the cavities is effected by conventional lithographic techniques. The diameter and depth of the cavities determine the solder bump size. The cavities are then filled with molten solder by passing the mold underneath an exit solder injection nozzle of the solder reservoir. Any excess solder is also removed in this step, thereby filling the cavities with a precise amount of solder. As solders do not wet glass, spherical solder balls are formed on melting (Ruhmer et al. 2006).

The filled mold is then aligned with the UBMs on the underside of the wafer to be bumped. They are both heated separately and the wafer is brought into contact with the mold so that the molten solder comes into contact with and wets the UBM pads. Due to the wetting of solder to the UBM, solder balls stay with the wafer. The solder bumps are thus transferred to the UBM as the mold is separated from the wafer, and the process is repeated to solder-bump the next wafer. The entire process takes place in a controlled reducing atmosphere ensuring clean UBM pads and solder. The process also precludes any post-bumping flux-cleaning steps.

The process is suitable for a variety of interconnection materials, such as eutectic Sn/Pb, Sn/Ag/Cu, Sn/Cu, etc., requiring only changes in the process parameters.

The multitude of techniques provide a choice either to bump singulated dies or full wafers. The singulated bumped dies are then attached to chip substrates or PWBs.

The footprint on the substrate is fluxed with water-white rosin for high-lead solders or water-soluble flux for eutectic tin–lead solder. The chip is then placed with a vacuum pick-up tool on the footprint. The flux generally acts as a temporary adhesive to hold the chip in place. This assembly is then subjected to a reflow using either a local heat source to bond single chips or an oven.

In the simultaneous bonding of multiple chip interconnections to the substrates or carrier, the high surface tension of the molten solder promotes significant self-alignment of the balls on to their respective pads. The self-alignment is such that chip pads and their counterparts on the substrates may be separated by as much as three times the average bump radius. Good interconnections are made as long as the mating surfaces touch each other and there is good wettability of the surfaces.

The assemblies are then cleaned with either organic solvent or water as appropriate and then tested for electrical functionality. The assemblies, if found defective, are reworked at this stage.

Encapsulation/molding. This step is required for protecting the silicon and its interconnect structure from physical, environmental, and handling damage during subsequent process operations as well as use environment. Encapsulation by molding is a common method for plastic packages used in most commercial electronics products. The molding polymers used for encapsulation are either thermosets or thermoplastics. Thermoset materials are those that are plastic or fluid at low temperatures and react irreversibley when heated to form a cross-linked network that cannot be remelted. Examples of this type of materials include some epoxies, phenolics, polyesters, diallyl-phthalates, etc. On the other hand, thermoplastic materials become soft and malleable when heated and can be repeatedly heated and cooled without change in their properties. Polyphenylene sulfides, polyetherimides, nylons, polysulfones, etc., belong to this category. New molding compound formulations such as modified epoxies, silicone-modified polyimides, etc., are constantly being developed to meet the emerging product manufacturing and reliability demands.

Three molding techniques are employed in the industry, namely, transfer molding, injection molding, and reactive injection molding. These are described in brief in the ensuing paragraphs.

In transfer molding, the mold consists of a top half and a bottom half. The lead frames with bonded silicon devices are loaded into the bottom half. The mold is closed and a clamping pressure is applied. The thermoset material preform, which is heated to about 90°C and held in a transfer pot, is conveyed under pressure through runners and gates in the mold by activating the transfer plunger. The mold is maintained at a temperature of about 175°C to ensure complete curing of the mold compound and the pressure in the plunger is maintained at about 170 MPa in order to ensure cavity filling. After a predetermined time, the mold is cooled and opened and the lead frames are ejected by the ejector system. The ejected lead frames are sometimes subjected to a post-cure operation at 175°C for 4–16 h to ensure total cure.

In injection molding, plastic pellets from a hopper are fed into a screw and go through several heating zones and exit the screw in the dry molten state. Approximately 300 kg/cm^2 pressure is applied so that the mold at the preset temperature and containing the lead frames is filled. The mold is then cooled and opened and the lead frames are ejected.

In reactive injection molding, the reactive components in liquid form are separately pumped into a mixing head in a fine stream and are mixed together. The liquid resin mixture is then pumped into the heated mold containing the lead frames and cured. Again, as in the previous cases, the mold is cooled and opened and the lead frames are ejected.

Some processes are better suited to thermoset materials and others to thermoplastic materials. The molding process can sometimes be modified to operate with either material. There are several merits and demerits to each of these molding techniques and a detailed discussion of these is beyond the scope of this book and the reader is referred to excellent references on the subject of plastic molding (Bryce 1991).

After the molding operation, the lead frames go through a deflashing operation to remove any resin bleedout on the surface. Deflashing is performed either with high pressure air mixed with fine abrasive particles, or with a high pressure water slurry containing an abrasive. This operation generally leaves a matte finish on the leads and is preferred for subsequent process operations.

Lead finish

Leads emanating from the plastic body are coated with material or materials to protect them from oxidative degradation and enhance their solderability to the PWB. A variety of lead finish materials are in vogue in the industry. The lead finish comprises operations such as electrolytic, electroless, or immersion plating or hot dipping.

Hot dipping is usually practiced mostly for the tin–lead eutectic or lead-free alloys. It consists of cleaning the lead frame followed by dipping in a warm flux, then dipping in a molten solder, removal of excess solder by air knives, and finally a water wash and drying. The initial cleaning could be a degreasing operation. Variations in the finish thickness in the hot dipping process are recognized.

Electroplating the leads with metals of choice provides the best coating thickness uniformity desired for many modern-day fine-pitch high I/O packages. Depending on the lead material, multiple finish layers may be necessary. For example, in the case of copper leads, first a layer of electroless nickel is applied as a diffusion barrier followed by a thin layer of immersion gold.

Electroless nickel immersion gold, nickel–palladium gold, immersion tin, etc., have been in use. Implementation of lead-free technology has given rise to the development of several new lead finishes. The different component lead finishes, their relative dimensions, and advantages are discussed in the section on surface finishes.

Trim and form

Subsequent to the lead finish operation, the commoning bar used for plating is trimmed carefully so as not to cause damage to lead surfaces and introduce any additional stresses. The package is held in a fixture, and using pneumatic presses the leads are bent with precision and accuracy to the requisite shape in a series of steps. It is important not to introduce additional stresses and also avoid any microcracks in the finish layers. Microcracks that are formed during the lead-forming operation permit ingress of moisture and other gases or vapors and promote oxidative degradation of the base alloy, which in turn affects the lead solderability. Lead forming is an extremely critical operation in terms of dimensions. Lead height, bend radii, and lead coplanarity have to be maintained to exacting standards especially for fine-pitch packages. A lead coplanarity to within 50–75 μ may be necessary for successful fine-pitch assembly. Two common lead forms are the J-lead and the gull wing lead. Gull wing leads are more prevalent in packages for portable electronic products. Figure 4–12 depicts an example of typical rectangular J-leaded and gull wing packages. The packages after the trimming and forming are finally marked, inspected, tested and are packaged in trays, reels, or tubes, as the case may be, for shipment.

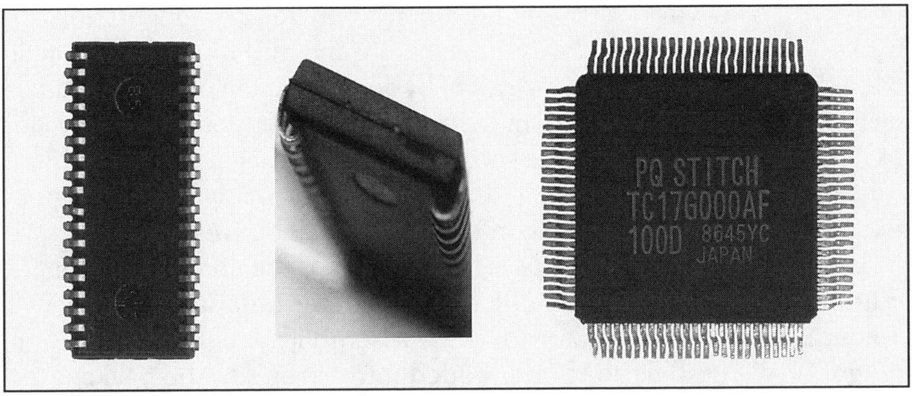

Fig. 4–12. Typical J-leaded and gull wing packages

Leadless packaging

These packages designated *leadless chip carriers* (LCCs) and are either ceramic or plastic in construction. They are made in square and rectangular formats. Ceramic parts generally have either an alumina or beyllia base. These chip carriers have gold-plated, semicylindrical grove-shaped termination called *castellations* on all the four sides of the package. Figure 4–13 shows an example of a leadless ceramic chip carrier LCC. Owing to the short signal paths, the terminations offer reduced interconnection resistance and improved power dissipation. They also have reduced inductive and capacitive losses. Ceramic LCCs are favored for high-frequency applications. The packages are hermetic and are offered in a range of termination pitches including 50, 40, 33, 25, and 20 mils. The range of I/O counts is 18–156.

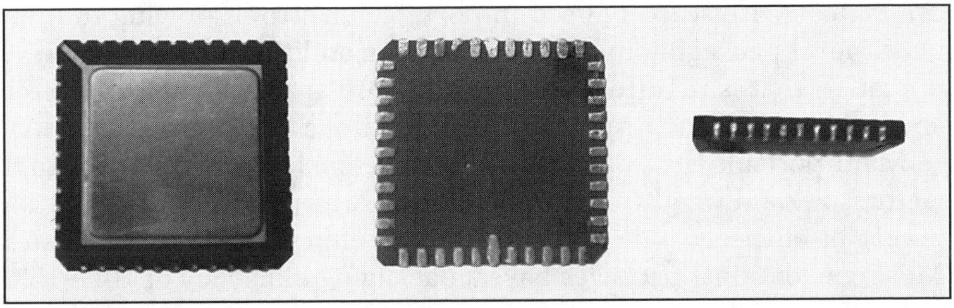

Fig. 4–13. Examples of leadless ceramic chip carrier (LCCC).

Leadless chip packages are further subclassified as type A, B, C, D, E, and F. Types A through D are square, while E and F are rectangular. The different types are necessitated by the different application requirements. The design is dictated by the mode of heat dissipation. Some constructions are lid-up while others are lid-down. With lid-up configuration, the backside of the silicon die faces the substrate and heat dissipation is facilitated through the substrate instead of by air convection or by an auxiliary heat sink. On the other hand, in the lid-down configuration the backside of the silicon is away from the substrate and the package is amenable to external heat sink attachment and air convection cooling. This configuration also limits the size of silicon device that can be enclosed.

Leadless packages offer the highest board packaging density, i.e. the number of packages that can be assembled per unit area of the PWB. Solder joint heights for many of these assemblies are in the 50–75 μm range. While castellations offer added joint strength, the package-to-board interconnection reliability is far below that obtainable with leaded packages for operation in a given temperature range. The package size is generally kept small enough so that the distance of solder joints from the neutral point of the package is maintained small enough to provide the required reliability. Ceramic leadless packages offer acceptable reliability when mounted on ceramic carriers owing to good CTE match with the carrier.

LCCC, when mounted on organic PWBs, may sometimes require enhanced solder joint height through the use of high-melt solder pedestals.

Perimeter leaded packages

Prior to the introduction of surface-mount technology in the early 1980s, package-to-board interconnection was effected by inserting the pins protruding from the package into the corresponding plated-through holes in the PWB, followed by soldering using a wave-solder process. Insertion mount components, alternatively referred to as *through-hole components*, are scarcely used in portable electronics owing to their board-level packaging inefficiency and the additional wave-soldering operation that is required. The packages are also of higher lead pitch and higher profile and do not lend themselves to a double-sided assembly.

Many portable electronic devices have defined requirements on form factor, size, and weight, and demand very high packaging efficiencies. Packaging efficiency is defined as the ratio of chip size to the package size. Most conventional packages have a packaging efficiency of 10%–15%. Low profile, high packaging efficiency, and light weight are essential for portable electronic appliances. In majority of cases, leaded packages are limited to gull wing packages. J-leaded packages are rarely used and hence do not constitute an integral part of this chapter. Low profile and high packaging efficiency impose constraints on CTE mismatch and lead compliance and hence have an impact on package-to-board interconnection reliability.

Owing to the form factor definition, space constraints, and high packaging density requirements of portable electronic products, thinner packages of low profile and lead pitch finer than 0.8 mm as well as higher packaging ratio have emerged. The products include personal computer memory card industry association (PCMCIA) cards, personal digital assistants (PDAs), cell phones, camcorders, digital cameras, etc. Thus evolved thin quad flat packs, thin small outline packages (TSOPs), very small shrink outline packages (VSSOPs), etc.

PCMCIA cards are of the size of conventional credit cards except they are thicker (85 mm × 54 mm × 3 mm). Data can be transferred from one computer to another with great ease. These packages enabled versatility, functionality, and flexibility in product design of many portable electronics and the eventual emergence of high-density packaging.

The predetermined form factor, however, imposed severe restrictions on packaging. The multilayer PWB has to be much thinner than the conventional boards (about 0.25–0.5 mm), and the first-level packages need to be of very low profile, especially if the assembly is a double-sided one. The packages are of very low profile, made of thinner silicon, and are finer in lead pitch. The packages have very small outline, implying that the package is not very much larger than the silicon device that is packaged.

TSOPs are mainly used for packaging memory devices and are generally of low I/O (under 100). They are rectangular in shape and are of two kinds: Type I and Type II. Type I TSOPs have leads emanating from the shorter dimension of the packages and are generally 0.5 mm in pitch, while in type II TSOPs the leads emanate from the longer dimension side of the package and are of a coarser pitch of either 32 or 50 mils (Viswanadham et al. 1993). The lead material of many of the earlier versions of these packages was alloy-42, and the lead shape is traditionally of the gull wing type. A typical illustration of TSOP I and TSOP II packages is shown in figure 4–14(a) and (b).

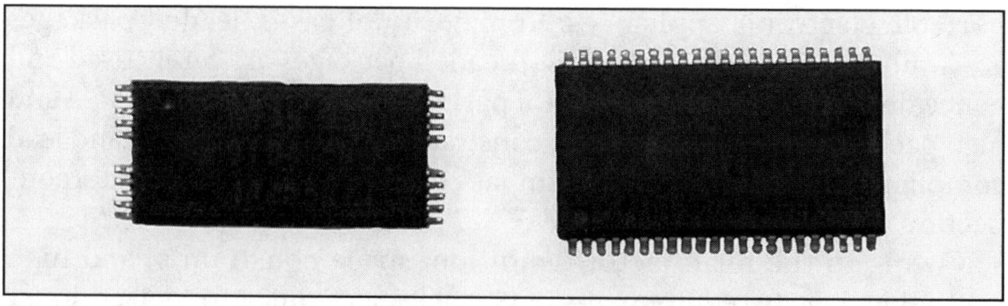

Fig. 4–14. (a) Typical TSOP I and (b) TSOP II packages

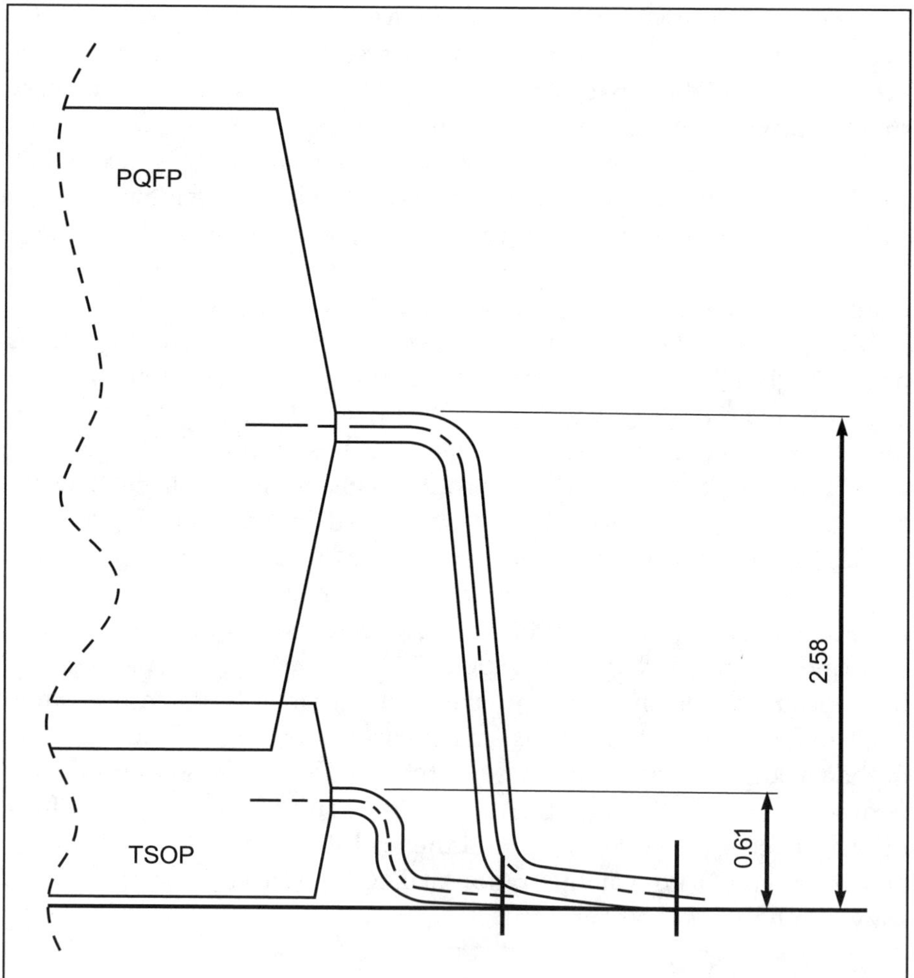

Fig. 4–15. Comparison of conventional quad flat package and TSOP illustrating the standoff height differences

In order to meet the demands of the form factor, low profile, and high packaging density, thin packages are designed with the minimum amount of molding compound surrounding the silicon covering the wire bonds. The silicon to plastic ratio is higher than in the conventional packages. As a consequence, the CTE of the package is lower. While conventional quad flat packs have a CTE of about 12 ppm/°C, TSOPs tend to have a CTE in the range of 5–7 ppm/°C depending on the size of the silicon. As a result, the global CTE mismatch between the PWB and the package is higher.

Owing to their lower profile, TSOP packages have very low standoff. Figure 4–15 shows a schematic of a comparison of a conventional quad flat package and a TSOP lead profile. These packages, owing to their low profile, form factor, small outlines, lower mass, and smaller volume, met many of the high-density packaging demands of the time. Their design, materials, and construction have also raised a number of reliability concerns and are described in the next sections.

Assembly concerns

As the packages are thin and small with an apparent low heat capacity, the leads get hotter sooner than the PWB during assembly reflow. Solder wicks up the lead, leaving insufficient amount of solder at the joint. This usually results in inadequate heel and toe fillet, affecting the overall solder joint reliability. Sometimes, the wicked-up solder gets lodged between the body of the package and the lead, thus reducing the lead compliance. Important steps in the package fabrication after molding operation, as discussed in an earlier section, involve lead frame plating with eutectic solder followed by lead forming and trimming operation. Lead forming involves two sharp bending operations close to each other, and the possibility of introducing microcracks in the solder plating cannot be ruled out. During storage, oxygen and moisture ingress oxidizes the underlying nickel of alloy-42, rendering it more difficult to solder subsequently. Also, lead trimming operation exposes alloy-42 metallurgy to air oxidation, leading to subsequent reduction in solderability and a weaker toe fillet. While the solder joint may look acceptable during visual inspection, little or no metallurgical bond between the lead and solder may be the result. Also, the lead trim operation exposes bare nickel in the toe region and adequate toe fillet may not form because of the oxidized nickel. (Viswanadham et al. 1994).

As the amount of molding compound surrounding the silicon is thinner, these packages are more prone to moisture absorption, and hence the propensity for the popcorn effect during assembly and rework is also higher.

Reliability concerns

Initial second-level thermal cycling reliability tests in the 0°C –100°C range indicated failures within a few hundred cycles in contrast to the conventional packages (Viswanadham et al. 1993). The requirement for 1,000 cycles was not generally met. Also, there was considerable variation in the performance depending on other factors such as board thickness, single-sided versus double-sided assembly, TSOP I versus TSOP II, etc. While the reliability of these packages may have been adequate for short-design-life products, they presented a general reliability concern since the same packages are used in several products with differing reliability and performance requirements.

The gull-wing lead design with very short or low standoff reduces the lead compliance significantly. During thermal cycling, the relative package-to-board movement due to CTE mismatch is very much restricted and, as a consequence, the solder joints experience greater stress. Also, alloy-42 is also a high modulus, low-CTE material.

Thus, while enabling high-functionality, high-density packaging, TSOPs as a class of packages have been found to have a relatively lower package-to-board interconnection reliability. A number of design and material modifications have been attempted and implemented to improve reliability. One of the design concepts is the lead-on-chip design (LOC), where the lead termination inside the package is located on the silicon device thereby providing an increased lead height. Figure 4–16 depicts a comparison of the traditional configuration and the lead-on-chip design. Use of copper alloy instead of alloy-42 as the lead metallurgy provided better lead compliance. However, copper alloys have a higher CTE compared to alloy-42 and silicon-to-lead metallurgy CTE mismatches inside the packages needed to be addressed.

Also, lead encapsulation after the board assembly, to reinforce the solder joints with epoxies, has been successfully implemented to enhance board-level reliability. Once the leads are encapsulated, any subsequent package replacement, repair, or rework is well-neigh impossible.

Considerable reliability enhancements were achieved by the aforementioned changes. Lead-on-chip design provided fractional improvement, while the copper alloy leads provided improvements by a factor of two. Lead encapsulation, however, provided by far the best improvement, with assemblies surviving in excess of 4,000 cycles of thermal cycling (Emerick et al. 1993).

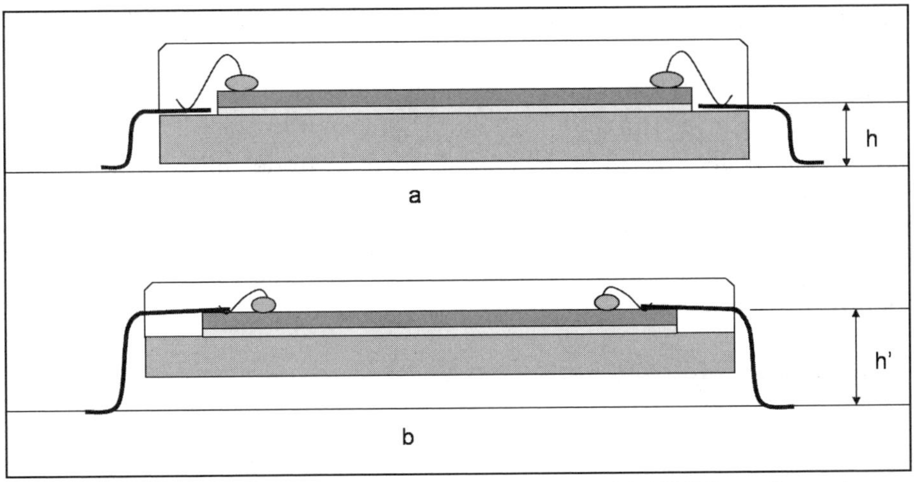

Fig. 4–16. Comparison of conventional and lead on chip TSOP configurations

The perimeter paralysis

The insatiable appetite for higher packaging density prompted package manufacturers to drive towards finer and finer pitch packages. Package pitches migrated from 1.25 mm to 0.5 mm and finer. A major effect in migrating to finer pitch assemblies is the impact on the assembly process. At finer pitches, the leads are extremely fragile and are prone to handling damage, resulting in bent, skewed, and nonplanar leads, etc. The assembly required stencils that could accommodate pitch diversity, with stepped down or stepped up stencils that could affect the optimal board real estate savings and printability of the paste with acceptable yields.

Also, assembly of finer pitch, high I/O devices also required higher routing densities on the conventional PWB technologies. Higher I/Os implied more board layers, finer circuit lines and spaces, better layer-to-layer registrations, better solder-mask clearances and registrations, etc., Conventional PWB technologies have approached their capability limits. Thus, the industry experienced what might be termed *perimeter paralysis* with perimeter leaded packages and required a paradigm shift.

Area array packaging

As indicated in the previous section, implementation of TSOPs, TQFPs, etc. has met some immediate needs of the portable electronics industry, but with limited package-to-board reliability. Also, these leaded packages utilized the existing lead-frame-based gull-wing package manufacturing infrastructure. The ultimate high-density enabler is, of course, the direct attachment of the silicon device on to the PWB. While flip-chip bonding has been in vogue since late 1960s, the main application, for obvious reasons, has been for very high or very low I/O packaging with ceramic substrates. The CTE mismatch between the organic PWB (18–20 ppm/°C) and silicon (2.8 ppm/°C), the need for underfilling for reliability enhancement, inability to easily rework, and the stand-alone dedicated tooling and equipment required have all been impediments in the rapid deployment of flip-chip technology for low-cost, high-function portable electronic products.

Ball grid array packaging

An innovative concept to second-level packaging is to bring the component terminations from the perimeter of the package to the underside of the package similar to the flip-chip attach. This approach will eliminate the lead fragility and handling issues associated with the fine-pitch packages and also enable opening up the terminations' pitch, thus enabling high assembly yield. Table 4–3 shows a comparison of maximum termination count possibilities for the area array approach for any given pitch and the perimeter only approach. In essence, it is to bring the area-array flip-chip attach concept from first-level interconnect to second-level interconnect. The package terminations are solder balls instead of copper or alloy-42 leads. Other advantages that accrue are the controlled collapse connections, self-alignment of the package, and consequently very high assembly yields.

The first of these packages brought into use are the 1.25-mm pitch IBM's solder ball connect packages on ceramic substrates. The package terminations are 90Pb/10Sn high-melt solder balls attached to the package pads by eutectic Sn/Pb solder. High-melt solder balls of the package when assembled on to the FR-4 multilayer board eutectic Sn/Pb solder maintain a noncollapsed fixed standoff for reliability. These area-array solder ball connect packages were appropriately called *ball grid array packages* (BGA) later and several variations of these packages evolved to suit the needs of the different segments of the electronics industry (Lau 1995).

Ball grid array packages can either be hermetic or nonhermetic. Hermetic packages use multilayer ceramics as the substrate with either metal or ceramic caps. The initial high I/O offerings of the BGA packages were of this type. There are several types of nonhermetic area-array packages that are constructed with organic laminates, which are rigid or flexible depending on the type of application. Two types of package designs, called *cavity-up* and *cavity-down*, are in vogue. In cavity-up design, the active surface of the chip and the I/O pads face up, while the back side is attached to the substrate with solder alloy or a die attach adhesive. Gold wire bonding constitutes the first-level interconnect. In cavity-down configuration, the active side of the chip and the respective I/O pads face down. First-level interconnects are either wire-bonded or flip-chip-attached depending on the design. Figure 4–17 depicts the cavity-up and cavity-down BGA configurations. Spherical balls of eutectic Sn/Pb, high Pb, Sn/Pb/Ag, and Sn/Ag/Cu alloys constitute the second-level interconnects to the PWB. The area array scheme offers several choices and opportunities for the ball configurations. These include full, perimeter, staggered, truncated, custom depopulated arrays, etc. Cavity-up wire-bonded designs are amenable to both full and perimeter array configurations. Only perimeter array configuration is possible with cavity-down wire-bonded configurations. Cavity-down flip-chip-attach designs can lend themselves to both perimeter and full array configurations.

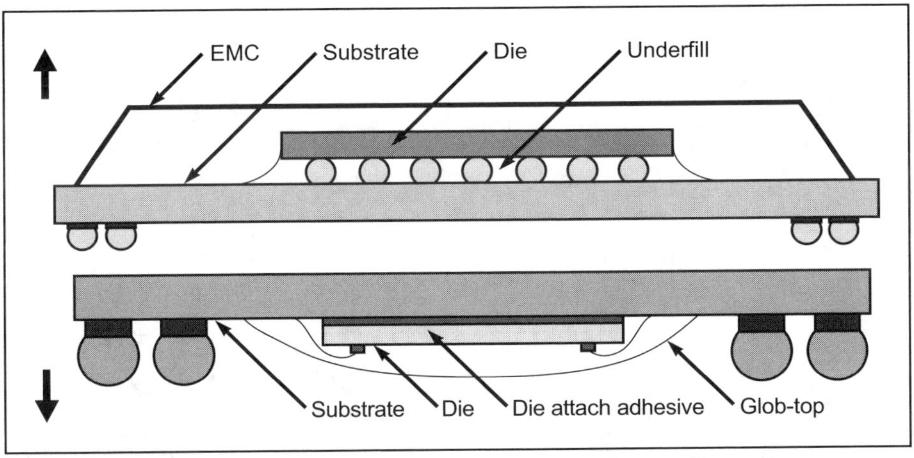

Fig. 4–17. Schematic of cavity-up and cavity-down BGA configuration

In general, wire-bonded packages offer limited I/O capability because wire bonding on the die is practiced only with two or three perimeter rows. With every additional row of perimeter pads on the die, one encounters progressively increasing wire lengths, increased propensity of wire sweep, shorting, parasitic effects, and decreased electrical performance. Thus wire-bonded packages are generally limited to low I/O packaging. Figure 4–18 depicts the different ball array configurations for common BGA packages. Flip-chip and cavity-down construction offers the facility for heat sink attachment to the backside of the chip either directly or through a thermally conductive adhesive or grease for effective thermal management. Figure 4–19 shows a package configuration with a heat sink.

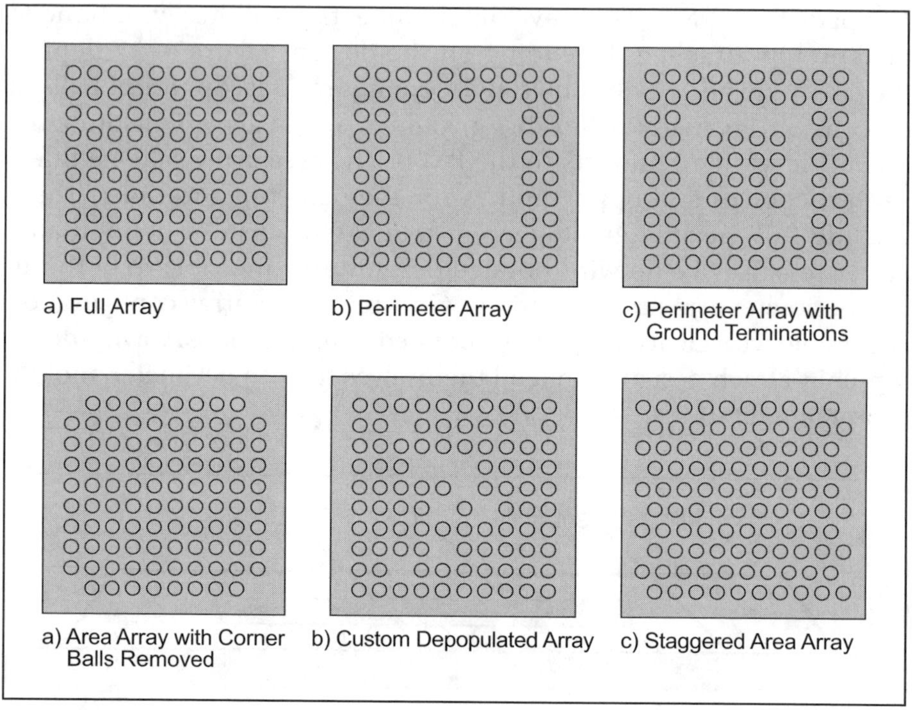

Fig. 4–18. Common ball array configurations for BGA packages

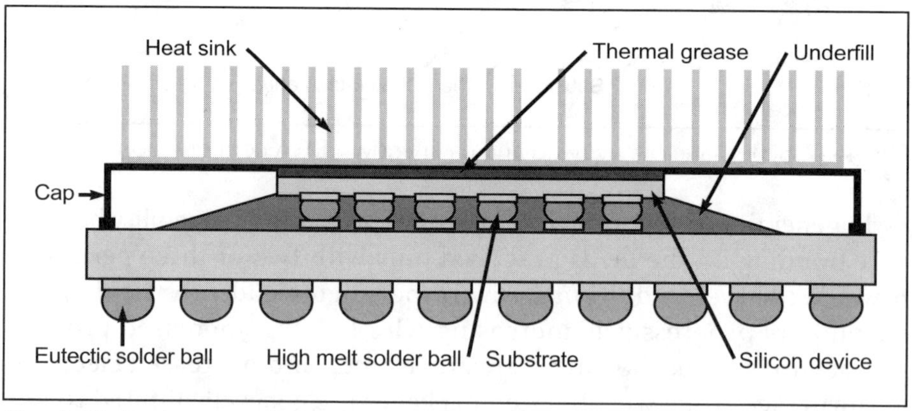

Fig. 4–19. Package configuration with heat sink

Ceramic BGAs offer high I/O capability and excellent electrical performance, which are essential for high-end computing and information processing. These packages are heavy, brittle, and of high thermal mass. Owing to their low CTE of ~7 ppm/°C, second-level thermal cycling reliability on organic PWB is limited. This is especially the case with bigger packages where the corner joint distance from the package neutral point is unacceptably large. To alleviate this reliability concern, solder balls are sometimes replaced with solder columns to provide the requisite compliance. These columns, of high-melt Sn/Pb solder, are either cast on the package itself or are attached as wires to the package pads with eutectic Sn/Pb solder. Thus, thermal cycling reliability is gained at the expense of electrical and thermal performance. Figure 4–20 shows a picture of a column grid array package.

Fig. 4–20. Ceramic column grid array package (bottom view) showing the configuration of the columns in the shadow

With the proliferation of portable and consumer electronics and the emerging need for high volume, lighter, and cheaper packages, plastic area-array packages such as super BGAs, plastic BGAs , and tape BGAs have come into increasing usage. Super BGA is a cavity-down, wire-bonded, high-performance, perimeter-array package with a built-in heat sink, as depicted in figure 4–21. Tape BGA, shown in figure 4–22, derives its design features from tape automated bonding (TAB) technology utilizing a two-layer (ground and signal) tape with T/C-bonded inner leads to the die pads. A copper stiffener is added to the structure to provide rigidity. The second-level interconnects are through high-melt solder balls. The package is moisture sensitive and very low in mass with very low I/O range and has a perimeter leaded structure and is well suited to low-performance applications (Keizer and Brown 1978).

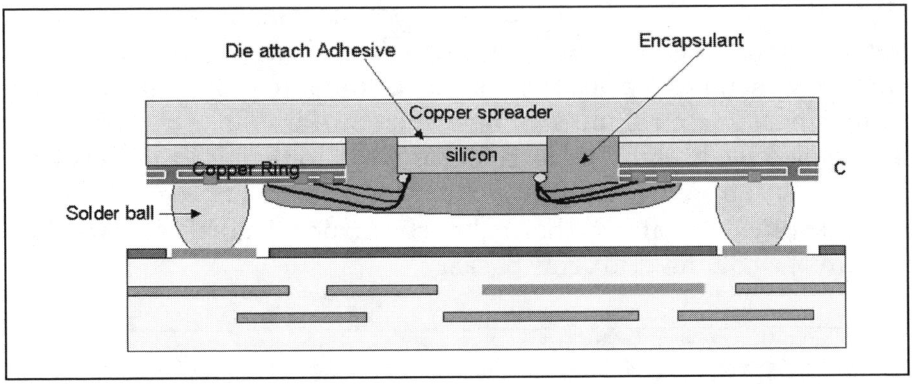

Fig. 4–21. Schematic of a super BGA

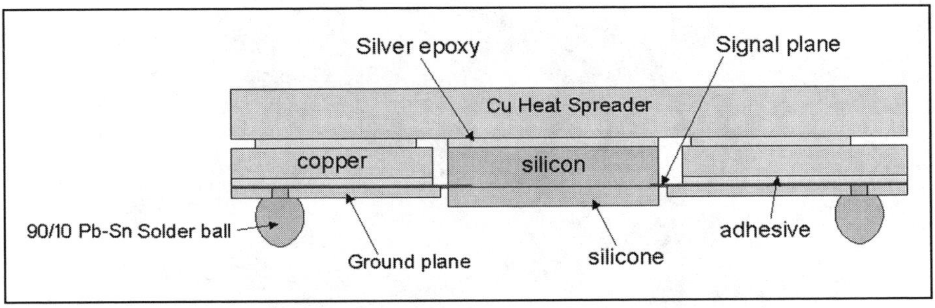

Fig. 4–22. Schematic of a tape BGA

Most of the BGA packages are 1.27 mm in pitch and the package efficiency is not as high as that of TSOPs. However, this technology enabled overcoming the perimeter paralysis associated with the fine-pitch leaded packages. BGA packages of 1.27 mm pitch contrast with the 0.5 mm pitch of QFPs. In addition, like the flip chip, these packages self-align in the assembly reflow process and are therefore much more forgiving to placement errors. Packages placed even up to 50% off the pad have been known to self-align; this feature increases the assembly yields and reduces the defect levels to single-digit ppm. Owing to the replacement of long leads with short solder balls, these packages offer better electrical and thermal performance and greater board packaging density.

However, there are a few disadvantages, too. Once assembled, the solder joints, except for the outer rows, are not amenable for inspection. Touch-up of solder joints, such as addition and removal of solder, is not possible. Removal and replacement of defective packages require special tooling, and mirrored packages on double-sided assemblies are extremely difficult, if not impossible, to replace. The advantages of BGA packages, however, far outweigh the disadvantages, and they have been well accepted by the electronics industry. Plastic BGAs (PBGAs) on organic laminate offer acceptable reliability, e.g., in excess of 1,500 cycles of thermal cycling in the −40°C–125°C temperature range, for most applications.

The ever-growing demand for high functionality in portable electronics places increasing emphasis on miniaturization and integration in packaging. Reduction in the ball pitch in area-array packaging, as well as increasing packaging efficiency, is perceived as a natural consequence. The need for high-density PWB technology to accommodate assembly of high density I/O packaging with finer lines, spaces, and smaller vias, as well as minimum number of board layers, was perceived and developed as discussed in an earlier chapter. Ideally, flip-chip (FC) on organic laminate would provide the highest packaging efficiency. Flip-chip assembly is generally not integral to the traditional surface-mount assembly process and requires additional dedicated stand-alone operation. In addition, assemblies of flip chip on laminate are associated with the largest CTE mismatch and hence are not reliable enough for most applications. The assemblies needed to be underfilled with appropriate materials and cured to achieve the desired reliability. Attendant with underfilling are several practical and infrastructure impediments. Underfilling the assembled packages is an additional stand-alone operation adding materials, assembly cost, and time. Once underfilled, the assemblies are not easily reworkable, and any defect discovered after underfilling and curing renders the assembly nonrecoverable.

Chip-scale packaging

The ever-increasing demand for high-functionality, high-density packaging, especially in portable electronic hardware, required innovative microelectronic packaging- and interconnection-related technologies. The key attributes for high-density package is thus a near-chip-size packaging, alleviation of the CTE mismatch concerns, elimination of the need for underfilling, use of traditional in-line surface-mount assembly process, and easy rework.

Chip-scale packaging (CSP) has evolved as a relatively new packaging technology innovation introduced in the industry around 1993. The concept was simultaneously proposed by Junichi Kasai of Fujitsu and Gen Mukarami of Hitachi (Mukarami 1994). This packaging concept can be viewed as an enabling technology for near-term miniaturization for the portable, high-function information processing industry. It can accommodate low to medium I/Os in the range of 20–350. The applications include SRAM, DRAM, Flash Memory, DSPs, logic ASICs, and a host of others. Figure 4–23 shows a schematic of a comparison of PWB real estate area requirements to support various packages and a bare silicon die. Obviously, almost 90% reduction in package area is attainable. CSP constitutes a low-cost alternative to flip-chip attach that does not require underfilling for many applications. Depending on the package format, CSPs can provide significant improvements in electrical performance compared to traditional packages. They are ideally suited for portable electronic products where size, weight, and robustness are important considerations.

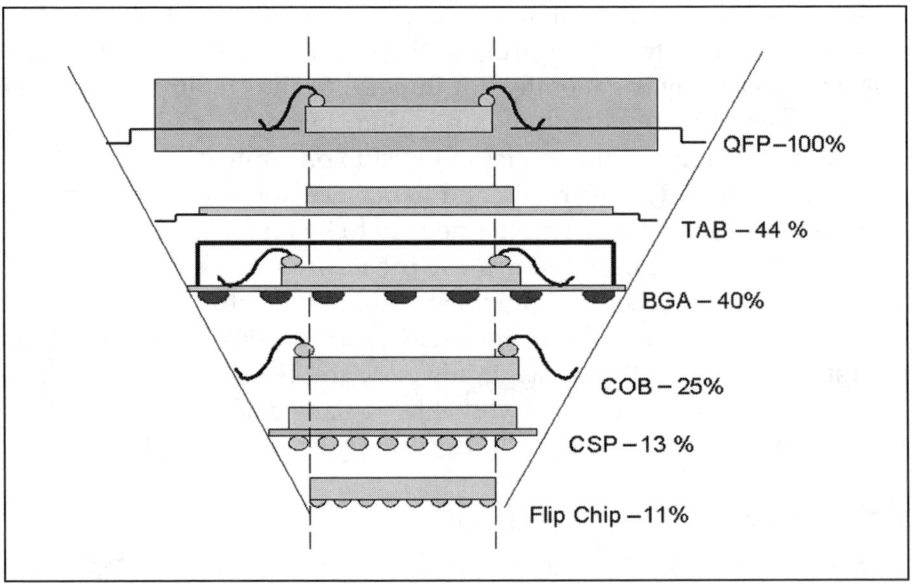

Fig. 4–23. Board real estate area requirements for different packages

CSP is generally a single-chip component technology that packages an IC device in a format that is either of the same size as or only slightly larger than the IC itself. They are semiconductor chip structures that have been made robust to facilitate ease of handling, chip assembly, and testing. Minimal size, no more than 1.4× the area of the silicon, and direct surface mountability are the attributes. The CSP nomenclature is also used for some packages with lead pitches of 1.0, 0.8, and 0.5 mm.

The main distinguishing feature from flip-chip attachment is that in the case of CSP there is still some packaging involved to enable attachment to the PWB. This is achieved by the incorporation of material layer(s) as interposers between the silicon die and the package substrate, which serve as compliant members for stress relief, space transformers, and/or mechanical protection for the silicon.

The number of variations of CSPs has been growing ever since their introduction, each with its own identification by the manufacturer, design, construction, or appropriate acronyms. There are well over 100–150 different styles. However, CSPs are generally classified into four groups based on the nature of the interposers and the construction as follows (Viswanadham et al 2001):

1. Lead frame-based packages
2. Flexible interposer packages
3. Rigid interposer packages
    a) Organic
    b) Ceramic
4. Wafer-level packages

It is nearly impossible to describe every package that is available in the industry. A brief description of each type is provided with representative examples to illustrate typical design, materials, and construction in the following sections. It is important to recognize that each of the descriptions provided is only one specific package configuration and does not represent others with the same design concept.

Lead frame-based packages

These are one of the first types of CSPs to be developed in the industry. Some of the many reasons for their evolution are the already existing package styles of thin small outline packages (TSOPs), use of copper and alloy-42 lead frame materials, traditional wire-bonding technique, and manufacturing infrastructure of the leaded packages. An initial purpose was also to provide a technique to stack components for memory enhancement. A variety of formats have evolved since then, and prominent among them are lead-on-chip (LOC), small outline no lead (SON), thin zero outline package (TZOP), bottom leaded package (BLP), land grid array (LGA) package, etc.

SON CSP

This package developed by Fujitsu was based on a multiframe lead-over-chip technology for low I/O pin count of less than 50 for memory devices. The inner leads are located on the top of the IC, and hence the traditional die paddle is not necessary. However, an additional lead frame may be used to hold the die. In the ultrathin small outline no lead package (USON), the die is face down and first-level interconnections are traditional wire-bonded. The second-level interconnections are plated flat lands, as shown in figure 4–24.

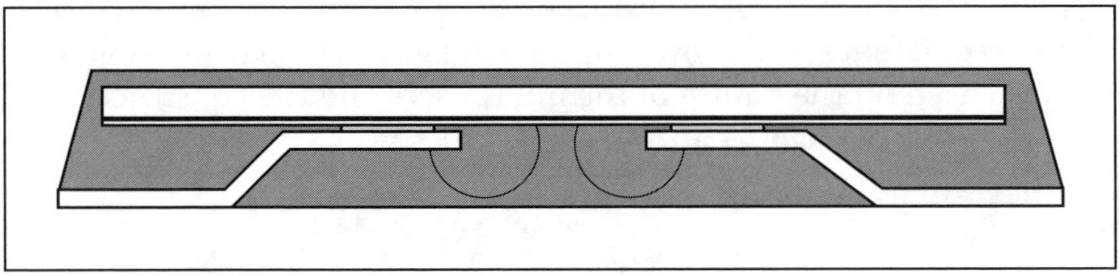

Fig. 4–24. Small outline no-lead package

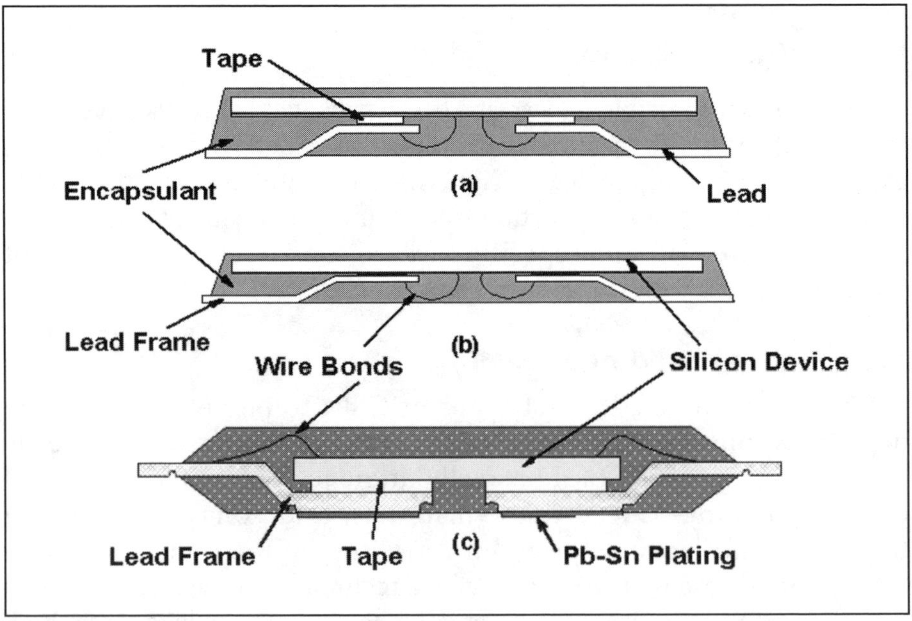

Fig. 4–25. USON, SOC, and tape-LOC package structures

In the small outline C-lead configuration, the SON is basically flipped with the bottom leads extended and wrapped around to the back of the package. In this configuration, the package is chip-face-up and has terminations on both the top as well as the bottom to facilitate package stacking, which is ideally suited for memory stacking. Another modification of the LOC configuration is tape LOC, where the leads are held to the silicon device with an adhesive tape in the BLP configuration. Figures 4–25a, b, and c depict the USON- BLP, SOC, and tape-LOC structures.

These packages are generally used for 16M flash ROM, SDRAM, etc. Figure 4–26 shows a four-package-stack three-dimensional packaging module. Single die packages are about 0.75 mm in thickness, and the two- and four-stack modules are, respectively, 1.5 and 3.0 mm in thickness. The SOC and SON package have been shown to have superior electrical performance compared to the corresponding TSOP structures owing to significant reduction in lead lengths (Hamano et al. 1997). The qualification and reliability tests include –65°C–150°C 100 cycle tests, pressure cooker test at 121°C, and 2 atm pressure for 168 h, with no functional failures.

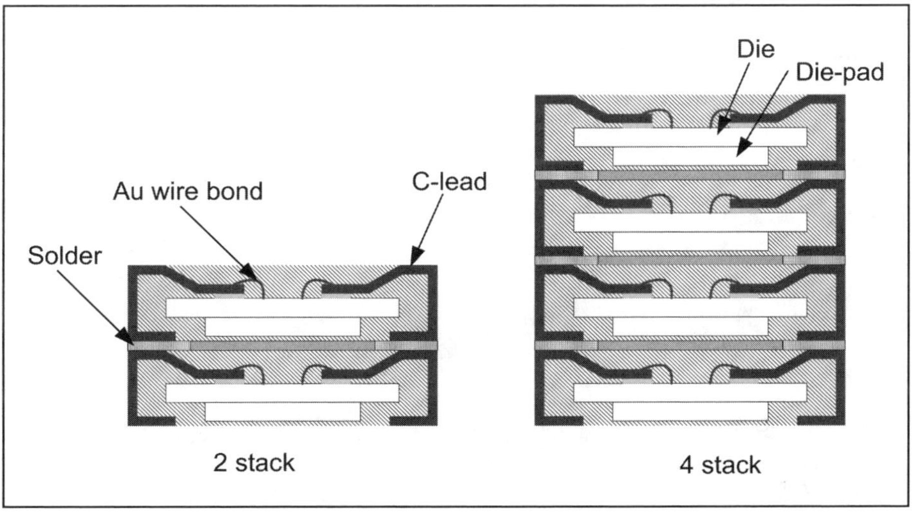

Fig. 4–26. Four-package stack module

Bumped chip carrier (BCC)

This package was developed as a low-cost replacement for low-pin-count shrink small outline packages (SSOP) and chip on board (COB) for communication products. It is a leadless package with 0.4mm wide mounting bumps at the bottom of the package (Yoneda et al, 1996). The terminations pitch is either 0.8 or 0.65 mm. Figure 4–27 shows a cross section of this package. The package construction starts with a 300-μm copper lead frame material. The copper lead frame is masked with an etch resist to expose the area where termination bumps are to be created. The lead frame is half-etched in those locations to form craters on the top side of the lead frame. This step is followed by electroplating operations to deposit sequentially, from inside out, 500 nm of palladium, 5 μm of nickel, 100 nm of palladium, and 5 nm of gold. The resist is then stripped away. The silicon device is mounted on the lead frame with a suitable die attach adhesive and cured. A modified wire-bonding technique is used by incorporating gold stud bumps in the craters followed by standard ball and stitch wire bonding to bond the die I/Os to the plated craters. A 0.6-mm wire loop with a 150-μm loop height is accomplished. Subsequently, the top side is encapsulated with an epoxy molding compound, allowing the epoxy to flow into the craters. The copper lead frame at the bottom is then chemically etched away exposing only the metallized bumps. Figure 4–28 shows schematically the steps in the fabrication process of the bump chip carrier.

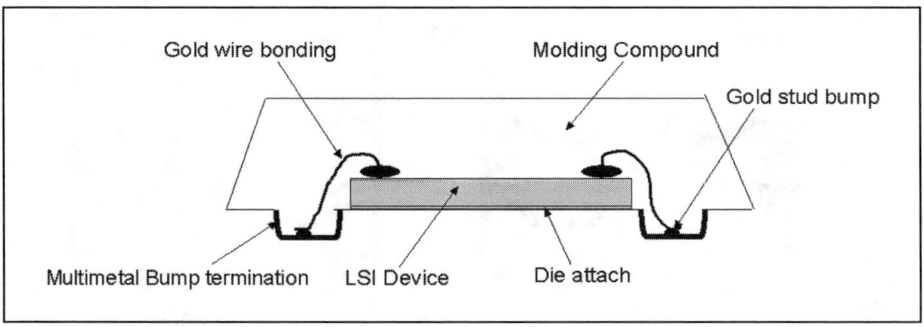

Fig. 4–27. Schematic of bumped chip carrier package

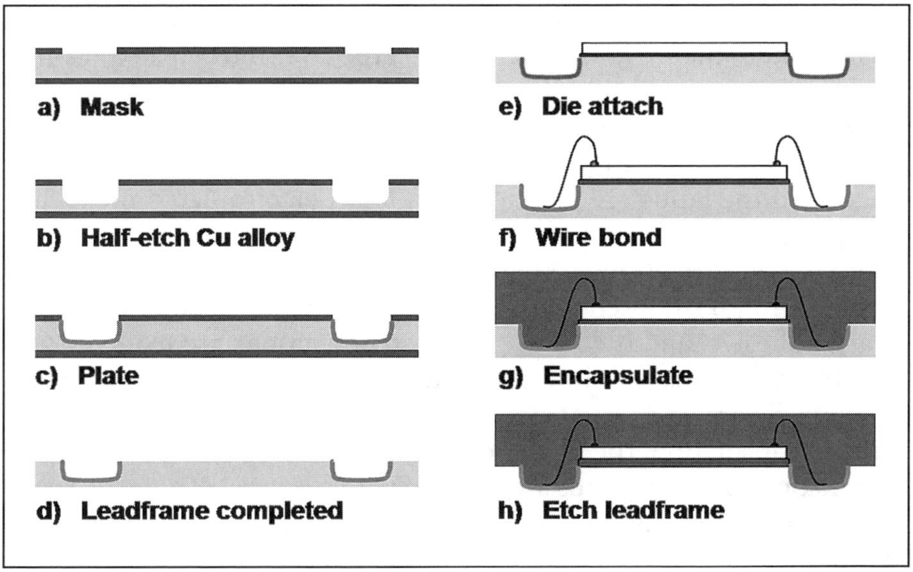

Fig. 4–28. Schematic of the processing steps in the fabrication of the bumped chip carrier package

BCCs have lower self inductance and capacitance compared to the SSOPs, and also show better heat dissipation characteristics. The wire-bond pull strength was around 10 g, much better than the 4 g specification. The packages have been shown to have good thermal cycling reliability in the –65°C–150°C range, surviving in excess of 200 cycles.

The low pin count (<50 I/O) BCC packages occupy 40% less real estate and 67% less volume compared to the SSOPs they replace. The packages have been in use as phase-locked loop synthesizers in many mobile phone applications.

Quad flat nonlead (QFN) packages

These are also custom lead-frame-based chip-scale packages and are of relatively higher pin counts, and were developed as substitute for the SSOPs. The concept utilizes low-cost lead frame along with conventional first-level wire-bonding interconnections. Standard manufacturing procedures of lead frame preparation, die attach, inner lead wire bonding, encapsulation, and termination formation are utilized.

The starting lead frame material is a 200-µm-thick Cu alloy with Ni plating. The nickel layer is arranged to form an area array pattern on the bottom side terminations. The Ni layer also serves as an etch resist. The lead frame is half etched to create 400 µm dia. and 50 µm height studs on top side and 500 µm dia. and 100 µm height studs on the bottom side. A 40-μm-thick build-up resin is coated on the top side and 100 μm vias are photoimaged and developed. On the layer thus created, 12-μm-thick copper is plated, and the desired circuit pattern with 60-μm lines and spacing is generated by the standard develop–etch–strip method. A design rule of maximum four lines per channel is employed. The Cu traces and the lands are finished with 3 μm electroplated nickel and 0.3 μm gold. The gold is used to facilitate gold wire bonding. The wire bonding pads are 200 × 100 μm rectangular in shape, are staggered, and are 140 μm in pitch. Subsequently, the die is attached with a die attach adhesive, and the first-level interconnections are made by gold wire bonding. This is followed by encapsulation with a suitable epoxy molding compound. The thickness of the molding is about 0.8 mm. After encapsulation, the back side is etched appropriately to form the terminations. And the lands are finished with a coating of eutectic solder to facilitate second-level assembly to the PWB. Figure 4–29 shows the schematic of the QFN fabrication (Kasai et al. 1996) and figure 4–30 shows a schematic of the construction.

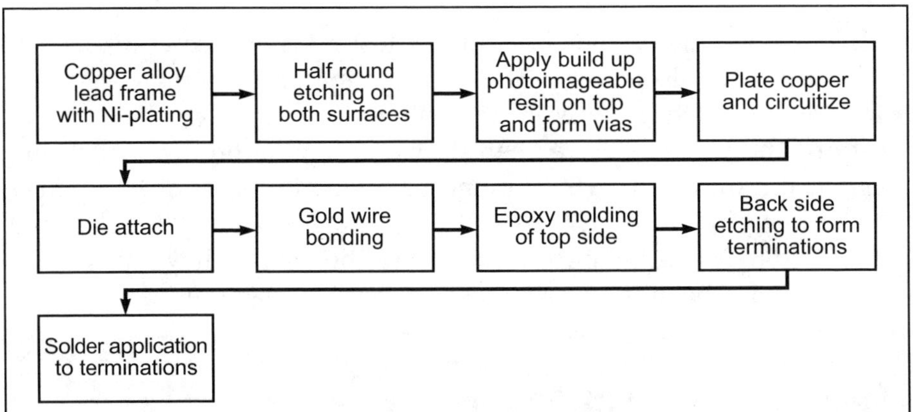

Fig. 4–29. Schematic steps in the fabrication of a QFN package

The same concept was utilized to create an area-array package by suitable etching of the back surface and attaching solder balls as the terminations for board assembly. This version is termed the micro BGA. The *micro BGA* version accommodates a higher pin count than the QFN package. These packages are very compact and are generally reliable.

The lead frame-based chip-scale packages thus utilized the existing material sets and manufacturing techniques with innovative design concepts to meet the demands of high density. Yet, the packages are limited to rather small pin counts owing to the perimeter wire bonding and the existing limitations of the manufacturing infrastructure. The second-level package to board interconnection reliability was not as high as with traditional leaded packages. It depended on the overall CTE of the package, which in turn is dependent on the size of the silicon, the lead frame material, and the amount of molding compound. Given the form factor of the chip-scale packages, the CTE can be rather low. It has, however, met the requirements in a limited manner since the packages are small in size and the solder joint distance from package neutral point (DNP) is small enough to meet the short product life requirements.

Fig. 4–30. Schematic of QFN showing the bottom and construction

Flexible interposer packages

Flexible interposer packages are the next evolutionary step in the development of CSP. As the chip is about the same as the package itself, the CTE is, by design, closer to that of silicon. There is an ever-growing demand for compact, high-function, high I/O devices in portable electronics. The global CTE mismatch between the package and the PWB is a concern for second-level solder joint reliability. It is important to alleviate the solder joint stresses under thermal loads by innovative package designs. Packages utilizing flexible interposers, some based on the tape automated bonding (TAB) technology, evolved as a possible solution. There are multitudes of package designs using this concept in the industry. However, a few package design types are described to illustrate this technology.

Tessera micro BGA

This is probably one of the earliest package structures and is based on the TAB technology. A compliant mounting tape is prepared from 25-μm-thick polyimide tape with copper on both sides with redistributed 25-μm I/O signal lines on one side and 5-μm ground plane on the other. The copper traces are gold-plated to facilitate bonding to the chip pads. The tape in reel form is then sliced into small strips and mounted onto a metal frame. Each strip contains a plurality of package footprints. Figure 4–31 shows one such frame. On to the frame is deposited, selectively, a 150-μm high-temperature silicone elastomer that is 50% filled with silica. An adhesive for die attachment is also dispensed at the appropriate location. The silicone elastomer layer serves as thermomechanical stress reliever, provides the necessary compressibility for testing packages in a socket, and also serves as a possible alpha-particle absorber. An automated pick-and-place machine is used to mount the silicon device. The first-level interconnects comprise 25-μm-thick, 25-μm-wide S-shaped gold-plated ribbon leads that are attached to the die pads by a single-shot thermosonic bonding. The solder spheres attached to the plated pads constitute the second-level interconnects. Figure 4–32 shows a schematic of the package fabrication steps (Fjelstad 1998). Thus, the module is a BGA package. The ball pitch can be 0.5, 0.75, or 1.00 mm. The total package height is 0.7 mm. With a fan-in configuration, the package footprint can be slightly smaller than the device itself. Four different package styles, namely chip size type, ring type, can type, and fan-in/fan-out types, are in vogue. Second-level package reliability has been shown to be equivalent to or better than conventional quad flat packs, surviving in excess of 1,000 cycles of 0°C–100°C. The dimensions of the S-shape ribbon are critical for the second-level reliability. The bend radii of curvature of the S, the height of bond, and the angle Θ are optimized to obtain the best reliability. Nonoptimal ribbon shape can lead to ribbon bond breakage in thermal cycling as shown in figure 4–33.

Fig. 4–31. Polymide strips with several package footprints

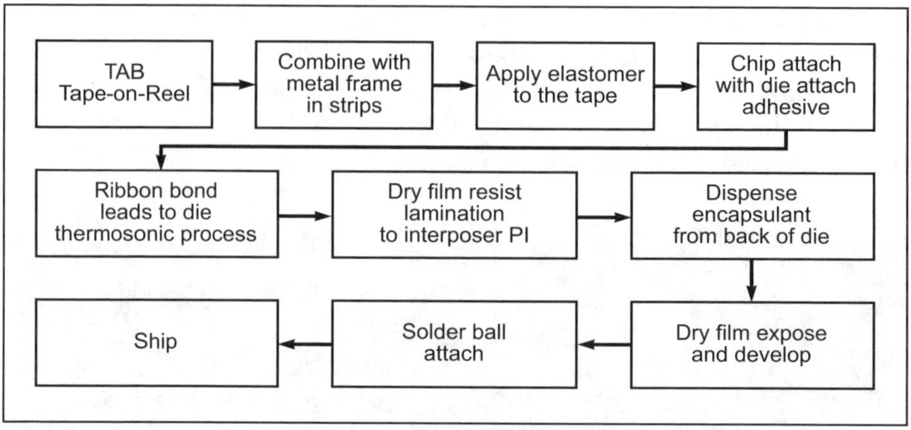

Fig. 4–32. Steps in the fabrication of the microBGA package

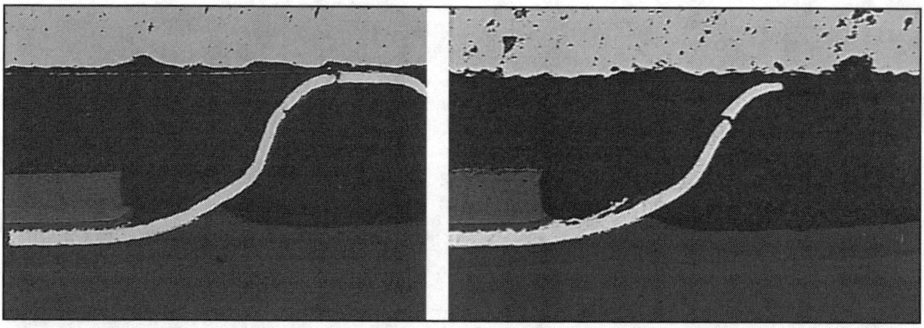

Fig. 4–33. Ribbon bond fracture in thermal cycling test in microBGA

The package has several advantages specific to portable electronic hardware. It is low profile, highly reliable, reworkable and die-sized for high packaging density, and has direct thermal path for heat dissipation, low inductance, and the potential for high I/O applications.

Micro-star BGA package

This flex-based tape carrier package also utilizes the TAB tape as an interposer, utilizes traditional wire bonding for the first-level interconnect and a grid array of solder balls for the second-level interconnect, and is over molded (Ano et al. 1997).

The interposer is a three-layer tape that is comprised of a copper layer, an adhesive layer, and a polyimide layer. The copper layer is patterned with 40 μm lines and spaces. The polyimide layer is drilled with an array of 200 μm diameter vias to facilitate second-level interconnect. The traces are plated with nickel and gold metallurgies for wire bondability.

The die attach is performed with a nonconductive die-attach adhesive and cured. The device I/O pads and the copper traces on the flexible carrier are connected by gold wire and T/S bonding. The die and the wire bonds are encapsulated with an epoxy molding compound. After encapsulation, the package is flipped and solder balls are attached to the bottom of the copper traces through the vias by an infrared reflow process. Figure 4–34 shows a schematic of the microstar package. The entire package assembly can be accomplished in a reel-to-reel format. The individual packages are singulated from the reel and are packaged for shipping. The package is only 2 mm larger than the silicon device and is 1.2 mm thick. The packages are supplied in 0.8–0.5 mm pitch, and the I/O count is in the range of 62–300, the most popular being 144 I/O package. The packages demonstrated excellent second-level thermal cycling reliability in excess of 1,000 cycles in the 0°C–100°C range and a JEDEC moisture resistance level of 3 or better. These CSPs are extensively used as ASICs and DSPs in many portable electronic products.

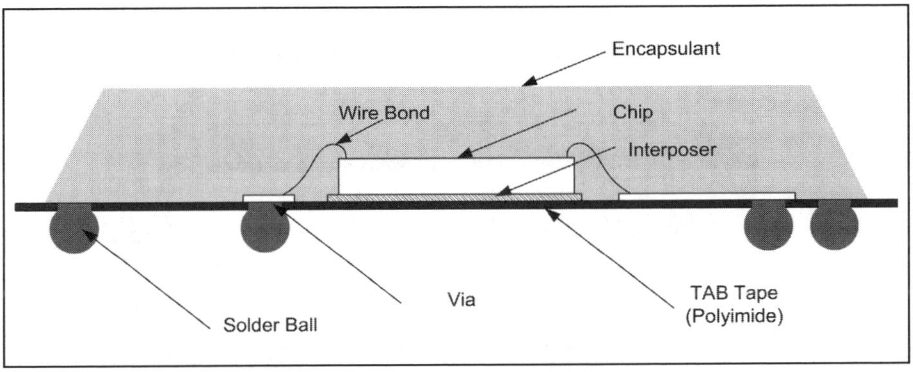

Fig. 4–34. Schematic of microstar BGA package

Enhanced flex CSP

This package is an offshoot of 3M's Microflex tape BGA technology. It is termed "enhanced" because of the incorporation of a copper lead frame to reduce the CTE mismatch between the package and the PWB. The Microflex TBGA technology incorporates a flex laminate with the capability of wire-bondable metallization for the first-level interconnects for tape BGA packaging. This capability was extended and utilized in the design of the enhanced flex CSP (Schueller 1997).

The three-layer flex consists of a 50-µm thick polyimide, a 25-µm thick thermoplastic adhesive layer, and a 127-µm copper layer. The copper layer is circuitized to provide 35 μm lines/spaces copper circuitry and tapered 375-μm diameter solder pad vias formed by chemical etching. The tapered side-wall vias are designed to provide a smooth profile, reducing stress concentration due to thermomechanical stresses. The copper traces are plated with 1.5-μm-thick Ni followed by 0.76-μm-thick Au for wire bonding. The flex circuitry and the 127-μm-thick copper lead frame are bonded together with a 25-μm-thick thermoplastic polyimide adhesive film. Together, this constitutes the lead frame/interposer. Slots are opened at the corresponding locations on the lead frame and adhesive film to access the wire bonding pads.

The silicon device is attached to the lead frame in the conventional manner with a die attach adhesive, and the wire bonding is effected with a gold wire. Subsequent to wire bonding, the package is over-molded with an epoxy molding compound. Solder balls are attached to the tapered vias by the infrared reflow process. The ball pitch is 0.5 mm. A schematic of the package is shown in figure 4–35.

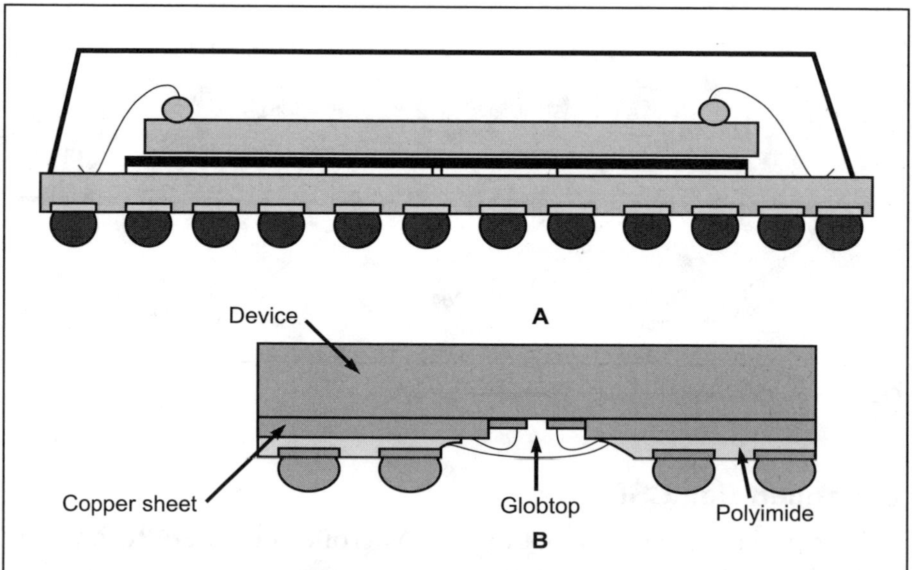

Fig. 4–35. Schematic of the enhanced flex CSP in the (a) cavity-up format and (b) cavity-down format

Both cavity-up and cavity-down formats are possible for this package, as shown in figure 4–35(a) and (b). The cavity-down version will have a lower pin count in the range 50–100 for applications such as DRAMs, while the cavity-up configuration has a relatively higher pin counts in the 100–300 range for high-performance microcontrollers and digital signal processors (DSPs).

Owing to the materials' choice and design, the package is very moisture resistant and has demonstrated acceptable reliability.

Rigid interposer packages

CSPs are also designed with a rigid substrate interposer. The rigid interposer can be either of ceramic or organic laminate material. The choice of rigid substrate depends on the functionality of the package required. Multilayer or thick-film ceramic interposer may be required for specific high-wattage, high I/O high functionality devices. Organic laminates generally include such materials as bis-maleimide triazine (BT), high-temperature tetra-functional FR-4, etc., and are generally well suited for moderate to low I/O applications, and compatible with organic PWBs.

Amkor chip array package

This package derives its name from the fact that it is manufactured in an array format. The interposer can be either organic or ceramic. The ceramic version is generally used for disk drive applications. The organic interposer is about 0.34 mm thick. The die thickness is 0.3 mm. The first-level interconnect is by gold wire bond on a bond pad pitch of 90 μm. The copper traces on the laminate are Ni- and Au-plated. It is a near-die-sized package (NDSP) with a lead count in the 28 to 128 range and is available in sizes of either 5 × 5 mm or 11 × 11 mm. The second-level interconnect is either solder ball or a land grid array at a variety of pitches in the 0.5–1.0 mm range in single or double row perimeter arrays. The package is overmolded with a resin and is about 0.7 mm in thickness. Figure 4–36 shows a schematic of the package. The package has a self inductance of 1.4–7.8 nH and an impedance of 170 Ω and is generally used for ASIC, memory, analog, and telecommunication applications. Since it is a plastic package, it is sensitive to moisture and is supplied as a level-3 package. In tests, the package survived without failure in the pressure cooker test at 121°C and 100% R.H. for 500 h; temperature, humidity and bias test at 85°C and 85% R.H. for 1000 hr, and thermal shock test of –55–150°C for 1,000 cycles.

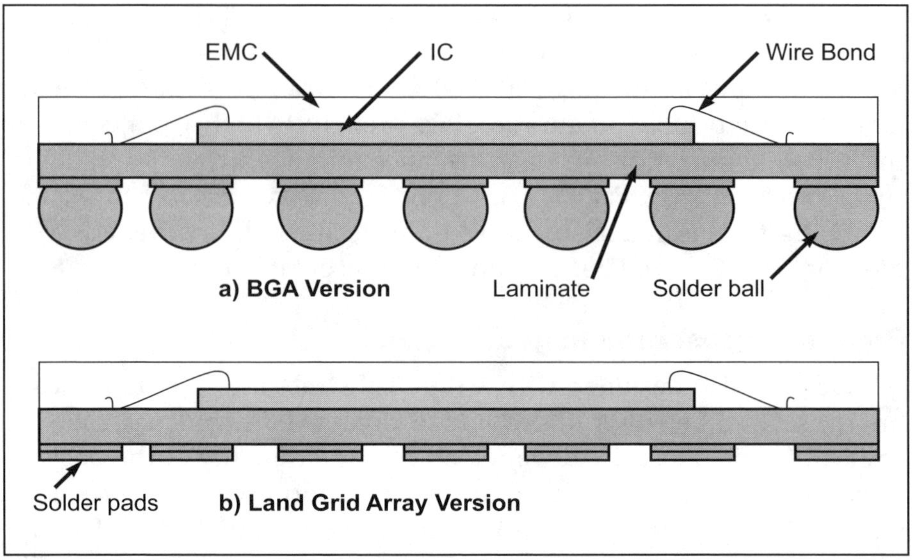

Fig. 4–36. Schematic of the construction of the chip array package

The slightly longer than IC carrier (SLICC) package

This package exemplifies a rigid interposer package that utilizes flip-chip controlled collapse chip connection (C4) licensed from IBM. The second-level interconnects are either 0.5 or 0.8 mm pitch solder balls. The package design employs both fan-in and fan-out patterns. The 0.2–0.3-mm-thick interposer is made of a glass-reinforced FR-4 or BT resin circuitized with 4 mil (100 μm) lines and spaces. Two layers of routing, one on each surface, are employed. Punched holes of 8 mil on 16 mil pads plugged with copper are employed. The top bonding pads are solder-cladded. The bottom interconnect pads are 0. 5 mm in diameter and 0.8 mm pitch and plated with electroless Ni and flash Au for solder ball attach. The silicon device is bumped with 76–127 μm 97Pb/3Sn high-melt solder balls at 10 mil pitch. After the C4 attach of the die, the interspace is underfilled with an epoxy and cured. The second-level interconnect is via 22-mil-dia. silver containing 62Sn/36Pb2Ag balls. The total package is about 28–100 mils larger than the die with a thickness of less than 1 mm. Figure 4–37 shows a schematic of this package with approximate dimensions.

Another variation of this package, called *just about chip size* (JACS) package, utilizes a semi-rigid FR-4 or polyimide laminate instead of BT resin. The high-melt solder balls for first-level attach are replaced by eutectic Sn/Pb solder balls.

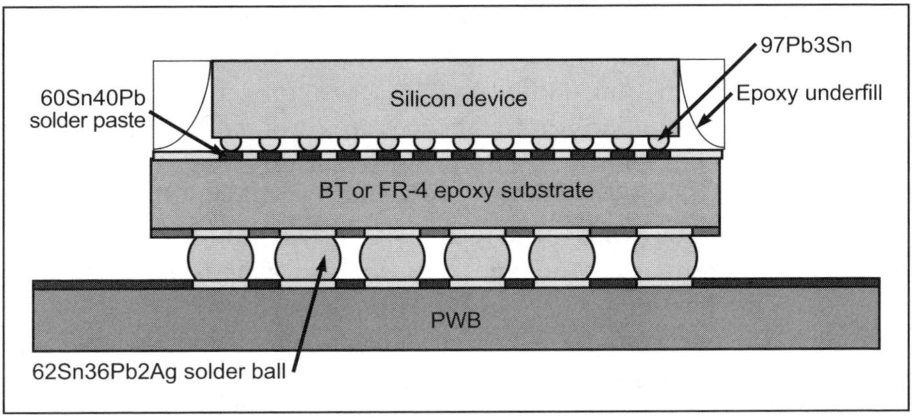

Fig. 4–37. Schematic of the SLICC package

These packages are of 150–200 I/O containing 3–4 perimeter rows and are assembled in a panel form. This package is better than its QFP counterpart in terms of capacitance, inductance, and parasitics. The package was demonstrated to pass traditional reliability tests such as 500 cycles of liquid-to-liquid thermal shocks at –55°C–125°C, 500 cycles of –40°C–125°C air-to-air thermal cycling, 1,000 h of temperature, humidity, and bias at 85°C/85% R.H., etc. The package is designated at level 3 moisture resistance. Interconnection failures during thermal cycling were reported to be on the package side after 1,500 cycles (Banerjee and Lall 1995).

Wafer-level packaging (WLP)

The package designs and constructions described in the preceding sections still require packaging of the singulated devices in an array or panel format. The panels are then singulated again and tested. The evolving trends in portable electronic packaging have been to reduce the overall price structure of packaging. Increase in wafer size and decrease in die size provide a higher device density on the wafer. A natural consequence is the development of the ability to perform all the packaging operations on the wafer itself that would enable singulated packages ready to assemble on the PWB. As a consequence, miniaturization, economies of scale, higher yields, and lower cost are achievable. As a concept, WLP has been in practice in a few companies involved in high-volume flip-chip attach. The device bond pads are provided with appropriate under bump metallurgies (UBM) and solder bumped at the wafer level, ready to be diced and flip-attached on substrates. WLP is thus a natural migratory path to a truly chip-size packaging with shorter interconnection, increased electrical and thermal performance, surface-mount compatibility, most of the time without the need for undefills for reliability enhancements, and ultimately at a significantly reduced cost per package. The I/O pads on the devices are rerouted in an area-array format using a compliant interposer redistribution layer.

It is important, however, to recognize that assembly of these devices onto the PWB requires advanced technologies such as high-density interconnect and micro-via technologies that would provide 100 μm pad sizes with 50–75 μm lines and spaces and 100 μm microvias. WLP is especially well suited for low I/O (~100) devices and also MEMS applications. One process for WLP consists of two major segments, namely preparing the wafer and preparing the complaint multilayer redistribution layer that is tested, and aligning the two and bonding followed by solder ball attach. A schematic of the typical WLP fabrication process is depicted in figure 4–38. Alternatively, the interposer distribution layer is fabricated on the wafer itself. A few examples of wafer-level CSPs are illustrated in the ensuing sections.

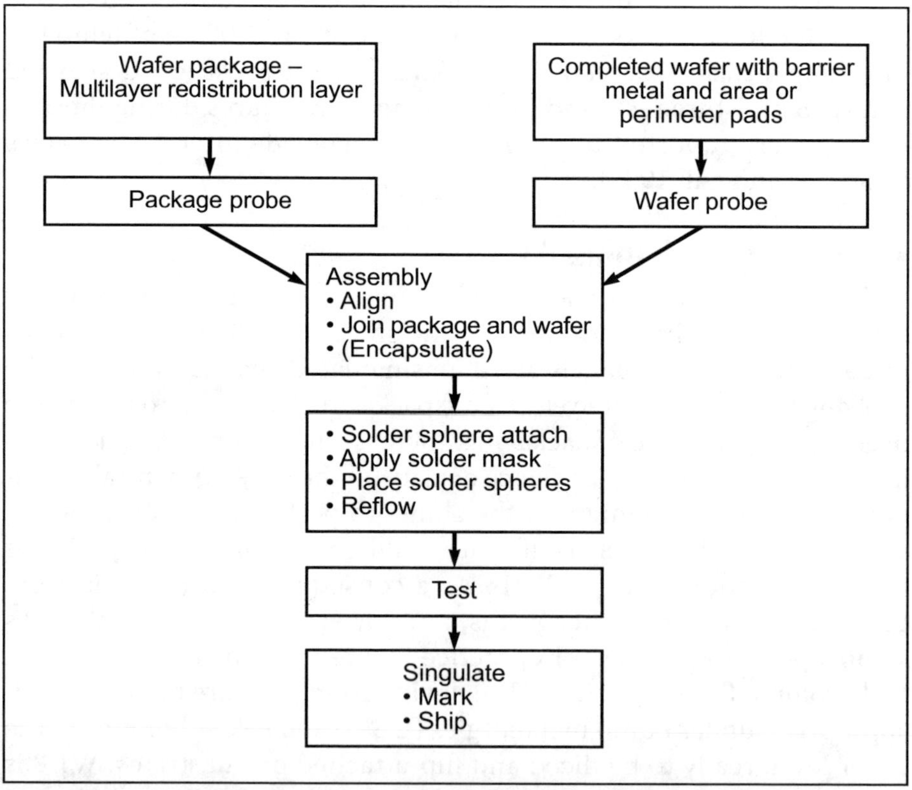

Fig. 4–38. Schematic of a typical wafer-level package fabrication process

Sandia mini BGA

This is a BGA where the processing is performed on the wafer itself. A 75-μm-thick dielectric such as benzocyclobutene or polyimide is first spin-coated on the wafer. The surface is patterned and etched so that the I/O pads on the die are exposed. Thin layers of barrier metals Ti/W and seed Cu layer are sputter-deposited on the dielectric. Then the layer is patterned, the interconnect metal Ni and Cu are plated to form the I/O redistribution pattern, and the barrier metals and seed metals are etched away leaving only the desired circuitry. Ti, W, and Ni act as diffusion barriers and Cu as the main circuit conductor. A second layer of dielectric is then spin-coated and vias are patterned and etched in order to expose the pads underneath for I/Os. Again Ti/W barrier metallurgies and Cu seed metallurgies are sputter-coated. The top layer is then patterned, and copper and solder are plated. The barrier and seed metal layers are etched away except in the area of the I/O. The plated solder is reflowed to form the solder bumps. Figure 4–39 shows a schematic of finished mini BGA with different layers identified. This package was designed to be a low-to-medium I/O in the range of 34–300, and its applications include DSPs, microcontrollers, and ASICs (Chanchani et al. 1995).

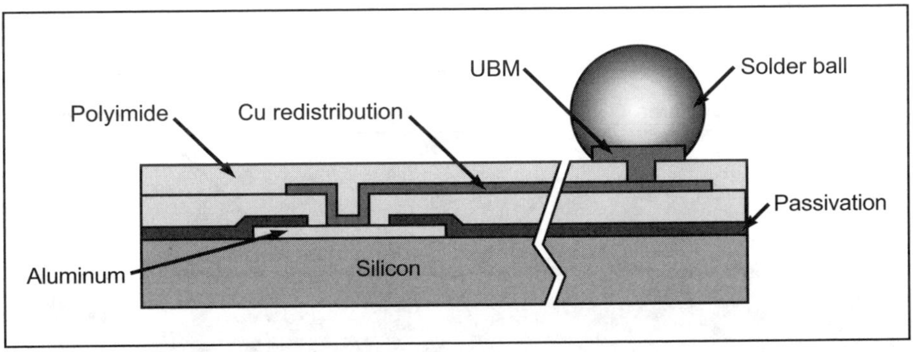

Fig. 4–39. Mini BGA chip scale package schematic

Shell case CSP

This package is also called *SlimCase, ShellPack*, etc. The concept involves encasing the silicon devices between two glass plates and offers the potential for optoelectronic applications. Also, owing to its extremely thin structure, it is amenable for embedding in smart cards.

A metallic layer of "pad extensions" is created on the die pads to link the original die bond pads to the wafer scribe line. A thin protective glass layer is bonded to the active side of a silicon wafer with an epoxy adhesive. Then an organic layer is applied to act as a compliant layer between the package body and the solder joints. Deep notches are then drawn between the die in the scribe lines, revealing the cross-sections of the pad extensions. A metal layer, patterned by lithography to individual leads, is deposited to contact the pad extensions at their cross-sections. A proprietary "T-contact" structure is formed. The leads connect the pad extension layer with the external surface of the package. A solder mask passivation layer is then applied. Subsequently, solder paste is screen-printed and reflowed to form the solder BGA on the compliant layer. The backside of the silicon is the ground and thinned down to 75 μm. The silicon between the dies in the scribe line area is etched away (back side), yielding individual ICs attached to the supporting glass layer. A second thin glass cover is bonded onto the backside of the etched silicon to achieve a complete protective enclosure for each die. The packaged wafer is finally diced on a UV-sensitive tape to yield individual chip-size packages. Figure 4–40 shows a schematic of a finished product. ShellCase packages exhibit superior electrical performance. The package thickness is only about 3 mm. Packages are offered in perimeter- and area-array formats. They are hermetic and are known to have good thermal performance, and the applications include ASICS, EPROMs, etc. (Badihi and Por 1995).

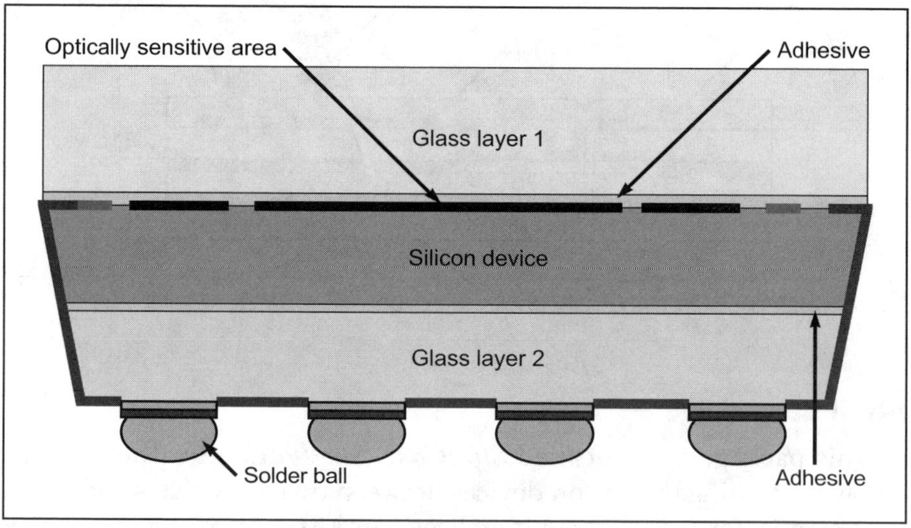

Fig. 4–40. Schematic of a ShellCase chip-scale package

Flip-chip packaging

Flip-chip attach has been described in detail in an earlier section as it pertains to first-level packaging. The practice is prevalent with ceramic packaging owing to its CTE compatibility. Flip attach directly on organic printed wiring board is practiced only on a limited basis. Several limitations have impeded its extensive use. In their early days, the standard surface-mount technology placement machines had limited capability of accurately placing the face-down chip on the board. Special dedicated equipment was required. The PWB technology was not advanced enough to accommodate high I/O, fine-pitch silicon devices. Thus flip-chip attach required dedicated stand-alone tooling. Also, the CTE mismatch between silicon and the PWB was too large, and the assemblies required underfilling with suitable encapsulant. Many of the underfills are not amenable for rework. Once assembled and underfilled, any defective chip cannot be reworked and the entire circuit card assembly has to be scrapped. Thus, flip chip on organic laminate has not been a favorite of many assembly shops. It is important to recognize that it is still practiced in the industry owing to some of its advantages.

Land grid array (LGA) packages

Land grid array packages are area-array packages, perimeter or full or partial array with no solder balls attached to them. Devoid of solder spheres, they constitute a lower profile option. Standoff can vary depending on the package and the volume of the solder paste applied in assembling them. Standoff is generally in the range of 0.6–1 mm. LGA package pads are generally finished with 0.1–0.5 μm electroless gold over nickel that acts as a diffusion barrier for the underlying copper pad metallurgy.

LGA packages are compatible with tin/lead or lead-free solder assembly process. Due to the absence of solder spheres, they have a lower mounted height compared to their BGA/CSP counterparts and are amenable to heat sink attachments, and are suitable for low-profile product applications especially in double sided assemblies. Owing to their small size, the corner solder joint distance from the package neutral point is small enough to provide adequate thermal cycling reliability in many applications. Also, the possibility of missing or damaged solder balls in BGAs and CSPs is not encountered. It is important to assemble LGAs on to PWBs with no-clean solder pastes since there is not enough standoff for cleaning. The assembly yields with LGAs are generally higher than with leaded packages as these too self-align even when placed 50% off the pads like CSPs and BGAs.

Typical LGA assemblies with appropriate solder joint volumes provide in excess of 3,000 failure-free thermal cycles in the 0°C–100°C range, which is more than the requirement for portable electronic hardware (Kujala et al. 2002).

Stacked silicon

As demand for high-function, high-density, and cost-effective portable electronic devices continues unabatedly, and at the same time the aerial real estate on the PWB is efficiently populated with chip-size devices, innovative methods of packaging to increase the packaging density are explored. Die stacking packaging thus evolved. It is the technique of mounting multiple silicon dies on top of each other within a single package. The initial efforts involved progressively stacking memory devices such as Flash and SRAM, which were followed by integration of logic, analog, mixed signal, etc. The scheme provided considerable savings on the real estate on the board at the same increasing the package performance owing to decreased interconnection lengths between circuits.

The designs consisted of pyramidal stacking of the chips, i.e., stacking smaller and smaller chips on top of each other to facilitate wire-bonding of the device either to each other or to the substrate. The designs are limited to one or two perimeter pads on the chips. A typical three-chip stacking is shown in figure 4–41. Use of adhesively attached spacers of appropriate thickness to allow for a minimum of 100 μm wire loops instead of 150–175 μm enabled stacking of chips irrespective of their relative sizes and shapes. Wire-bonding within such tight spacing was indeed a challenge. The traditional ball bond on the device and stitch bond on the substrate was sometimes replaced by stitch bonds at both terminations to reduce or eliminate wire loops and reduce the overall profile. Figure 4–42 shows the schematic of chip stacking with spacers.

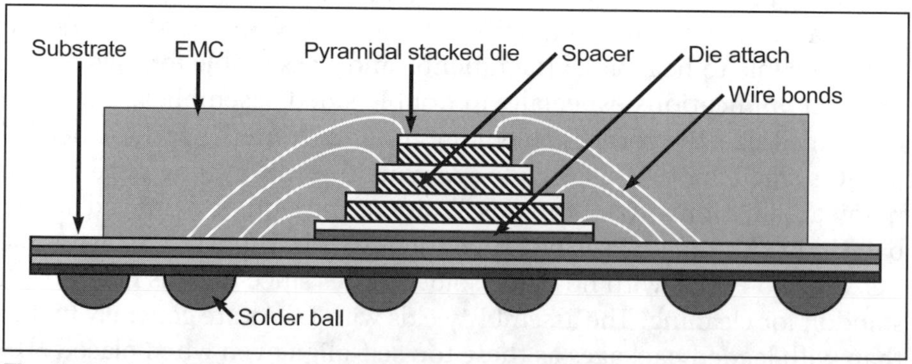

Fig. 4–41. A typical three-die stacking arrangement

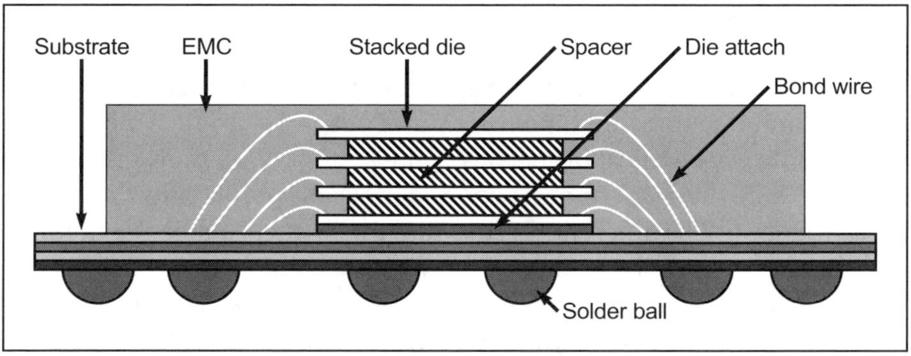

Fig. 4–42. Schematic of chip stacking with spacers

When high I/O area-array chips are also involved in the design, a combination of flip-chip and wire-bond scheme is employed. The flip-chip device is bonded first to the substrate and the wire-bonded device is mounted back to back on the first chip as shown in figure 4–43.

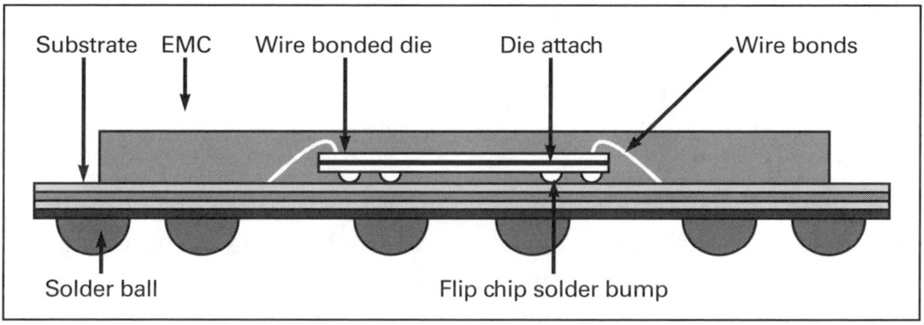

Fig. 4–43. Schematic of die stacking with flip chip and wire bonding

As more dies are designed into the package subsystem, maintaining the total package height and meeting the standards become a formidable challenge. For multiple die stacking, thinner chips in the 75–150 μm range are employed. Six-die-stacked and four-die-stacked packages with 1.4 and 1.2 mm height, respectively, have been demonstrated (Khan et al. 2008). Consumer and cell phone applications predominantly utilize two-die stacking for logic and memory architecture. High-density DRAM and Flash memory modules have four-die stacks, and some capability has been demonstrated for stacking even eight dies. Stacked die provide low-profile packaging options with advanced wafer thinning. It is also a lower package cost option with lower substrate consumption.

While die stacking has become mainstream packaging technology wherever applicable, it is imperative that thermal and mechanical reliability aspects require careful evaluation and assessment in each case. Package construction with die of different sizes and thicknesses, spacer materials and their dimensions, properties of adhesive and the thickness, and the laminate material dimensions and properties induce inhomogeneities in the mechanical properties of the system. Variations in the local CTE, modulii, thermal conductivity, and other properties affect assembly performance under different thermal and mechanical loading conditions and should be carefully evaluated using finite element modeling (FEM) techniques and verified through testing (Mitchell et al. 2004).

An important aspect in the die stacking technology is that the individual semiconductor devices are tested good, implying that known good die (KGD) is a prerequisite for stacking. A single defective die in the stack discovered after assembly renders the whole package unusable. The possibility of incompatibility among semiconductor chips from different suppliers can cause testability issues and difficult-to-isolate problems when encountered.

As the number of dies stacked increases, it is recognized that process complexity and expense factors escalate exponentially and alternative technologies are explored. One such alternative is package stacking instead of die stacking, which is described in the next section.

Stacked packages

In view of the complexities recognized in the concept of multiple die stacking in a single package, an alternative and complementary approach involving stacking individual packages on top of one another emerged. This is variously called *package stacking* or *package on package* (POP) (Sjoberg et al. 2007). Package stacking option permits combining packages of select functionalities from different suppliers and affords flexibility. In addition, each package is individually tested for functionality and reliability. A majority of these packages are BGAs with perimeter array of balls. The bottom package substrate extends beyond the periphery of the encapsulation with rows of pads on the perimeter to enable attachment of the top package. The top package is a perimeter-array BGA package with a corresponding row of balls. The solder balls on this package are large enough to clear the top of the bottom package. The bottom package is first placed on the solder paste-screened PWB, and the bottom package is then either dipped in a liquid flux or a specially formulated paste and then placed on the top package. Both packages are reflowed simultaneously along with the rest of the bill of materials for the product. Figure 4–44 depicts a schematic of a POP configuration.

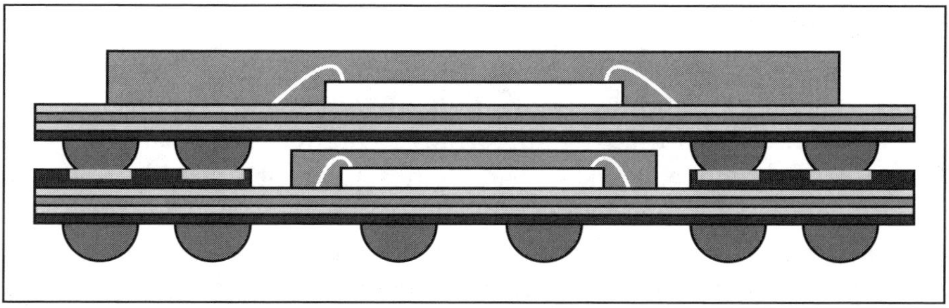

Fig. 4–44. Package on package stacking for increasing the density of packaging

While the concept of POP is conceptually simple, it is important to be cognizant of the implication in assembly and reliability. As the packages are thin with thin substrates and very thin encapsulation, there is propensity for moisture absorption. The physical and mechanical properties of the individual constituents of the package such as the adhesives, the laminate, the encapsulant, silicon, etc., and the package design can introduce package warpage, resulting in poor solder joints. It is important to understand, evaluate, and assess the effect of these factors on product performance.

Stacked silicon in stacked packages

A logical extension to the above concepts of silicon and package stacking is to combine these two concepts to achieve higher packaging efficiencies. As a consequence, POP each with a plurality of die stacks emerged. Figure 4–45 shows a schematic of a POP with multiple chips in each.

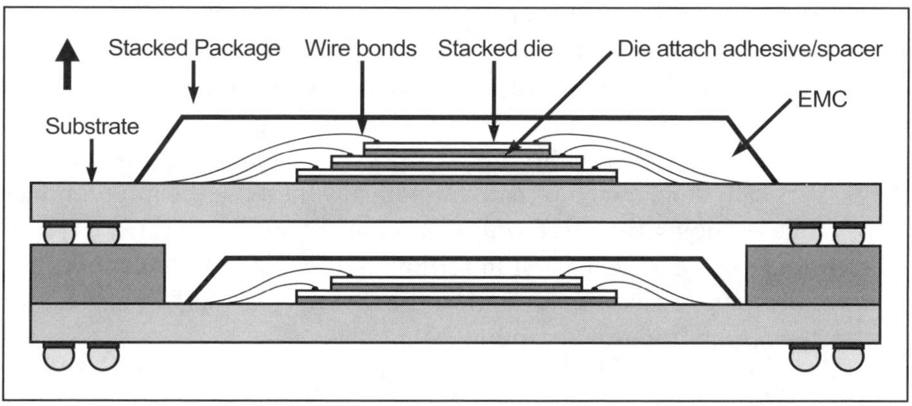

Fig. 4–45. Schematic of package on package with multiple chips

Emerging trends

In the packaging arena, wafer-level packaging provides as small a package as can be fabricated. Efficiencies are to be obtained by increased integration at the silicon level, higher I/O in the range of 300–500, and reduced package pad pitches. Current assembly practices provide acceptable yields with 0.5 and 0.4 mm pitch packages. The trend for the near term seems to be migration to 0.3-mm-pitch packaging with ball diameter in the range of 50 μm, which may place severe constraints on the board technologies. PWB circuit lines and traces may have to be in the 18–20 μm range and 75 μm via diameters and solder mask registration better than 20 μm (Katahira et al. 2007). Routability of the high I/O packages on the PWB might require stacked microvias and extensive use of via-in-pad structures. With smaller pads, registration of the vias in the pads can be a challenge. With small pad sizes, the ball mounting pads may have to be solder-mask defined instead of non-solder-mask defined.

Through-silicon via

The need for higher integration and smaller footprint cannot be met by package stacking and Si stacking alone. Embedded applications for portable communications and driver assistance systems for automotive applications require reliable integration of mixed technology subsystems such as sensors, actuators, analog, or memory. Approaches such as multichip modules (MCMs) that utilize global bussing to connect subsystems sacrifice too much to electrical path losses and effectively limit performance and low-power uses. For such applications, 3D stacking of multiple, mixed technology dies is seen as an attractive alternative.

Through-Si via interconnection between dies was first proposed more than 20 years ago (Stuby and Falls 1972; Warabisako 1983). It took the maturation of important process technologies such as deep silicon etching (Laermer and Schilp 1996), wafer thinning, etc. to realize the potential for high-density and high-speed signal transmission between chips. Vertically stacking chips can dramatically increase the interconnections while reducing the interconnection path at the same time. Vertical interconnections can also assist in removing the generated heat because in addition to their electrical function the through-Si vias can also act as heat pipes (Boetcher and Ostman 2001).

Making these through-Si vias and bonding chips to each other, then, is at the heart of 3D stacking dice. It important to note that through-Si vias (TSVs) are independent of the other circuitry made during the device fabrication line. Some ways of making vias are as follows (Denda 2007):

1. Via first (Vias made prior to the wafer process classified as "via first").
   a. Before front end of line process (FEOL)
   b. After FEOL
2. Via last
   a. Via from top: the via is made from the top of the thick Si chip (Takahashi 2003). Typically, the via is deep-etched by reactive ion etching (RIE), and the inside of the via is covered with sputtered SiO_2 and filled with Cu. A barrier layer is typically used between the SiO_2 and Cu, and solder is attached to the top bump surface. The wafer thickness is reduced by thinning to expose the backside Cu bump.
   b. Via from back: the via is made from the back of the Si chip after thinning (Hitachi-Renesas, Tanaka et al. 2002).
3. Poly Si filled via: Instead of Cu via, poly Si is used.
4. Resin insulation: Since SiO_2 insulation needs high temperature and good deposition system, which adds to the cost; resin insulation has been proposed by Toshiba (Sekiguchi et al. 2006).
5. Laser drilled: Since high-speed RIE is slow and still maturing, laser drilling can also be used. However, since thin Al is not effective in stopping laser drilling, a thick Ni bump is formed on Al metalization to define laser drill stops.

Preliminary reports on the reliability of the 3D stacked modules are promising (Tanaka 2002 et al.). Applications such as CMOS image sensors on board have been demonstrated using TSV technology (Charbonnier et al. 2008).

Stud bump bonding flip chip attach

Although solder bumping is the predominant method of attaching dies to the substrate, there are alternatives that are driven by finer pitch (<100 um) and smaller standoff for portable and consumer electronic applications. These alternatives can be divided into primarily two types: (a) metal stud bump adhesively bonded with nonconductive polymer attach and (b) metal stud bump metallurgically bonded on solder or metal leads. The common element in both is the presence of the stud bumps on the die, which are simply ball bonds that have been suitably shaped for coplanarity.

Stud bumping is a low-cost bumping technique since it does not require special UBM and bumping is achieved via wire-bonding machine technology (fig. 4–46). The process involves making ball bond on standard pads using Au or Cu wire and clipping the wire just above the ball bond (fig. 4–47). Then, to ensure a tight distribution of bump heights, the ball bond is coined. Since this operation is to be carried out for each target pad on the die, stud bump flip-chip technology is especially suitable for low-cost, primarily perimeter-array, low standoff applications. After the coining operation, the die is flipped over and attached to the substrate using a nonconductive adhesive, solder, or metal–metal bonding (Tsunoi et al. 1995; Zakel et al. 2001). Often, the flip chip is underfilled after the bonding operation (or during the bonding operation) and the underfill performs a dual role of holding down the die (adhesive) and providing support to the bumps during thermomechanical exposure.

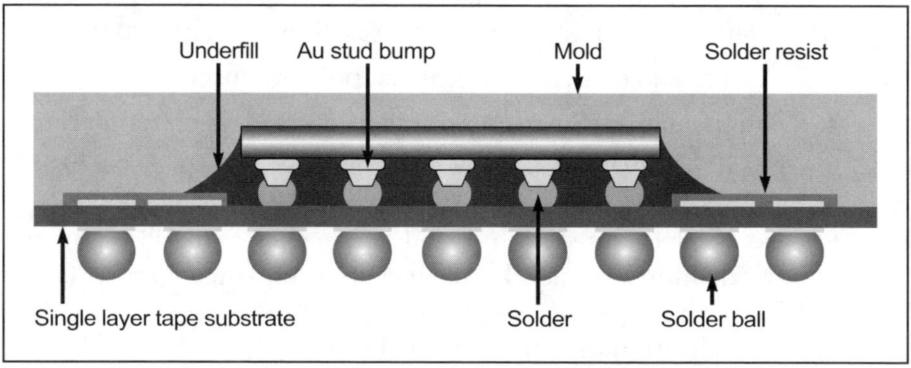

Fig. 4–46. Schematic of a stud-bump-bonded flip-chip CSP with Au solder attach

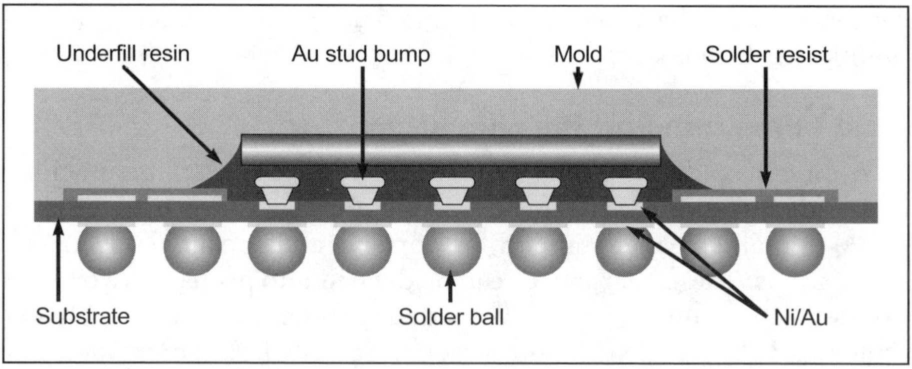

Fig. 4–47. Schematic of a stud-bump-bonded flip-chip CSP Au to Au bonding

Each of these interconnection options for stud bump flip chip comes with the potential for an ultrathin package with pitches better than 50 μm. Although the technology has been demonstrated with acceptable thermomechanical fatigue performance, the process window optimization is crucial to ensure the reliability for each of these options. Reliability under hygrothermal loading and prolonged high-temperature exposure has been found to be lacking due to moisture effects for adhesively bonded stud bumps and Au–Sn bonded stud bumps, respectively. Further cost savings are also possible when Cu wire is used for the stud bump instead of Au stud bump. For portable electronic applications that do not require as high a thermomechanical reliability or electromigration resistance as server applications, stud bump bonding offers low cost and extremely fine-pitch interconnection option suitable for high-speed signals.

References

Ano, K., T. Ohuchida, K. Murata, and M. Watanabe, 1997. Reliability study of the chip scale package using flex substrate. Paper presented at the Proceeding of the Chip Scale Packaging Symposium, Surface Mount International Conference, San Jose, Sept.: 44–47.

Badihi, A., and E. Por. 1995. Shell Case—A true miniature integrated circuit package. Paper presented at the Proc. of the Int. Flip Chip, BGA, Adv. Packaging Symp., San Jose, CA: 244–252.

Banerjee, K., and P. Lall, 1995. Development of SLICC package. Paper presented at the Proc. of NEPCON East95, Boston, MA. 441–451.

Bischoff, A., F. Adlinger, , and W. Heraeus, 1984. Reliability criteria of new low cost materials for bonding wires and substrates. Paper presented at the Proc. 34th Elec. Comp. Conf., IEEE: 411.

Boetcher, L., and A. Ostman, 2007. Through silicon via interconnection enabling 3D level SIP. Paper presented at the Proc. Technical Program of SMTA-International: 15–22.

Bryce, D. M. 1991. *Thermoplastic Troubleshooting for Injection Molders, Processing Series*. Brookfield, CT: Society of Plastic Engineers.

Chanchani, R., K. Treece, , and P. Dressendorfer. 1995. Mini ball grid array (mBGA) technology. Paper presented at the Proc. NEPCON West'95, Anaheim, CA: 938–945.

Charbonnier, J., D. Henry, F. Jacquet, B. Aventurier, C. Brunet-Manquat, and G. Enyedi, et al. 2008, Wafer level packaging technology development for CMOS image sensors using through silicon Vias. Paper presented at the 2nd Electronics System Integration Technology Conference 2008 (ESTC 2008), September 1–4 : 141–148.

Darken, L. S. 1948. Diffusion, mobility, and their interrelation through free energy in binary intermetallic systems. *Trans. Met. Sci. AIME*, 175: 184–201.

Day, J., C. Koehler, and T. Tessier. 1997. A laminate based flip chip scale package. Paper presented at the Proc. Surface Mount International Conference, San Jose, CA: 12–16.

Denda, S. 2007. Process examination of through-Si via technologies, *IEEE Polytronic*: 149–152.

Emerick, A., J. Ellerson, McCreary, R. Noreika, and P. Viswanadham. 1993. Enhancement of TSOP solder joints reliability using encapsulation. Paper presented at the Proceedings of the 43rd Electronic Component and Technology Conference, Orlando, FL. June 1–4: 187–192.

Fjelstad, J., T. DeStefano, B. Faraji, C. Mitchell, , and Z. Kovac. 1995. MicroBGA packaging technology for integrated circuits. Paper presented at the Proc. of NEPCON East, Boston, MA: 221–227.

Fjelstad, J. (1998). Novel approaches to compliant area array chip connections. Paper presented at the Proceedings of SEMICON Europa.

Fried, L. J., J. Havas, J. S. Lechaton, J. S. Logan, , G. Paul, and P. Totta. 1982. A VLSI bipolar metallization design with three level wiring and area array solder connections. *IBM J. Res. Develop.*: 26: 362–371.

Hayes, D., and D. Wallace, 1996. Solder jet printing for low cost wafer bumping. Paper presented at the ISHM Proceedings: 296–301.

Hamano, T., T. Fujisawa, M. Seki, K. Mitobe, S. Orimo, M. Sato, and J. Kasai., 1997. SOC package and 3D package module for memory devices. Paper presented at the Proc. IMAPS Advanced Technology Workshop on Chip Scale Packaging, Austin, TX.

Kasai, J. 1996. Low cost chip size package solutions for ASIC and memory application. Paper presented at the advanced IC packaging technologies: Array and chip scale, Semocon Europa'96.

Kasai, J., K. Tsuju, Y. Yoneda, , S. Orimo, R. Nomoto, H. Sakoda, , and M. Onodera. 1996. Development of microBGA and QFN by lead frame terminal formation method, Paper presented at the Proc. 1st Pan Pacific Microelectronics Symposium, Honolulu, HI: 23–30.

Katahira, T., P. W. Scanlon, and K. S. Ho. 2007. 0.3 mm pitch CSP/BGA development for mobile terminals. Paper presented at the Proc. Technical Program of International Microelectronics and Packaging Society Conf.: 393–401.

Kawanobe, T., K. Miyamoto, and Y. Inaba 1981. Solder fabrication by electrochemical method for flip chip interconnnnection, Paper presented at the Proc. Electronic Components Conf., Atlanta, GA: 149–155.

Keizer, A., and D. Brown, (1978). Bonding systems for microinterconnect tape technology, *Solid State Technol.*: 59–64.

Khan, N., S. W. Yoon, A. G. K. Viswanath, V. P. Ganesh, , R. Nagarajan, D. Witarsa, et al. 2008. Development of 3D stack packaging using silicon interposer for high power application. *IEEE Tran. Adv. Packaging* 31(1): 44–50.

Khoury, S. L., D. J. Burkhard, D. P. Galloway, and T. A. Scharr. 2007. A comparison of copper and gold wire bonding on integrated circuit devices. *IEEE Trans. Comp., Hybrids, Manuf. Technol.* 13: 673–681.

Koopman, N., and P. A. Totta, 1988. Controlled collapse chip connection, ASM World Mat'l Congress.

Kujala, A., T. Reinikainen, , and W. Ren. 2002. Transition to Pb-free manufacturing using LGA packaging technology. Paper presented at the Proc. IEE Electronic Component Technology Conf.: 359–364.

Lau, J. H., 1995. in *A Brief Introduction to BGA Technologies in Ball Grid Array Technology*, ed. J. H. Lau, New York: McGraw Hill Inc. ch. 1: 1–64.

Laermer, F., and A. Schilp, 1966. Method of anisotropically etching Si, U.S. Patent 5,501,893.

Miller, L. F. (1969). Controlled collapse chip joining, *IBM J. Res. Dev.* 13: 239–250.

Mitchell, D., B. Zahn, and F. Carson, 2004. Board level thermal cycling reliability and solder joint fatigue life predictions of multiple stacked die chip scale package configurations. Paper presented at the Proc. Electronic Components Technology Conf.: 699–703.

Mori, S., H. Yishida, and N. Uchiyama, 1988. The development of new copper ball bonding wire. Paper presented at the Proc. 38th ECTC, IEEE: 539–545.

Mukarami. G. (1994). Package industry roadmap to the future, *Proc. Semicon West.*

Nguyen, A. S., Danker, N., Santhiram, and C. R. Sherwin. 1992. Flow modeling of wire sweep during molding of integrated circuits. Paper presented at the ASME Winter Annual Meeting: 27–38.

Nguyen, L. T., and F. J. Lim. 1990. Wire sweep during molding of inegrated circuits. Paper presented at the IEEE Electronic Components and Technology Conference: 777–785.

Nguyen, L. T., D. McDonald, A. R. Danker, and P. Ng, 1995. Optimization of copper wire bonding on A1–Cu metallization, *IEEE Trans. Comp. Packag. Manuf. Technol.* 18: 423–429.

Numata, H., Sato, N., Shirakawa, T., Matsuo, M., Yoshikawa, H. Yanagida, M., et al. (2006). Novel low cost integration of through chip interconnection and application to CMOS image sensor. Paper presented at the Proceedings of the ECTC: 1367–75.

Onuki, J., M. Koizumi, and Iraki (1987). Investigation of the reliability of copper call bonds to aluminum electrodes, *IEEE Trans. Comp. Hybrids Manufact. Technol.* 12: 550–555.

Peltola, A., S. Akkanen, and V. Halkola. 2002. J. *Surf. Mount. Technol.* 15 (2).

Ruhmer, K., E. Laine, and P. Gruber. 2006, Lead free and low cost solder bumping technology for electronic packaging, *CircuiTree* 19(8): 10–18.

Schueller, R. D. 1997. New chip scale package with CTE matching to the board. Paper presented at the Proc. Int. Electronics Manuacturing. Technology Symp., Austin, TX: 205–215.

Sekiuchi, M., H. Numata, N. Sato, T. Shirakawa, M. Matsuo, H. Yoshikawa, M. Yanagida, H. Nakayoshi, K. Takahashi, 2006. Novel low cost integration of through chip Interconnection and application to CMOS image sensor. Paper presented at the ECTC: 1367–74.

Singer, P. H. 1984. *Hybrid Wire Bonding Advances,* Semiconductor International: 66.

Singh, I., J. Y. On, and L. Levine. 2005. Enhancing fine pitch high I/O devices with Cu ball bonding. Paper presented at the ECTC 2005: 843–47

Smigelkas, A. D., and E. O. Kirkendall. 1947. Zinc diffusion in alpha brass, trans. *AIME* 171: 130–34.

Sjoberg, J., T. Castello,, D. A. Geiger, D. Shangguan. 2007. Process development and reliability evaluation of in-line package –on-package assembly. Paper presented at the proceedings of the technical conference SMTA International (2007).

Spencer, T. H. 1982. Thermocompression bonding kinetics: four principles, *Int. J. Hybrid Microelectron*: 404.

Stuby, K. P., and W. Falls, 1972. Hourglass shaped conductive connection through-hole interconnections in Si substrate," US Patent 3,648,131.

Takahashi, K., M. Umemoto, , N. Tanaka, K. Tanida, , Y. Nemoto, , Y. Tomita, et al. 2003. Ultra-high-density interconnection technology of 3D packaging, *Microelectronics Reliability* 43: 1267–279.

Tanaka, N., T. Sato, Y. Yamaji, , T. Morifuji, M. Umemoto, and K. Takahashi, 2002. Mechanical effects of Cu through-vias in a 3D-die stacked module. Paper presented at the IEEE ECTC: 473–79.

Toyozawa, K., K. Fujita, S. Minamide, and T. Maeda. (1990). Development of copper wire bonding application technology. *IEEE Trans. Comp. Hybrids Manuf. Technol.* 13: 667–72.

Viswanadham, P., M. Stennett, and R. Haggett. 1993. Second level assembly and reliability aspects of thin small outline packages by advances in electronic packaging. Paper presented at the ASME International Electronic Packaging Conference, Vol. 2: 1127–134.

Viswanadham, P., D. Coplin, and R. Haggett, 1993. Solder joint reliability of TSOPs–an overview. Paper presented at the Proc. 43rd Electronic Component and Technology Conference, Orlando, FL: 863–870.

Viswanadham, P. 1994. Reliability aspects of fine pitch assembly, in *Handbook of Fine Pitch Surface Mount Technology*, Ed. J. H. Lau, Van Nostrand Reinhold: 598–636.

Viswanadham, P., K. Ewer, R. Aguirre, and T. Carper, 1998. Package to board interconnection and reliability of ball grid array packages over extended temperature range. Paper presented at the Proc. Pan Pacific Microelectronics Symposium, Mauna Lani, HI. and 4th National IPC BGA Symposium in Minneapolis, MN, 1998.

Viswanadham, P., K. Kulojarvi, , V. Halkola, , and J. Ahlstedt, 2000. Reliability aspects of mobile communication products. *Future EMS International*, No. 3: 61–64.

Viswanadham, P., T. Chung, and S. Dunford, 2001. Chip scale package technology, in *Handbook of Area Array Packaging*, Eds. P. Totta and K. Putlitz, Kluwer Publishing: 702–61.

Viswanadham, P., S. Dunford, , and T. Carper, 1999. Board level reliability evaluations of 40, 32, and 30 mil pitch ball grid array packages over –40–125°C . Paper presented at the Proceedings of the. 4th Pan Pacific Microelectronics Symposium, Island of Kuaii, HI, Feb. 2–5: 362–375.

Wallace, D. B. 2005. Method and apparatus for forming microdroplets of liquids at elevated temperature, U. S. Patent 5,415,679.

Warabisako, T. 1983. Stacked semiconductor integrated circuits and manufacturing method, Japanese Patent S59-222954.

Winkler, S. L. 2005. Multicomponent IC packaging, Paper presented at the Proceedings of the Technical Conference SMTA International.

Yano, Y., T. Sugiyama, S. Ishihara, Y. Fukumi H. Juso, Miyata, et al. 2002. Paper presented at the Proceedings of Electronic Components Technology Conference: 1329

Yoneda, Y. 1996. Development of BCC (bump chip carrier). Paper presented at the Technical Digest 3rd VLSI Packaging Workshop of Japan, Kyoto, Japan: 71–72.

Yoshida, A., B. Miles, , V. Perelman, H. Y. Wok, , and R. Grover, 2004. Paper presented at the Proceedings of the Technical Conference SMTA International.

5

Interconnect Technologies—Second Level Packaging

Introduction

Interconnection is the process of effecting mechanical and/or electrical attachment of any two parts of an electronic assembly. The variety of interconnecting technologies in practice include (a) wire wrap, (b) alloy interconnections, (c) interconnections with thermally conductive, electrically conductive, or nonconductive adhesive attachments, and (d) simple mechanical insertion attachments amenable for multiple insertions and extractions, such as in connectors. The choice of a particular interconnection scheme depends on the design, functionality, and reliability requirements. This chapter provides an overview of the different interconnection schemes pertinent to portable electronic products.

Since most modern portable electronics can be characterized as light weight, small size/volume, high functionality, low cost, with high thermal, mechanical, and electrical reliability, they are manufactured with thinner printed wiring boards (PWBs) and fewer layers. There are a few ceramic packages, and the preferred packages are low-standoff plastic packages with relatively large chip to package ratio and hence high packaging efficiency.

Wire wrap

Wire wrapping is perhaps one of the oldest methods of interconnections in electronic assemblies. It was used in the fabrication of radios, radars, telephone exchange equipment, sonar, etc. In fact, the Apollo guidance computer was wired using this technique. The techniques consists of wrapping tightly the two terminations to be joined with a 28–30 gage AW6 silver- or tin-plated soft copper wire using a wire wrapping tool that resembles a soldering tool. The wire is generally coated with a fluorocarbon polymer called Kynar (a product of Pennwallt

Corp.). The connections thus made are very robust. The interconnections are dry, highly repairable, and vibration and corrosion resistant. Owing to the nature of the process, it is slow and hence is more amenable to small, low-volume electronic assemblies. The technique has become obsolete owing to the evolution of advanced joining methods and is only of historical interest.

Alloy interconnections

Alloy interconnections are by far the most widely used joining methods in the electronics industry and are accomplished with solders. The word solder itself is derived from the French word *solidare* meaning to *make firm*. Thus, any metal or metal composition in molten state used to attach or patch two metal surfaces has come to be regarded as a solder. Historically, alloys of tin (Sn) and lead (Pb) have been in use as joining materials for more than 5,000 years, first in Mesopotamia and later in Egypt, Greece, and Rome (Wolters 1977). And, over a period, the word solder has become synonymous with eutectic or near-eutectic tin–lead alloy.

Anatomy of interconnections

Wire wrap interconnection. As mentioned previously, this is a physical, mechanical binding involving the wrapping of the two termination leads tightly with another conductor, thus providing an electrical interconnection. This interconnection is rarely used in electronic packaging because it is labor intensive and incompatible with high-volume manufacturing.

Insertion, pin-in-hole, or through-hole interconnection. Through-hole interconnection process involves first the insertion of the component leads that are either shaped or formed into the corresponding plated through holes of the PWB. The component leads are generally of copper alloys or alloy-42 and are coated with tin, tin-based alloys, or gold for easy solderability. The PWB through holes are also coated with either tin-based alloys or an organic solderability preservative for wettability and formation of the bond during soldering. The assembly is then passed over a wave soldering machine for effecting the interconnection. During the wave solder operation, the board first traverses over an inclined conveyor where the assembly is sprayed with a foam flux containing an activator that cleans the component leads as well as the barrel of plated-through hole, and then passes over a turbulent and laminar flow solder waves. Molten solder wicks up the leads and wets the two surfaces, thus effecting a metallurgical bond. The turbulent wave facilitates the solder wicking penetration into fine-dimensioned crevices, and the laminar wave allows flux outgassing, prevents oxide entrapment, and provides a uniform solder distribution. The time over the wave is long enough to permit solder penetration but short enough to prevent too much thermal exposure that might result in either component or board damage. The temperature and the dwell time depends on the type of the solder used, the complexity of the board such as thickness, number of layer, etc., as well as the nature of the component involved. For tin–lead solders, the temperature is in the range of 230 to 260°C depending on the alloy used. The assembly then passes over a hot-air knife for blowing off excess solder followed by an optional solvent or aqueous cleaning, and is then cooled. The cleaning step is superfluous when no-clean fluxes are used.

An acceptable solder joint comprises the formation of concave fillets both at the top as well as the bottom and when the joint interior is devoid of any entrapped voids and a metallurgical bond is formed between solder/copper barrel and solder/lead interfaces. Figure 5–1 shows a typical through-hole or insertion-mount interconnection.

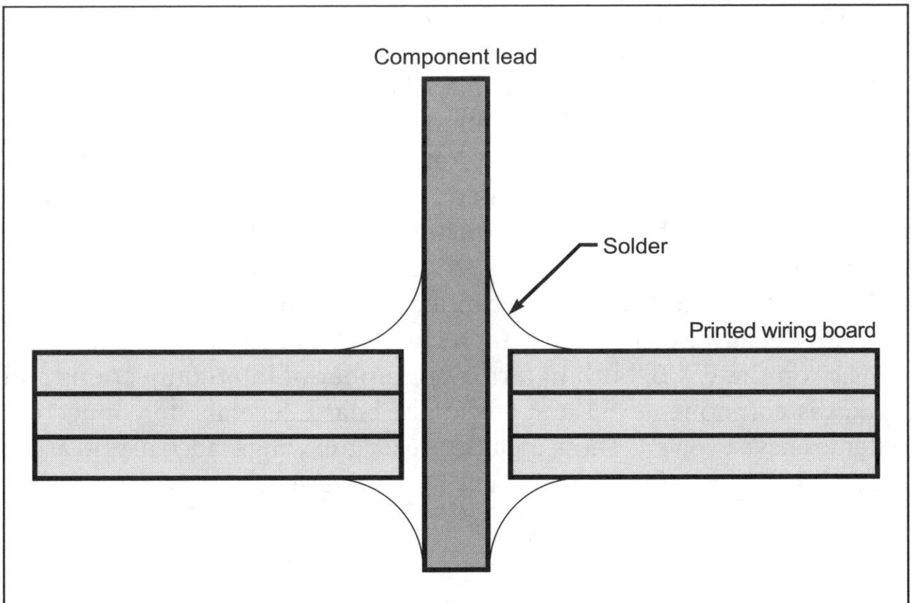

Fig. 5–1. Schematic of a typical through-hole interconnection

Convex profiles of the solder fillet suggest poor wetting of the lead or board pads, and the joints are likely to be less reliable. Also, one might occasionally see poor hole-fill giving rise to the absence of the top fillet altogether. Poor hole-fill can occur due to a number of reasons: (1) poor wetting, which may be due to less active or inadequate fluxing; (2) a discontinuity or break in the barrel copper plating; (3) extremely thin copper plating where the copper is dissolved in the solder exposing the laminate thus limiting the solder wicking; (4) a high aspect ratio of the plated-through hole, etc., and (5) an out-of-control wave solder process. In such a scenario, the cause of the poor hole-fill needs to be investigated, established, and remedied as appropriate. Figure 5–2 depicts some examples of poor hole-fill.

The lead pull strength of such an interconnection with dual-in-line package is generally in the range 6–8 lb. Thus, through-hole, pin-in-hole, or insertion interconnection is by far the most robust alloy interconnection and is therefore preferred for many high-reliability applications such as military electronics. Although most consumer, portable, and business electronics migrated to surface-mount technology, total surface-mount technology is adopted only in a few applications such as cell phones. In instances where multiple insertions and extractions are involved, such as sockets for displays, printers, power input, etc., as well as in laptop computers, gaming controllers, etc., pin-in-hole (through-hole) technology is still employed for reliability. Alternatively, the surface-mount connectors are anchored with additional stress-relieving reinforcements.

In general, pin-in-hole interconnection technology is fraught with several disadvantages such as single-side assembly only, limited routability (lines/channel), and wiring density, wider component lead pitch of 100 mils, heavier boards, etc., and has been replaced whenever possible by surface-mount interconnection technology.

With the advent surface-mount technology to increase the packaging density on the PWB, both in terms of number of interconnections and packages/components per unit area of the board, several interconnection schemes have evolved. These include J-lead, gull wing lead, ball grid array, etc., which are briefly discussed in the ensuing sections.

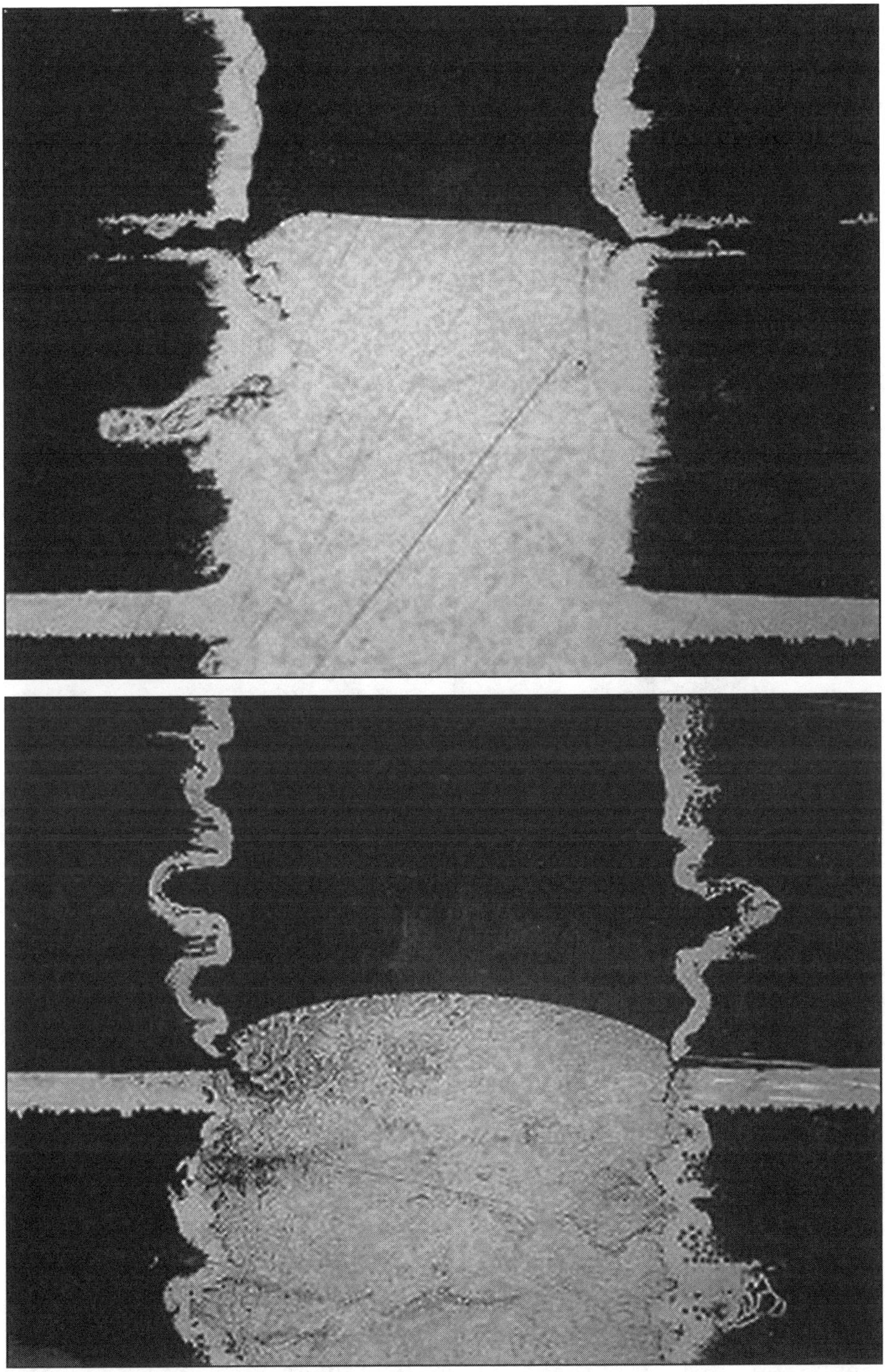

Fig. 5–2. Typical examples of poor hole-fill in plated-through hole

Leadless interconnection

This class of packages is known as leadless chip carriers and the package has only perimeter terminations. They are available both in ceramic and plastic versions. As the name implies, the package-to-board interconnection scheme is devoid of any lead. Instead, they have gold- or silver-plated half-round cylindrical grooves on the sides of the package which are connected to small pads on the underside of the package. The package is placed on the solder-paste-applied footprint on the PWB and soldered. During soldering, in addition to the footprint connection, the solder wicks up the grooved surface forming side fillets. Figure 5–3 shows packages with leadless interconnection terminations.

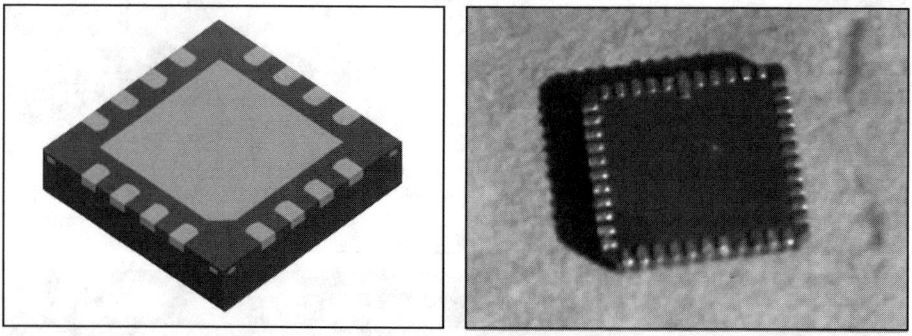

Fig. 5–3. Schematic and picture of leadless chip carrier terminations

This interconnection scheme provides considerable real estate savings on the board and is amenable to high-density packaging, the only limitation being that they are offered in 50-mil pitch format.

J-leaded interconnection

In this interconnection, the component leads are formed in the form of the letter "J". The lead material is usually alloy-42 and is rather stiff. They are more tolerant to shipping and handling stresses. The leads are vertical and the compliancy of the package mainly depends on the lead height and lead stiffness. They also have a higher profile than gull wing packages. The packages are mostly manufactured in 50-mil pitch format. As indicated in an earlier chapter, the leads are plated with a tin/lead solder. The joint has two major fillets, namely, the toe and the heel on outer and inner sides, respectively, of the U-bend. Figure 5–4 shows an example of a J-leaded interconnection package (top view and bottom view). The integrity of these toe and heel fillets constitutes the interconnection strength. The heel fillet is accessible for optical inspection, while the toe fillet is not. It is important to note that the lead bending operation is performed after the lead plate operation. The likelihood of inducing microcracks or delamination of the plating during multistage lead bend operation cannot be ruled out. In such an event, the underlying metallurgy, namely Ni, may get oxidized and become less solderable at those locations. Even though the toe and heel fillets may look acceptable upon visual inspection, the actual metallurgical bond may be less than satisfactory. These nonoptimal solder joints are likely to result in early failures during thermal stress excursions either during accelerated testing or under the field conditions. Figure 5–5 shows an example of J-lead failures under thermal loading.

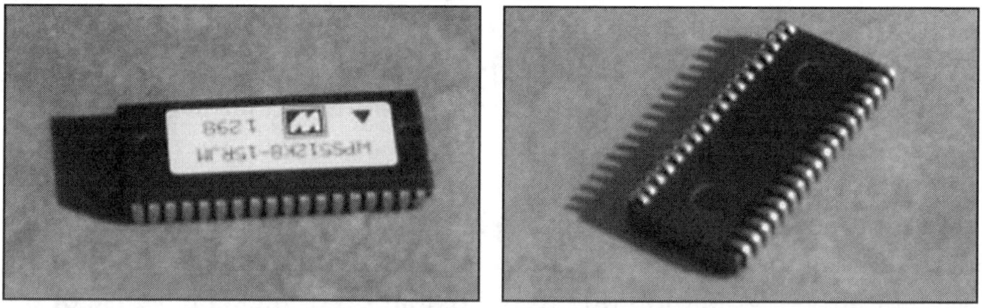

Fig. 5–4. Examples of a J –leaded interconnection package: (a) top view; (b) bottom view

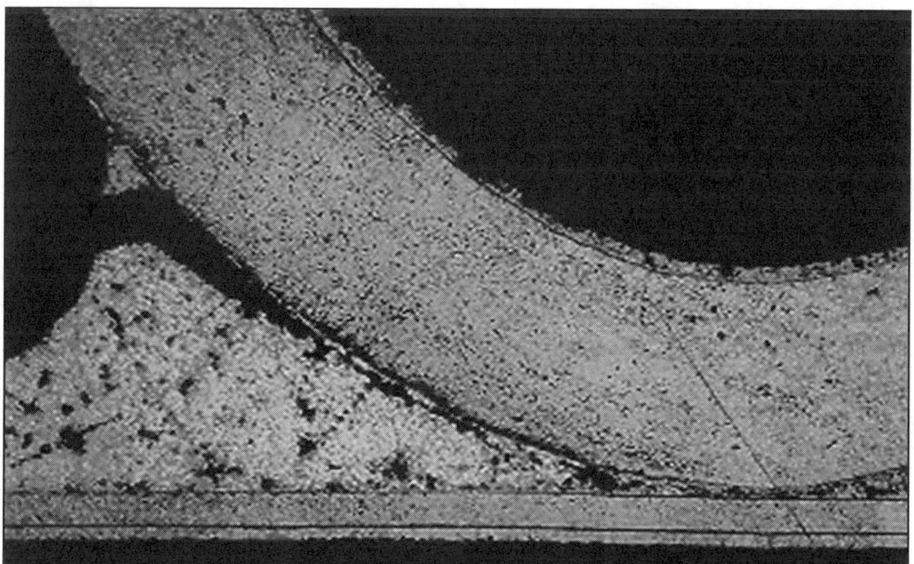

Fig. 5–5. J-lead failure due to poor metallurgical bond

Gull wing lead interconnection

This interconnection configuration is by far the most compliant of all the interconnection schemes. The lead is shaped like the wing of a seagull bent down and out, and hence the name. It is used in small outline integrated circuits (SOIC), quad flat packs (QFP), tape automated bonding (TAB), plastic leaded chip carriers (PLCC), thin small outline packages (TSOP), and many others. The configuration is amenable to infrared, convection, and vapor phase mass reflow techniques as well as hot bar attach, laser soldering, etc. In contrast to a J-leaded package, the toe fillet is accessible for inspection while the heel is not, except for the corner joints. Figure 5–6 shows a schematic of the anatomy of gull wing leaded interconnection in an SOIC. During package manufacturing, leads are trimmed after the plating operation, and as a consequence the underlying alloy, such as that of copper or nickel, is exposed and easily oxidized during storage. This affects the solderability at the tip, and, hence, consistent toe fillet formation is not assured. This is especially important in dealing with low-standoff, low-CTE TSOP interconnections of type I fine pitch stiff alloy-42 leads where the joint reliability can depend on good toe fillet. These interconnections can also suffer from solderability degradation due to the oxidation of nickel because of the microcracks developed during lead bending and formation similar to J-lead shown earlier.

Gull wing leaded packages are amenable to high pin count and finer pitch, and packages with pitches up to 12 mils (300 μm) have been reported. Attendant with the capability of migrating to fine pitch is the lead fragility. Lead fragility of the fine-pitch packages raises several concerns. These include lead bending, nonplanarity, lead crushing, etc., during shipping and handling and/or during assembly. Additionally, solder paste screen printing limitations, potential solder bridging during assembly, limitations in the ability to work defective components, etc., contribute to low yields. Despite all the difficulties, gull wing packages with proper handling and process optimization have enabled high manufacturing yields and acceptable reliability down to 0.4 mm pitch packages.

Gull wing leaded interconnections occupy a larger real estate on the board surface in comparison to J-leaded interconnections but they provide the best thermo-mechanical reliability.

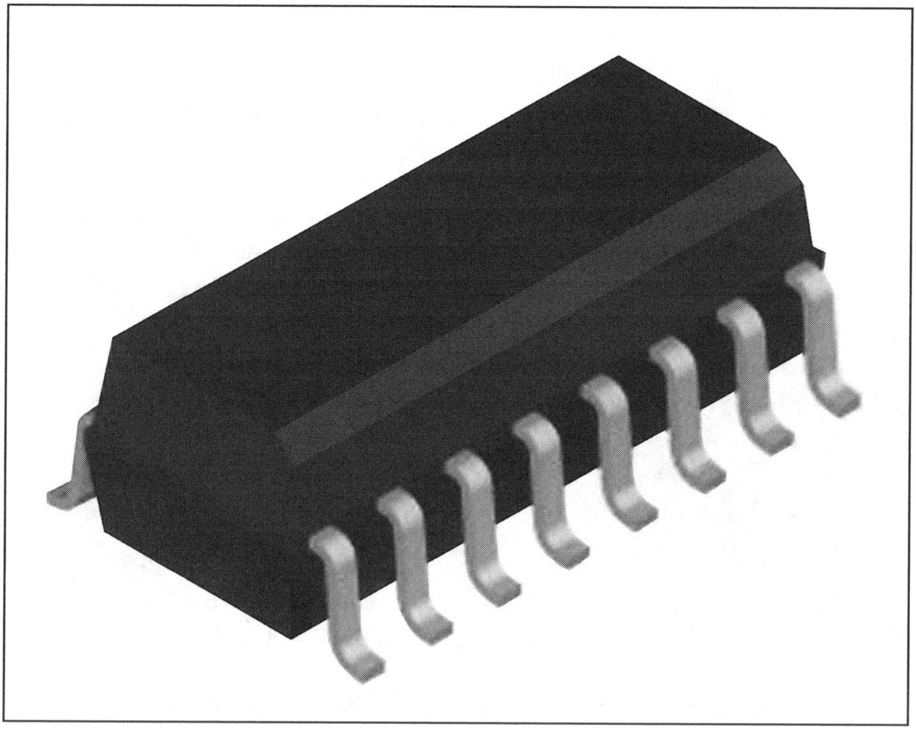

Fig. 5–6. Schematic of the gull wing shaped lead configuration in a small outline integrated package (SOIC)

Area-array interconnection

The difficulties associated with fine-pitch perimeter-leaded interconnections led to the adaptation of area-array packaging, thus overcoming the perimeter paralysis. Area-array packaging was originally developed for first-level packaging of high I/O semiconductor devices on ceramic substrates and was known as the as *flip-chip attach*, also popularly known as *controlled collapse chip connection* (C4) (Miller 1969). This concept was extended to second-level package-to-board interconnection and has come to be known as *solder-ball connect* or *ball-grid array* (BGA) technology. The interconnection scheme does not have emanating leads from the package perimeter. The device I/Os are routed through the multilayer ceramic or organic substrate terminating as circular pads on the underside of the package to which solder balls are attached. The package-to-board interconnection consists of attaching these packages on to the corresponding area-array pads on the PWB. A BGA interconnection is depicted schematically in figure 5–7.

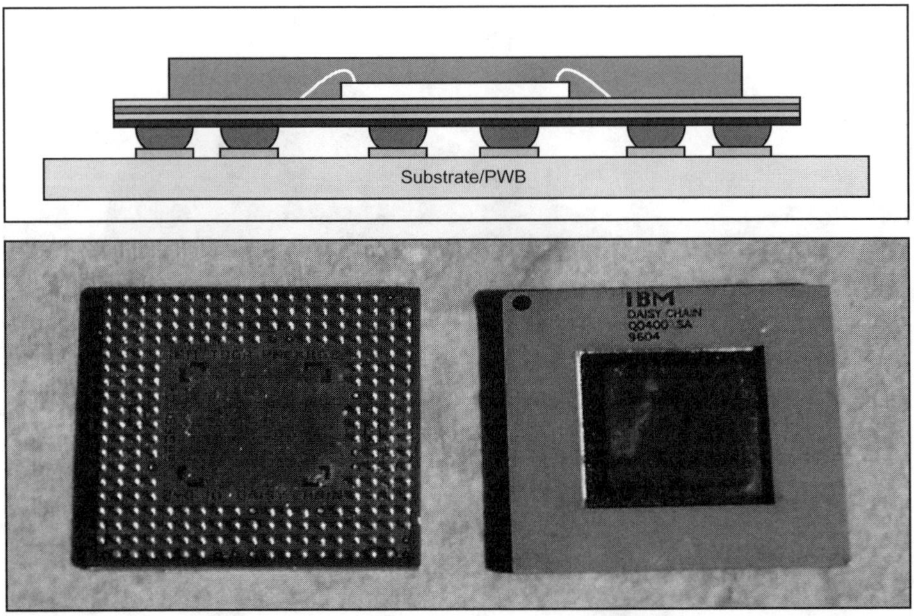

Fig. 5–7. Schematic and photograph of a ball-grid array interconnect scheme

The initial offerings of these packages were of 1.27 mm pitch, which subsequently migrated to finer pitch interconnections, and are variously known as micro-BGAs, micro type BGAs, fine-pitch BGAs, chip-scale packages, etc. Thus, area-array packaging interconnection facilitated the highest packaging density on the PWB, also pushing the limits of the capabilities of conventional PWB technology. Higher device I/Os imply larger number of board layers that enable routing the I/O, and this led to the development of high-density interconnect (HDI) board technology described elsewhere in the book.

Solder alloys and interconnections

Several interconnection alloys have been in use in the packaging industry. The choice of an interconnection alloy depends on several factors, the primary among them being as follows:

1. Product application environment
2. Design life
3. Metallurgical compatibility
4. Corrosion resistance
5. Component and package sensitivity to assembly
6. Reliability requirements

Among the interconnection alloys that are employed for second-level packaging, the following are considered the more common ones and are discussed in some detail.

1. Tin/lead eutectic or near eutectic alloy
2. Tin/lead/silver
3. Lead-free alloys
   a. Binary alloys
   b. Multicomponent alloys

Tin/–lead eutectic or near-eutectic alloys

Tin–lead solders are by far the most common alloys for joining metals because of their compatibility with other metallurgies of relevance in electronic packaging. These include copper and its alloys, nickel and its alloys, gold, silver, palladium, platinum, etc. An important physico-chemical aspect in the elucidation of behavior of alloys is the phase rule which relates the number of phases such as solid, liquid, or vapor (*P*), the number of components (*C*), and the minimum number of independent variables (such as temperature, pressure, concentrations, etc.) or degrees of freedom (*F*) required to define the system as originally proposed by Willard Gibbs (Findlay, Campbell, and Smith 1951). For a system where temperature, pressure, and concentration are the only variables, the Gibbs phase rule is given by

$$P + F = C + 2 \qquad (5\text{–}1)$$

A pictorial equilibrium representation of the system is called a *phase diagram*. In most of the phase diagrams involving metals, pressure is assumed to be the constant, i.e., atmospheric pressure, and hence the phase rule takes the form

$$P + F = C + 1 \qquad (5\text{–}2)$$

The behavior of the tin/lead system is pictorially represented in figure 5–8.

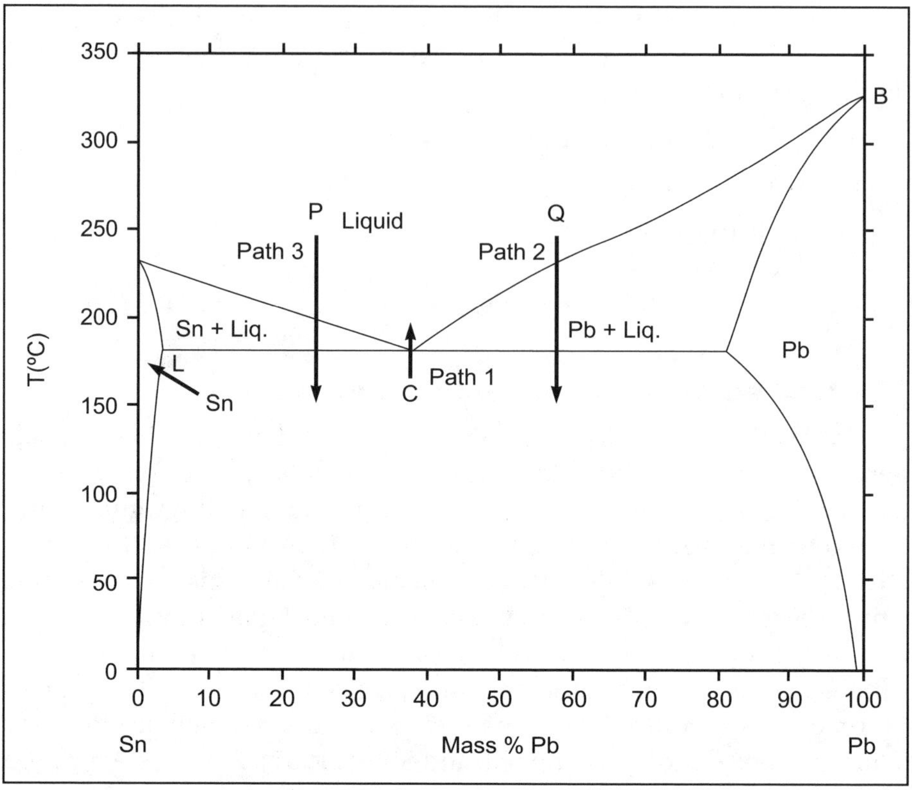

Fig. 5–8. Tin/lead phase diagram

The horizontal axis represents the composition in either weight percent or atomic percent, and the vertical axis represents the temperature in degrees Celsius or Kelvins. Point A represents the melting point of Sn. As lead (Pb) is added to the melt, the melting point is lowered and follows line AC. Similarly, point B represents the melting point of lead, and as tin is added to the melt, the melting point is lowered and follows line BC. Point C is thus the intersection of lines AC and BC and represents the lowest melting point for this system. This temperature is called the eutectic ("well-meltable") temperature and the composition of the alloy is termed the *eutectic composition*. The regions left of AL and right of BM are single-phase solid solution regions α (Pb with small amount of tin dissolved in it) and β (Sn with small amount Pb dissolved in it).

Since, electronic assembly involves making the interconnection by melting the alloy and subsequent cooling, it is important to understand the cooling characteristics of the alloy in question. Let us consider the cooling behavior to the left of the eutectic composition. At point *P*, the system is completely in the liquid phase. One phase and two components means we need two variables, namely, temperature and composition, to define the system ($F = C - P + 1; 2 - 1 + 1 = 2$) completely. As the cooling is continued, once we reach the line, the first solid phase appears and in that region the solid α phase and liquid coexist. Thus $C = 2$, $P = 2$ and hence the degree of freedom is 1. Thus either temperature or composition is adequate to define the system. In the region below the line is also a two-phase region until the line LCM is reached, at which point complete solidification of the system occurs consisting of α and β phases. And $F = 2 - 2 + 1$, representing a univariant system.

When one starts in the liquid phase at eutectic composition and cools the system, when the system reaches point C, the system consists of α phase, β phase, and a liquid phase of eutectic composition. Thus, one has three phases and hence $F = 2 - 3 + 1 = 0$. Thus the eutectic temperature of the system is uniquely defined with zero degrees of freedom. As the system is cooled further, the entire system is in solid state consisting of α and β phases only.

Thus phase diagrams aid in understanding the system behavior as a function of temperature and composition.

Utilizing a soldering alloy of eutectic composition in electronic assembly provides a system that melts at the lowest temperature as well as without any mushy two-phase regions where a part of the interconnection is in a liquid state and another part is in a solid state. While that is the most desired scenario, it may not always be achievable, and slight deviations from eutectic compositions are the norm as will be discussed elsewhere.

The tin–lead interconnection

The two surfaces in the interconnection are the copper pad on the PWB and the lead metallurgy of either copper or nickel alloy generally plated with Sn, Sn/Pb, or Au.

As the solder is brought to temperature 20°C–25°C above its melting point, flux dissolves any surface oxides and exposes pure metal surfaces. On the PWB surface, as the solder begins to wet, copper and tin reacts to form a thin layer of copper/tin intermetallics. At the reflow temperature, the predominant intermetallic phase observed in a cross-section is Cu_6Sn_5. Thermodynamically, copper forms two intermetallic phases with Sn, namely, Cu_6Sn_5 (η phase) and Cu_3Sn (ε phase): η phase forms on the Sn-rich side and the ε phase on the Cu-rich side. The activation energy of the ε phase is about twice as that of the η phase. The thickness ratio for these two, ε/η, is about 1:500, suggestive of the appearance of only the η phase at lower temperatures. After prolonged exposure for a few hundred hours at elevated temperatures of about 150°C –170°C, growth of both these two phases is observed. Figure 5–9 depicts the copper/solder interface indicating the predominance of the Cu_6Sn_5 intermetallic phase. The growth of these phases is parabolic and is given by

$$\delta = k\, t^{0.5} \tag{5–3}$$

where δ is the thickness, t the time, and k is the growth constant given by

$$k = k_0 \exp\,(-Q/RT) \tag{5–4}$$

where T is the temperature in Kelvin, Q the activation energy, R the gas constant, and k_0 a constant.

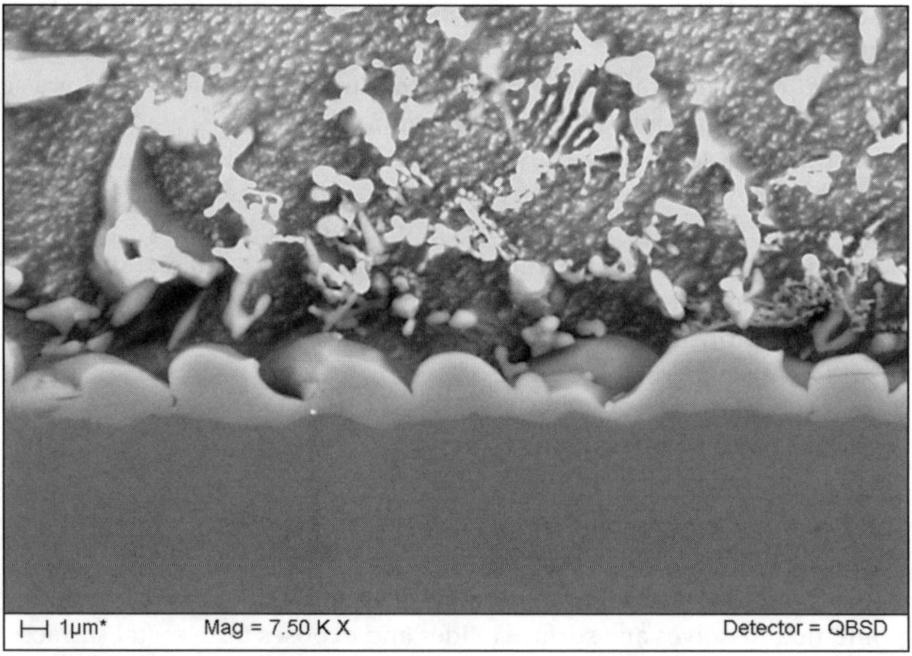

Fig. 5–9. The Cu intermetallic phase in the shape of scallops

Near the interface, formation and growth of the Cu–Sn intermetallic phase depletes Sn from that region and renders it Pb-rich. Thus, in Sn–Pb interconnection, the composition of the material adjacent to the interface on the solder side is likely to be different from the bulk region of the joint. This variation in the composition can have an effect on the modulus, strength, ductility, etc., and thus have a reliability impact. Pb-rich regions are generally considered to be less reliable.

It is important to recognize that the intermetallic growth is diffusion-limited and ceases after a particular thickness is reached at a given temperature.

Intermetallic phase formation at the interface of the two surfaces being joined is a necessary condition for a metallurgical bond. Thus, absence of the intermetallic layer is an indication of poor interconnection. However, intermetallic phases are generally more brittle than the constituent elements, and an excessively thick intermetallic layer in an interconnection is more prone to fracture. Intermetallic phases can grow as contiguous layers, spiky needles, nodules, rods, plates, etc., depending on the composition of the particular intermetallic phase and the growth conditions such as temperature and its gradient, time, liquid or solid matrix, etc.

In the alloy-42 leads, the intermetallic phase at the lead–solder interface consists of Ni/Sn intermetallic plates, mostly Ni_3Sn_4 on the nickel-rich side. The bond formation is just as rapid as in the case of copper and tin. The growth rates of copper and nickel intermetallic phases are different. The growth of the intermetallic layer at elevated temperature, like the Cu/Sn intermetallics, is also diffusion-limited. It is generally recognized that the Ni/Sn intermetallics are more brittle than the Cu/Sn intermetallics. Thus, interconnections with Ni/Sn intermetallics are less forgiving under mechanical loading and can result in brittle fractures at the interface. The intermetallic growth and morphology depends on whether it is a solid–solid interaction or solid–liquid interaction.

Surface Finishes

In some instances, copper pads are coated with electroless nickel followed by a layer of gold. Copper and gold have a great affinity to each other with high mutual solubility. When copper and gold are in contact with each other, copper quickly diffuses into the gold layer. In order to prevent diffusion of copper into gold, a layer of nickel is provided as a diffusion barrier. Thus, for PWBs in which a very high degree of surface planarity is required for fine-pitch assembly and connector pads, the choice of finish is gold on a layer of electroless nickel. For connector tabs, hard gold is plated over nickel. For soldering purposes, immersion gold is used over nickel to prevent oxidative degradation of nickel and preserve nickel solderability. The thickness of nickel layer is in the range of 2.5 to 4 μm. The thickness of immersion gold is in the range 0.5–2 μm. It is important to note that the electroless Ni usually contains 6%–12% P by virtue of the sodium hypophosphite reducing agent used.

During the interconnection process, the surface gold rapidly dissolves in the molten solder and the metallurgical bond between solder and nickel is made up of Ni/Sn intermetallics.

The presence of gold in solder introduces some nuances in the properties and behavior of the interconnection and these are dependent on the thickness and amount of gold involved. According to the phase diagram of Sn/Au, Au forms a number of intermetallic phases such as AuSn, $AuSn_2$, $AuSn_4$, etc. Presence of Au in a joint is known to cause embrittlement of the interconnection. A 3 wt% of Au in a solder joint is generally considered as an upper limit in terms of any impact on interconnection reliability (Glazer, Morris, and Morris 1991). A variety of physical properties are influenced by gold content in solder. Tensile strength increases with gold content up to 5%, and then drops off, while shear strength drops after 2%. Ductility decreases with gold content, and after 8% Au elongation is almost negligible. Hardness increases gradually up to 4%, and then rapidly with higher Au content. Also, voiding in solder joint is known to increase after 4% Au. Thus given these dependencies, a 3 wt% upper limit of Au in a joint does not seem unreasonable as a guideline. However, the following aspects need to be considered in evaluating the effect of gold intermetallics on the joint reliability even within the suggested limits.

1. Location of the intermetallics: If the intermetallics are uniformly dispersed in the bulk of the joint, it may add strength to the joint. On the other hand, if the intermetallics are localized adjacent to the intermetallic bond, and are located in the high stress crack propagation region, it may be deleterious.
2. The interconnection geometry: If the interconnection structure is compliant, the stresses are accommodated by the compliance of the flexible members of the structure, and hence the solder joint may not experience undue stresses. This is the case with structures like TAB. Such interconnections can withstand much higher gold concentration in the joint. On the other hand, in structures like TSOP interconnections with low standoff, high CTE mismatch, and stiff alloy-42 leads, even low gold concentrations in a joint can adversely affect reliability. Thus, reliability is very much dependent on the nature of the interconnection structure.

In cases where there is very thick gold, the intermetallic layer sequence is AuSn, $AuSn_2$, and $AuSn_4$ starting from gold side to the solder. Reliability evaluation in such interconnections needs to take into considerations the intermetallic morphology, thickness, and the physical properties along with the mechanical and thermal loads to which the system is exposed.

Lead/tin/silver solder

Historically, origin of this solder can be traced to the hybrid circuit technology. Hybrid circuits utilize thick-film technology where circuit lines/traces and lands are made of silver-containing polymers, which are also called *silver inks*. When one utilizes a eutectic Sn/Pb solder as the interconnection alloy to assemble the components, molten eutectic solder scavenges some of the silver from the lands. This silver leaching exposes the underlying ceramic and renders electrical interconnection impossible. In order to prevent this silver leaching, silver is deliberately added to eutectic solder to an extent of 2%. The 2% limit is arrived at considering the 3% solubility of silver in molten eutectic solder at the reflow temperature.

Silver addition lowers the melting point of the solder from 183°C to 179°C. The electrical conductivity of 62Sn36Pb2Ag solder is 6.8×10^4 Ω^{-1} cm^{-1} compared to 6.9×10^4 Ω^{-1} cm^{-1} for the eutectic solder. Thus, the electrical conductivity is unaffected. The shear strength of silver containing Sn/Pb solder is about 10% greater than the Sn/Pb eutectic at room temperature and about 20% greater at elevated temperatures (at about 100°C). Thus the material yields at slightly greater stress than the Sn/Pb alloy. This also implies a slightly greater elastic deformation than plastic deformation. The relative magnitudes of the deformation depend on the package–board combination and their relative stiffness.

Fatigue life improvements due to the use of silver-doped eutectic solder are not very significant (~10%). Thus, they are not viewed as providing any significant advantage to the printed wiring assembly industry. Presence of silver is reported to reduce solder wicking and is attributed to the surface tension modification, and this attribute can be advantageous in fine-pitch low-standoff surface-mount assembly involving TSOP, TSSOP, and VSOP packages. In these cases, eutectic Sn/Pb solder wicks up the leads, depleting solder from the heel and toe areas of the interconnection at the same time rendering the lead even stiffer.

Silver forms Ag_3Sn intermetallics, and the effects are not expected to be any different than gold intermetallics in the range of 1%–3% Ag concentration and hence is considered not a significant risk.

An often expressed concern in regard to presence of silver in electronic packaging is electromigration of silver. It is most often observed in thick-film circuits on bare alumina. The propensity of dendritic growth depends on several factors that include applied potential, temperature, humidity, presence of moisture on the film, potential nucleation sites, etc. The migration propensity is reduced by alloying it with metals like palladium. A 30% addition of Pd reduces silver migration to almost zero. Since, the amount of silver under discussion in the case Sn/Pb/Ag is only 2%, it is not considered a potential problem. While this is generally true, it is important to be aware of its possibility.

Lead-free alloys

With the advent of the implementation of Waste in Electrical and Electronic Equipment (WEEE) and Reduction of Hazardous Substances (ROHS) directives and guideline, the electronic packaging industry in general and the portable electronics industry in particular have also been required to migrate to lead-free interconnection materials by July 2006. While the range of compositions studied for Pb-free solder is vast, many of the more commercially available formulations are tin based. Several binary and ternary alloys have been investigated. Some of these pertinent to portable electronics are described.

Tin/bismuth alloy contains 42 wt% tin and 58 wt% bismuth and has a eutectic temperature of 138°C and has been known to have reasonable fatigue and shear strength properties (Wild 1975). Owing to its low melting temperature, this alloy is less stressful on the PWB as well as the components. However, it is important to note that flux activation at this temperature is lower than in the case of tin/lead solder assembly. Since its mechanical properties are similar to the eutectic tin/lead solders, it has found applications in the high-end computers. A characteristic feature of bismuth-rich alloys is that they expand on cooling. Also, there exists a ternary Sn/Bi/Pb alloy of composition Sn16/Pb32/Bi52 with a melting point of 96°C. Accidental contamination of tin/bismuth alloy with Pb can result in the formation of this low-melting eutectic and can pose a reliability risk. The Pb contamination can be from the tin/lead-plated lead finish and the hot-air-solder leveling surface finish of the PWB. When the assembly experiences temperatures in excess of 96°C either during accelerated thermal cycling test or during the use conditions, the low-melt eutectic forms snow-ball-like spheres and falls out, thus destroying the interconnection integrity. (Singh 1977; Wild 1971). Addition of small amounts of indium up to about 2% is shown to prevent the formation of the low-temperature ternary eutectic (Hua, Zequn, and Glazer 1998). Tin/bismuth solders are also susceptible to brittle fracture under shock loading, and there may be need for composition modification to reduce the strain rate sensitivity (McCormick et al. 1995).

Sn/Bi solders are also considered as bump metallurgy for flip-chip applications in place of eutectic Sn/Pb solder. In the case of very fine pitch flip chip or CSP attach on to the substrate with copper pads, the interfacial reaction can significantly deplete tin, making the interconnection rich in bismuth. The attendant changes in the composition and microstructure of the interconnection need to be carefully assessed.

Tin/silver/copper and related interconnection alloys

In the last decade or so, several multicomponent lead-free alloys have been evaluated for use as interconnection materials to replace the eutectic tin/lead solder. The evaluation considerations included the physicochemical properties such as thermo-mechanical fatigue, creep, microstructures, tooling and process compatibility, compatibility with the components, reliability, and, of course, cost. No drop-in replacement could be immediately arrived at. However, Sn, Ag, and Cu-based alloys have come to be regarded as possible choices with some modifications to the assembly parameters. There have been several studies to determine the eutectic Sn/Ag/Cu ternary composition and temperature since it is complicated by the variations in the microstructure due to cooling rate. One study (Snugovsky et al. 2004) reports the composition and eutectic temperature as 95.6%Sn, 3.5%Ag, 0.9%Cu, and 217.2°C, respectively (Loomans and Fine 2000). Another study based on calculation of phase diagrams (CALPHAD) and results from other alloys suggests that the eutectic composition is probably Sn3.63Ag0.8Cu (Park et al. 2003). Thus, the true eutectic is close to the published compositions with a melting point close to 217°C.

Commercially available Sn/Ag/Cu solders have a wide range of compositions and are not strictly eutectic.

In portable electronics, reliability against mechanical loads such as drop and bend is just as important as thermal shock and cycling loads. Two of the significant aspects that influence interconnection reliability are (1) microstructure of the bulk joint and (2) composition and microstructure of the interfacial layer. Under thermal cycling, stress and strain rates are relatively low and the fatigue crack failures are generally ductile in nature and propagate in the regions adjacent to the interface in the solder. On the other hand, under mechanical drop the strain rates are relatively high, and failures are characterized by brittle fractures generally in the interfacial intermetallic layer.

The reflow temperature in Pb-free assembly is generally around 240°C compared to 215°C–220°C for Sn/Pb eutectic alloy. It is important to recognize that, irrespective of the starting alloy composition, the interconnection metallurgy is more likely to be in considerable variance from the starting alloy composition due to some degree of dissolution of the pad metallurgies such as copper, nickel, silver, etc. and component termination metallurgies such as nickel, iron, palladium, etc. Ni/Au component terminations and ENIG PWB surface finish contribute Au to interconnections in Pb-free solders. Two intermetallic phases are formed: the binary Ni_3Sn_4 and a ternary $(AuNi)Sn_4$ Thus, the solder joint itself can be a complex multicomponent system. The interface between these two intermetallic phases (IMCs) is rather weak and constitutes a low-energy path for crack propagation. Subsequent aging of the assemblies indicates transgranular cracks in the ternary intermetallic layer (Glazer 1994).

In addition, minor elements are also deliberately incorporated to enhance wettabilty, microstructure, and/or performance. Thus, multicomponent alloys have come into use. Table 5–1 shows some of the lead free solders in usage.

Table 5–1. Some Pb-free solders in usage

Sn93.6/Ag4.7/Cu1.7	Sn95.4/Ag3.1/Cu1.5
Sn95.2Ag3.9/Cu0.9	Sn95.4/Ag4.1/Cu0.5
Sn95.2/Ag3.8/Cu1	Sn95.8/Ag3.5/Cu0.7
Sn95.5/Ag3.5/Cu.1	Sn95.5/Ag3.9/Cu0.6
Sn96.2/Ag3/Cu0.7	Sn95.5 /Ag3.8/Cu0.7
Sn96.5/Ag3/Cu0.5	Sn/Ag4.0/Cu0.5
Sn96.2/Ag2.5/Cu0.8/Sb0.5	Those containing antimony
Sn97/Cu2/Ag0.2/Sb0.8	Those containing In and Ga
Sn95.2Ag2.5/Cu0.8/Sb0.5	Those containing Bi and Sb Etc.

Effects of deliberate minor element additions to the base Sn/Ag/Cu alloys have been studied and reported in the literature. The studies are too numerous to include all of them within the scope of this chapter. However, typical examples to illustrate some of the effects are provided. The elements added include Ni, Bi, In, Sm, Y, Mn, Ce, etc. (Liu and Lee 2006; Liu et al. 2007). In some cases, the result is a modification of the microstructures, while in others the changes are in the interfacial microstructure. It has been reported that addition of Ni reduces Cu_3Sn growth, which is more brittle than Cu_6Sn_5 . It has been reported that addition of 0.2% In and 0.004% Ni to Sn1.0Ag0.1Cu alloy shows a 20% increase in drop performance even after 150°C thermal aging. Lower In content also aids in reducing the voiding (Amagai et al. 2004). With a base alloy Sn1.0Ag0.5Cu, the effects on reliability due to several minor elements were reported. The elements included Mn, Bi, Ce, Y, Ti, and several others. These elements either used alone or in combination exhibited marked improvements in the mechanical drop reliability of the interconnection. Of these, 0.13% of Mn addition exhibited significant improvements in drop-test reliability. Thermal aging at 150°C for 700 h, drop testing, and microstructures were evaluated. The interfacial layer thickness was not significantly affected by aging. The melting point of the base alloy is not significantly affected by the addition alloys. Manganese appears to migrate towards the interface layers and forms crystals of $MnSn_2$ in the interface region. Failure location shifted from the interface area to bulk solder. The intermetallic layer thickness as reflowed was 0.86–1.4 μm, and after aging the thickness was in the range of 1.56–2.5 μm. The tin-silver copper (SAC-Mn) alloy is reported to have outperformed both Sn/Pb eutectic and SAC alloy. Thermal aging seems to improve subsequent drop reliability. The exact role of Mn in the reliability enhancement was still not fully evident (Liu and Lee 2006).

In one study, the effect of Au (in the range 0.1%–5%) on Sn3.8Ag0.7Cu on the interconnection morphology was examined. Addition of Au results in a quaternary eutectic microstructure with $AuSn_4$ in equilibrium with Ag_3Sn, Cu_6Sn_5, and β Sn phases at 204 ± 0.3°C. It was found that in the range of 1.0%–2.0% Au addition, the interface Cu_6Sn_5 structure, instead of having a scallop type microstructure, gradually transforms to a fine-grained $(CuAu)_6Sn_5$ layer with finely dispersed β Sn islands (Park et al. 2003). Figure 5–10 shows the interface microstructure change due to the addition of Au.

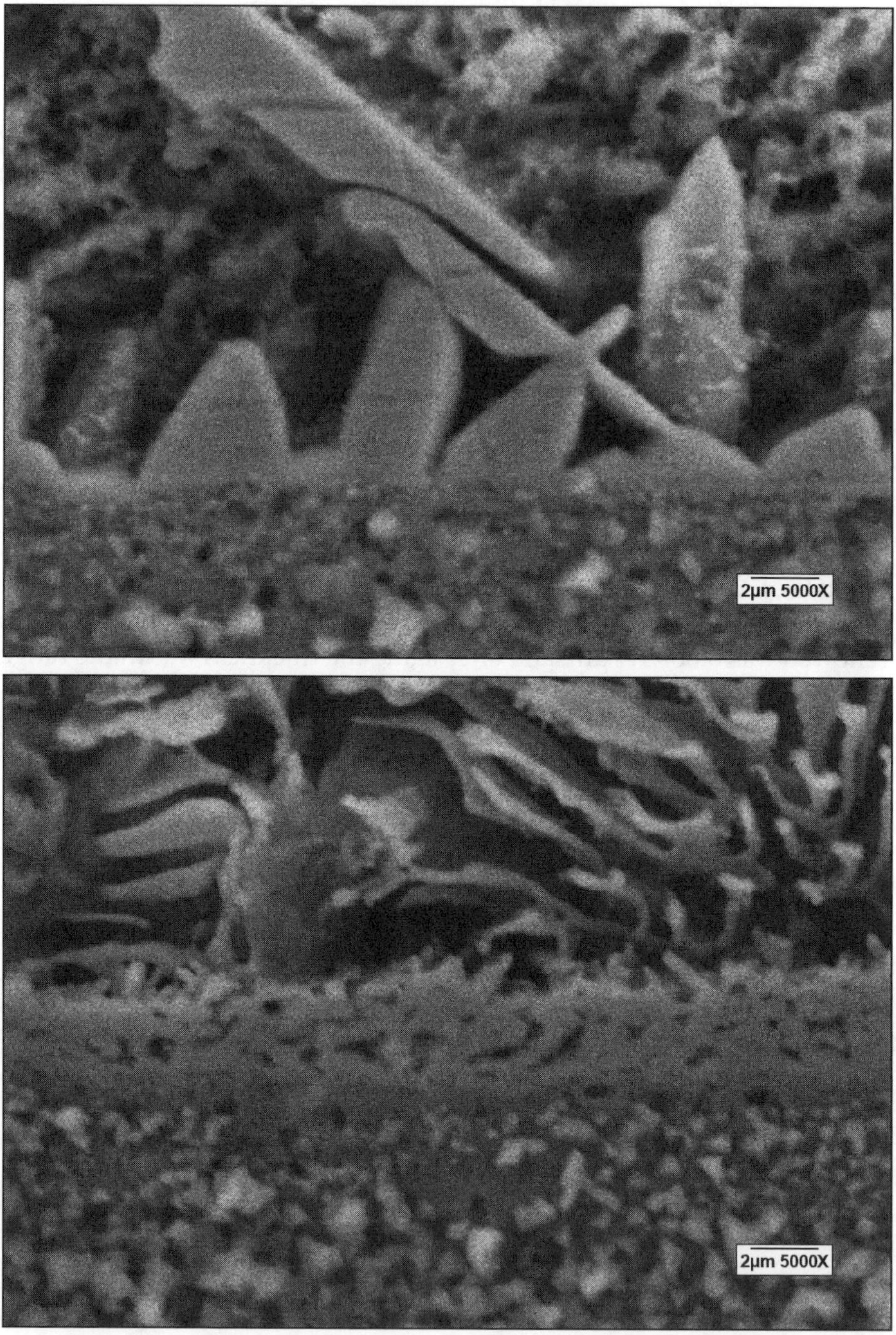

Fig. 5–10. Interconnection interface morphology change due to addition of Au: (a) without Au addition; (b) with 1.0% addition of Au to SAC solder

Also, it has been reported that the change in the interfacial microstructure to a contiguous layer of the ternary intermetallic Cu/Au/Sn phase, in contrast to the scallopy Cu/Sn intermetallic structure, significantly improves the cyclic bend reliability of the interconnection as shown in figure 5–11 (Kabade 2003).

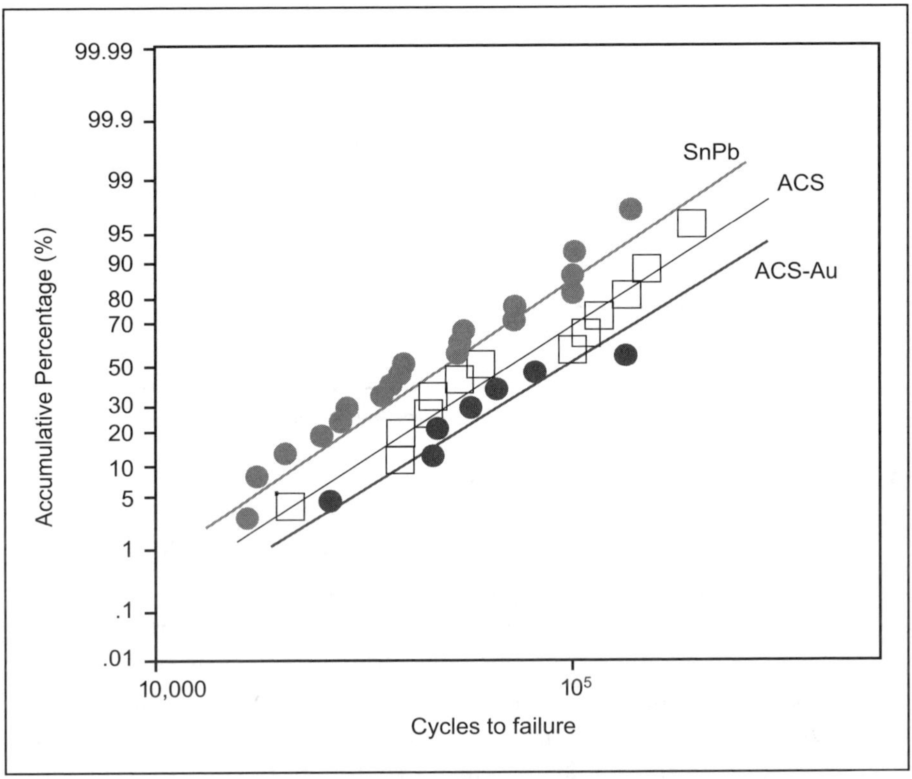

Fig. 5–11. Cyclic bend test Plot of percent fails vs. number of cycles

Effects of supercooling interconnects

Microstructure, as indicated earlier, is an important factor in the thermomechanical reliability of the package-to-board interconnection. Low-cycle fatigue cracks are generally intergranular and therefore a fine-grain joint structure is preferred; hence rapid cooling rates of about 4°C/s are used to achieve a fine-grain joint structure. Slower cooling promotes grain growth and results in coarse grain or lamellar structures. Sn/Ag/Cu solders are known to exhibit supercooling. These materials, when cooled from a liquid state, continue to remain in a metastable liquid state past the freezing temperature and then suddenly solidify. The temperature difference between the true freezing temperature and the actual freezing point is a measure of the extent of super cooling. Figure 5–12 shows a schematic of the supercooling behavior.

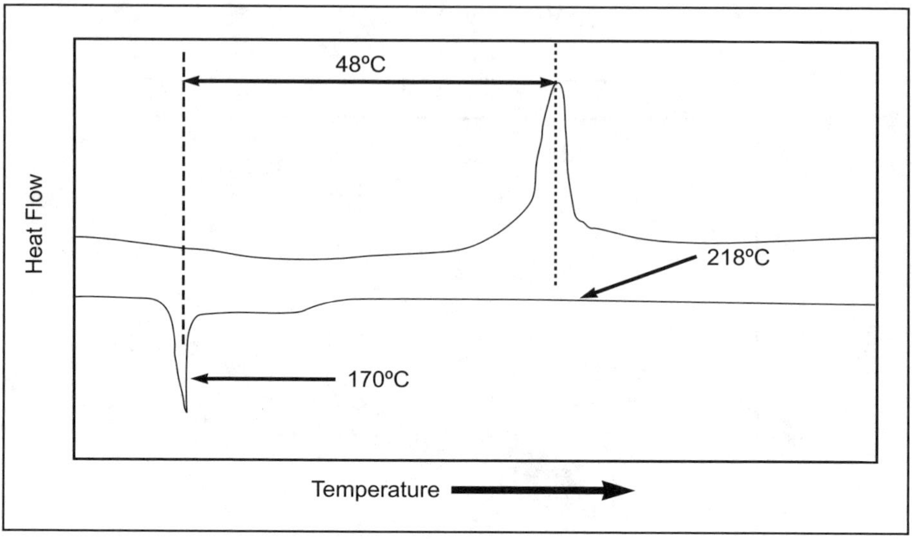

Fig. 5–12. Schematic of the super cooling phenomenon

The extent of supercooling in Pb-free solders can be as much as 50°C to 60°C. One consequence of supercooling is the continued intermetallic growth. Secondly, intermetallics continue to form and grow in energetically favored crystallographic orientations manifesting themselves as rods, plates, bars, etc., and are frozen in place in the solid alloy. Cooling rates depend on the thermal characteristics of the system parts. Low heat capacity light parts may cool faster than high heat capacity ceramic parts. Thicker multilayer boards may take longer to cool. Even for a given package-to-board combination, given the properties of the constituents of the component, different parts of the package may cool at slightly different rates. Thus, it is not surprising to find different joints under the same package having different microstructure morphologies.

Figure 5–13 depicts a partially etched CSP solder ball showing the intermetallic structure inside the ball. The plate-like structures comprise the Ag/Sn intermetallics, while the rod-like structures are Cu/Sn intermetallics. Figure 5–14 depicts a row of partially etched solder joints under an area-array package. It can be seen that not all the joints are identical in their internal structure.

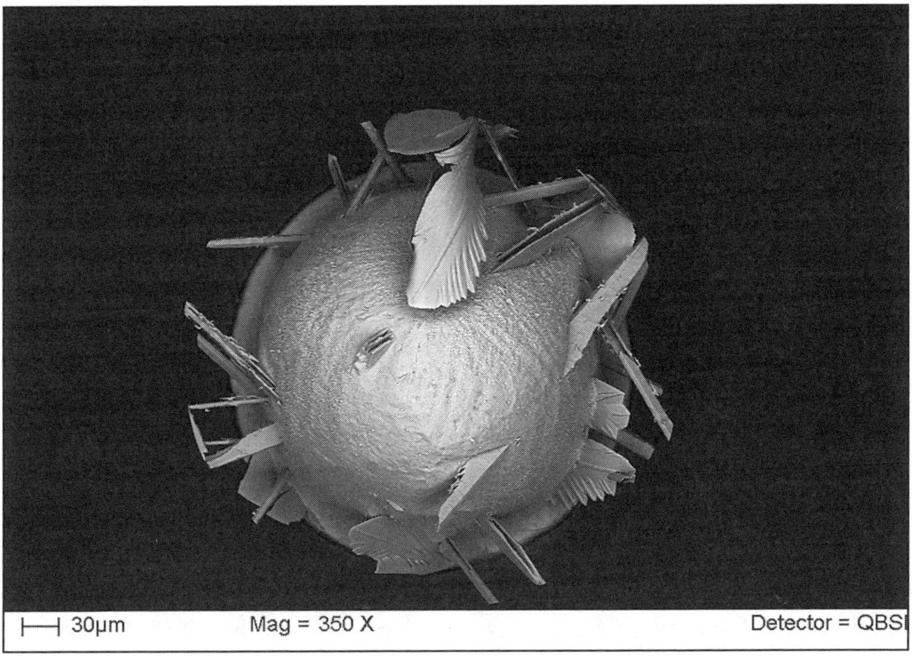

Fig. 5–13. Partially etched CSP solder ball showing the presence of different intermetallic morphologies. The leaf-like dendrites are composed of Ag_3Sn phase and the rods are composed of Cu_6Sn_5 phase

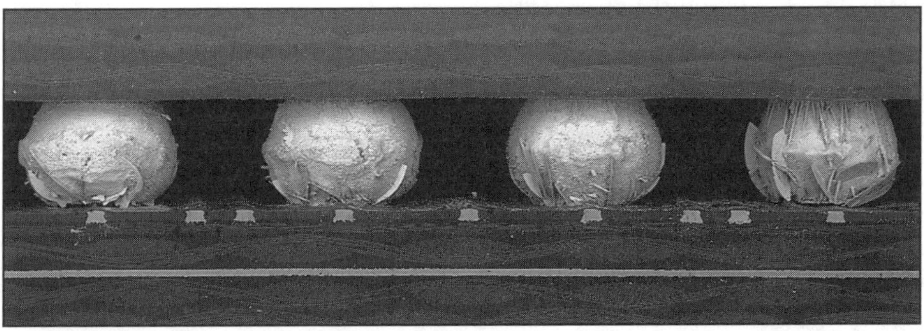

Fig. 5–14. Row of partially etched interconnections under an area-array package showing the variation in the IMC morphology from joint to joint (Dunford et al. 2004)

The interconnection microstructure morphology depicted is characteristic of Sn/Ag/Cu alloys and can have implications in product reliability. The fine β Sn grain structure is interspersed with Ag_3Sn and Cu_6Sn_5 intermetallics and, depending on their location, can influence the fatigue crack propagation. Much of the rigidity of SAC alloy is attributed to the presence of Ag/Sn and Cu/Sn intermetallics. Lowering the Ag content in the solder has been observed to reduce the Ag/Sn intermetallics. Ag concentration of 2% or lower in the SAC alloy has been shown to improve the microstructure and give a more ductile and compliant solder to tolerate high-strain-rate drop stress. However, these exhibit poorer creep and thermal fatigue performance, which can limit their range of potential applications in the electronics industry (Kang et al. 2005). Also, the non-eutectic composition will have a pasty range and can result in higher defect rates in assembly.

Steps to reduce the supercooling of the alloy are necessary to reduce differential cooling.

Addition of Al as a minor element in SAC is shown to reduce the number of hard Ag_3Sn and Cu_6Sn_5 IMC particles, and forms larger, softer, and non-stoichiometric AlAg and AlCu particles, reducing the overall yield strength and increasing creep rate. This results in a significant reduction in yield strength, and also causes some moderate increase in creep rate. For high Ag/SAC alloys, adding 0.1%–0.6% Al to SAC is most effective in softening, and brings the yield strength down to the level of SAC105 and SAC1505 (containing 1.5% Ag and 0.5% Cu), while the creep rate is still maintained at SAC305 level. Incorporation of Ni results in formation of ternary IMC $(Ni,Cu)_3Sn_4$ at the expense of Cu_6Sn_5 particles. This softens SAC alloys, although only to a smaller extent. Thus, addition of Al and Ni reduces the modulus and elongation at break (Huang, Hwang, and Lee 2007).

Thus, Sn/Ag/Cu alloys with several minor element additions have been shown to improve the microstructure and modify the thermomechanical and mechanical reliability. To a large extent, ternary SAC alloys have been shown to behave equivalent to or better than Sn/Pb alloys. While a eutectic composition is preferred for assembly simplicity, it is neither possible to attain with a multicomponent system nor is it necessary as long as the pasty region is narrow, within a few degrees range.

Whiskers

With the advent of WEEE and ROHS directives, the electronics industry along with other industries has been migrating to lead-free materials alongside eliminating hazardous substances in their products. Component terminations and lead finishes have migrated to lead-free balls and coatings such as immersion tin, Ni/Pd, Ni/Pd/gold finishes, etc., respectively. PWB finishes migrated predominantly to ENIG, immersion silver, immersion tin, OSPs, etc. Replacement of Sn/Pb finish with tin has resulted in the re-emergence of an old phenomenon of tin whiskers and associated potential risks. The fine conductive filamentary outgrowths from the tin surface, namely whiskers, can break off and cause electrical shorts and result in permanent or transient system failures. A perspective of the tin whisker phenomenon, its potential mechanisms, and remedial measures as relevant to portable electronics is provided in the ensuing paragraphs.

Tin occurs in two allotropic modifications, namely white tin and grey tin. Disintegration or transformation of white β tin into grey α tin due to volume increase (by 26%) is termed *tin pest*. This transformation is associated with long incubation period below 13°C, and is fastest at about −30°C. Similarly, mechanical tensile loads can accelerate the transformation. Minor element additions such as Sb or Bi are known to substantially reduce this phenomenon. Tin can be plated in three different finishes, namely, matte, bright, and satin. Matte tin film has a large columnar grain structure (1–10 μm size) and is relatively stress free. Bright tin film is made up of small grain (<1 μm) columnar structure and is associated with inherent compressive stress. The third is satin tin with a grain structure of 1 μm. Matte and bright finishes are the most common types.

Historically, during the pre-World War II period, electroplated cadmium was the material of choice for electronic component finish. However, failures due to conductive filamentary growth were reported in 1946 (Cobb 1946). Later, in 1948, similar failures due to Cd filaments in channel filters of multichannel transmission lines were noted by Bell Telephones. As a result, Cd was subsequently replaced by tin, only to discover that tin too is prone to whisker growth. As a consequence, long-term investigations were initiated by the Interconnection Technology Research Institute (ITRI), the European Space Agency (ESA), Northern Electric, Ericsson, etc., to elucidate the phenomenon and explore remedial measures.

According to the dictionary, a whisker is usually defined as one of the long projecting hairs growing on the sides of the mouth of a cat or other animals. As such, any filamentary growth of a defined dimension emanating from a surface is termed a whisker. Plated tin occurs in two modifications: bright tin and matte tin. Whiskers are 1–5 μm in diameter; their length can range upwards of even 5,000 μm and their ambient growth rate can be 0.1 A /s. Tin films plated on steel surfaces are known to produce whiskers when subjected to compressive forces (Fisher, Darken, and Carroll 1954). Applied compressive forces up to 7,000 psi accelerated the whisker growth from <1 to 10,000 A/s suggesting that compressive forces within the material extrude tin as whiskers and the extruded material undergoes several recrystallization events due to the occluded energy. Once the propensity of tin to grow whiskers was recognized, several mitigation measures have been proposed and practiced. Addition of Pb to Sn reduced the whisker potential significantly and hence was adopted as the remedial mitigation strategy. Thus, the propensity for whisker growth in Sn/Pb electronic assemblies is not common. Another mitigating strategy involved annealing between 150°C –and 200°C (Glazunova and Kudryavtsev 1963). It should be noted that it may not always be possible to anneal at 150°C–200°C owing to the temperature sensitivity of the components involved. Whisker incipiency and growth is a distinct form of recrystallization. For pure tin films, post-plating annealing at 150°C for 1 h can reduce subsequent whisker growth. Another remedial measure is to use an underlay of nickel or silver. Still another method is to use a proprietary process that can provide films with favorable combinations of crystal orientations. Whisker growth can have an incubation period before manifestation and this can be from a few seconds to a year.

It is generally recognized that whisker growth is primarily due to stresses in the tin film. These stresses can be external or internal. External stresses can result from the mechanical forces of bending, twisting, torque, tensile force, etc., imposed on the product during fabrication or usage. Internal stresses can be inherent to the plating processes, grain size, crystal orientation, etc. The following is a list of some of the stress sources:

- Tin plating formulations
- Plating rates and grain size
- Copper contamination in Sn plating baths
- Crystallographic orientations
- Plating thickness
- Nature of the substrate
- Nature of the under layer
- Nature of the intermetallic phases formed with tin
- Manufacturing stresses imposed on the product
- Stresses imposed during product usage
- Temperature
- Humidity
- Bias voltage

The relative magnitudes of these stresses vary considerably and several of the mechanisms can be operative simultaneously. Thus, it can be a formidable task to estimate the magnitude of the risk. Owing to the multitude of the variables involved, individual results of investigations can differ from each other considerably.

Proprietary methane-sulfonic acid and mixed-acid plating formulations with special additives can produce tin films with favorable combinations of crystal orientations. Variations in the plating parameters with the same plating chemistry can give rise to different grain structure: one more prone and another less prone to whisker growth. It has been reported in one study that matte tin films plated onto a variety of lead frame materials when subjected to 5 V bias consistently produce whiskers (Romm and Abbott 2003). Another study indicates that, while pure tin plating does not exhibit whiskers, controlled Cu addition in excess of 30 ppm to the plating baths produces tin films that are prone to whiskers (NIST 2001). Nickel and silver underlayers are known to inhibit whisker growth.

There are several mechanisms that have been proposed to explain the whisker growth phenomenon. These include concepts of intermetallic formation, recrystallization, dislocations, diffusion, oxide defects, etc. either acting alone or in combination.

An integrated theory was recently advanced to explain the mechanism of whisker formation (Galyon and Palmer 2005). It is generally agreed that whisker formation is due to internal stresses in system. The stresses that are operative can be classified as macro and micro stresses. Macro stresses are long-range and influence lattice dimensions, and the gradients drive and determine the net mass transport. On the other hand, micro stresses have little or no influence on the lattice dimensions or the mass transport. Both these affect localized rearrangements, such as recrystallization, and influence grain growth.

Intermetallic phases and grain boundary networks are important features in whisker formation. Their formation and growth is a pathway to relieve the stress, and is accompanied by a net mass transfer. It is the resultant of the available energy in the system and the energy barriers in the system that needs to be overcome. Intermetallic phases are the primary reason for the stress generation at interface of the plated tin film and the substrate. Grain boundaries offer a path way for the atoms to move from high stress regions to low stress regions. Table 5–2 gives the salient physical properties of the metals and intermetallic phases involved in the systems.

Table 5–2. Properties of Sn, Cu, Ni and related intermetallic phases

Element or Intermetallic Phase	Density/ gm/cc	Coefficient of Thermal Expansion (ppm/C)	Atomic or Molecular weight	Young's Modulus GPa	Poisson's ratio	Molar Volume/ cc/g-mole
Copper	8.9	16.6–17.6	65	124	0.33–0.36	7.11
Tin	7.28	23	118	42	0.36	16
Nickel	8.9	13	58.7	214	0.31	6.59
Cu_6Sn_5	8.27	16.3	980	85.6	0.309	118.5
Cu_3Sn	8.9	19.0	313	108.3	0.299	35.16
Ni_3Sn_4	8.65	13.7	294	133.3	0.330	33.98

Source: Annotated Tin Whisker Bibliography and Anthology by Galyon and Palmer, 2005.

In the case of tin plating on a copper substrate, copper and tin interdiffuse into each other with different rates. Copper diffuses much more rapidly into tin than tin into copper. As copper diffuses into tin, it forms the Cu_6Sn_5 intermetallic phase. Since the combined volume of the Cu_6Sn_5 and the displaced tin atoms is greater than the original volume of tin atoms in that space, a compressive stress is generated in that region. The expansion of the intermetallic phase is also subjected to the restraining effects of the adjacent copper and tin layers on either side. The differential diffusion rates of tin and copper into each other promote a Kirkendall voiding on the Cu side of the interface (Smigelskus and Kirkendall 1947). As the concentration of voids increases, it causes shrinkage in the Cu layer thereby creating a tensile stress in the region.

The plated tin can have co-deposited contaminants that can cause internal stresses in the layer. An example is Cu contamination. A few parts per million copper contamination can create internal stresses in the tin plating. The CTE of Sn is higher than that of copper. This mismatch results in stresses at the interfaces during thermal excursions.

Thus, immediately under the tin layer is an intermetallic and tin zone where compressive forces prevail. Underneath the IMC zone is the Kirkendall vacancy zone, a shrinkage zone with tensile stress zone due to the overlying IMC zone, and below the vacancy zone is the copper substrate that is relatively inactive. The expansive forces due to the Cu_6Sn_5 intermetallics constitute the driving force for the upward migration of Sn atoms forming the whiskers.

The migration of tin atoms away from the compressive zone is generally localized to a specific whisker grains. These whisker grains are oriented (along [210]) differently than the neighboring or surrounding Sn atoms (which are oriented along [321]). There is generally an incubation period for the formation of these, which can vary from a few seconds to almost one year. This is essentially a recrystallization event. This has been attributed to the weakness of oxide layers on certain grains. Weak oxide grains differentiate themselves from the neighboring grains by their oblique grain orientations in contrast to parallel and columnar as-plated grains. It was proposed that a weak oxide layer enables localized relief of internal stresses by allowing whisker growth through the cracks in the oxide layer (Tu 1994). Figure 5–15 shows the schematic of the Sn/Cu interface layer structure identifying the stresses.

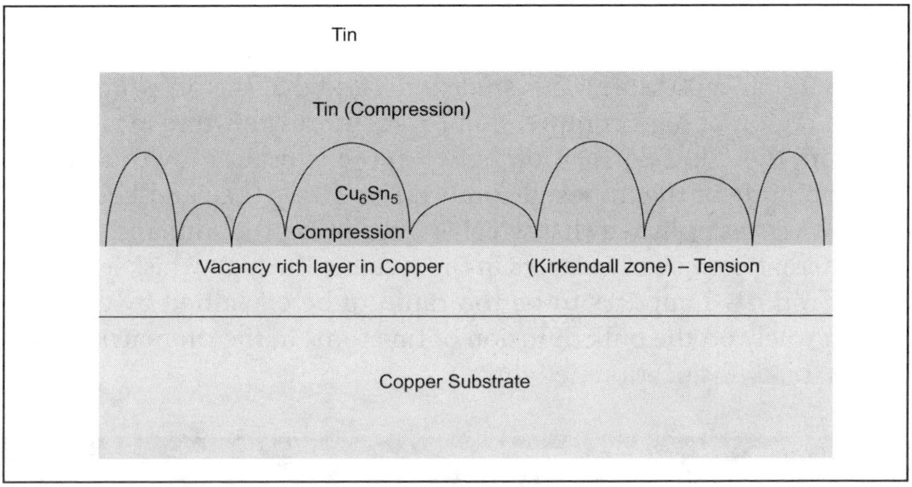

Fig. 5–15. Schematic of Sn/Cu Interfaces and associated stress zones for whisker formation (Courtesy of Steve Dunford, Nokia, Inc.)

Whisker shape, lengths, and diameters vary considerably owing to the relative magnitudes of the stresses involved, the type and thickness of plating, the underlayers, etc. Figure 5–16 shows examples of Sn whiskers.

Other alternative mechanisms involving the theory of dislocations have also been proposed over the years. Spiral prismatic dislocations are believed to move to the surface via a climb mechanism thereby adding what is termed a Burger's vector thickness. One such layer is added for each complete loop of spiral, thus building up the whisker (Amelinckz et al. 1957). In another study it was postulated that tin atoms move through a screw dislocation at the center of the whisker and deposit themselves at the tip. In still another study it was reported that whisker grows from the base and not from the Sn atoms deposited at the tip (Koonce and Arnold 1953, 1954). It was also postulated that whiskers grow from dislocations at the base as a result of a diffusion-limited process. Frank–Reed dislocation sources produce loops that expand by climbing to a boundary. These loops in turn glide to the surface and deposit their half plane of extra atoms on to the surface. The rotating edge dislocation stays in the same plane after each revolution, but additional atoms are added to the whisker base. The driving force is believed to be the oxidation which creates a negative surface tension in the region where the whisker is formed (Eshelby 1953; Frank 1953). In a U.S. Steel study, it was observed that a Sn-plated steel mounted on a metallographic clamp grew significant number of whiskers only in a few days. Subsequently, growth measurements as a function of pressure indicated that the induction period approached zero as the clamp pressure is raised. A growth rate of 10,000 A/s was observed at a pressure of 7,500 lb/in^2. This study indicated that the tin atoms move from a region of high compression to a compression-free location. In addition, there is a correlation between the whisker growth rate and the amplitude of the imposed compressive stress (Lee and Lee 1998). Thus, it was established that whisker growth is predominantly due to the inherent compressive forces in the system. The high whisker growth that is reported appears to be too rapid to be explained by theories relying solely on the bulk diffusion of tin atoms in the tin matrix, grain boundary, or dislocation networks.

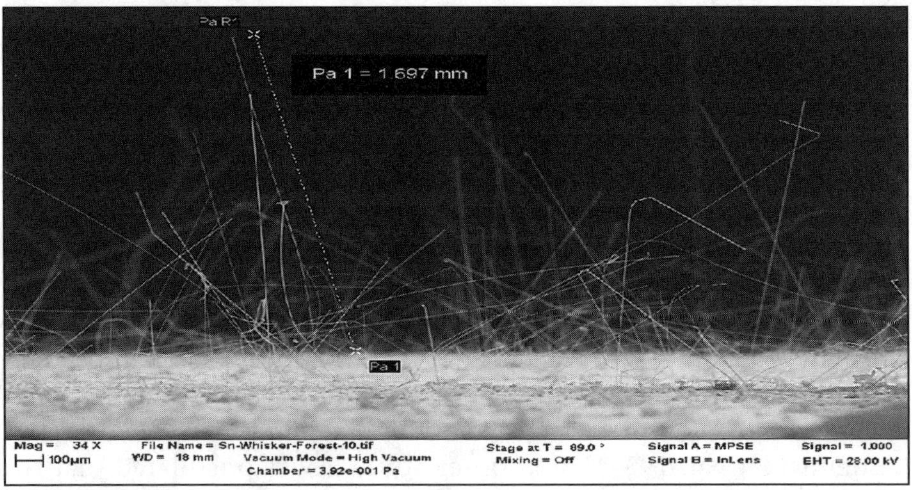

Fig. 5–16. An example of whiskers (Image provided by Robert Champaign and Bob Ogden, Raytheon Corp.)

When electroless nickel-immersion gold (ENIG) is used as surface finish, or the lead material is such as alloy-42, the whisker propensity is considerably lower. If Ni is present, the stress levels are generally tensile. Focused ion beam experiments indicate that whisker grains have roots close to intermetallics projecting from the Cu/Sn interface. Thus, Ni acts as a barrier for copper diffusion into Sn. Annealing of Sn on nickel at 175°C indicates that the boundary moves into nickel, whereas in the case of Sn over copper the boundary moves into Sn. Thus, Sn moves at a greater rate into Ni than Ni into tin, while Cu moves readily into tin, and whisker mitigation due to Ni is more because of an inter-diffusional relationship with Sn than the blockage of copper atoms.

Whisker growth phenomena is not limited to tin on copper only, but has been observed on several metallurgies, like brass, zinc, cadmium, germanium, indium, etc.

Several methods have been tried to slow down, minimize, and mitigate the risks associated with whisker growth. These are based on the particular stage of the product in its life cycle, the designed life of the product, as well as the magnitude of the risk and the cost to mitigate. These can be categorized as follows:

1. Use of matte tin that is 10 μm or thicker. While this may not provide 100% protection against whisker growth, it is superior to bright tin.
2. Use of specially formulated proprietary tin plating formulations that have been verified to provide stress-free tin films with favorable combination of crystal orientations.
3. Use of material sets that are not prone to whisker formation. Under this category come the use of Pd/Ni/Au terminations, ENIG for PWB finish, etc.
4. Relieving the inherent stresses by annealing at 150°C for 1 hr or reflowing, etc. This may relieve the stresses, but stresses developed subsequently during manufacturing and assembly and usage conditions can still cause whisker growth.
5. Use of underlayer materials such as nickel, silver, etc., which have been known to abate whisker growth. These underlayers are applied immediately below the tin layer and above the copper substrate.
6. Use of tenacious conformal coatings such as parylene, which have been known to prevent whisker growth. It should be noted that most protective coatings only delay the inevitable. Depending upon the product design life, it may provide adequate protection.
7. Use of anticorrosion coatings on the exposed Cu regions, which can potentially eliminate whisker growth.

Product classifications and acceptance criteria for whiskers

Electronic products for the purpose of whisker propensity are classified into various categories by the Joint Electronic Device Engineering Council (JEDEC). High reliability and critical applications, such as in medical, military, and space, pertain to Class I. High reliability business applications, high-end servers, disk drives, long-product-life and low-downtime products, etc., are Class II. Consumer products with relatively short product life of five years or less, which suffer no major problems when whiskers break off and cause problems elsewhere in the system, are designated as Class III products. Thus, most portable electronic devices fall under class III. A three-step testing methodology was indicated by the JEDEC and is shown in table 5–3.

Table 5–3. JEDEC-recommended test conditions for whisker growth

Stress Type	Test Condition	Duration of Test
Air-to air thermal cycling	-55 + 0/-10C to 85 + 10/-0 C	Up to 1,500 cycles
Low temperature/humidity storage	30 ± 2C/60 ± RH	Up to 4,000 hrs
High temperature/humidity storage	55 ± 3 C/85 ± 3% R.H.	Up to 4,000 hrs

Acceptability or pass/fail criteria are different for the three classes of products. Pure tin and high tin content alloys are generally forbidden for class I product applications. For class II product applications, the maximum whisker length allowed is 40 μm. For class III products, the maximum whisker length permitted is 60 μm with a minimum component lead pitch of 0.05 mm, and 50 μm where the operating frequency is 6 GHz or greater. Often, product manufacturers may choose to develop their own acceptance criteria depending on the nature of the product, its operating environment, the product warranties, etc., to ensure reliability.

Portable electronics involving medical applications will need special consideration even though the risk due to whisker formation may not be a big concern for consumer-grade portable electronics.

Another independent acceptance criterion was a 5 V bias test. Matte tin films plated onto a variety of lead frame materials when subjected to 5 V electrical bias consistently produce whiskers (Romm and Abbott 2003). Absence of detectable whiskers in this test may indicate lack of whisker propensity.

In spite of extensive whisker studies reported in the literature, and the several theories that are put forward, considerable variation in the results is not uncommon. There can be more than one mechanism operative. The variables involved are too numerous to be addressed and to arrive at an unambiguous mechanism. The current scenario may be summarized as follows. The exact mechanism of whisker growth is not fully understood. No agreed-upon acceleration factors exist that enable accurate projection of risk-free product life. There does not seem to exist a known method to assure the elimination of whiskers over the product life.

However, it is comforting to note that there exist several screening and mitigation methods to address the potential risks and still produce quality and reliable products while on-going investigations aid in a better understanding of the phenomenon of whiskers. It is prudent to refer to the latest version of JEDEC or IRC or other standards documentation for the most up-to-date test and acceptance criteria.

Conductive adhesive interconnection

Use of conductive adhesives as a replacement for metallurgical surface mount interconnects has been attempted several years ago and has found limited application. The thrust for non-lead interconnects with conductive adhesives and lead Pb-free alloys is due to environmental factors (Buckley 1994). Considerable effort has been expended in finding suitable materials to replace eutectic Sn/Pb solder. No immediate drop-in replacement material that would satisfy the requirements of low resistivity or joint resistance, adequate joint strength, high-volume manufacturability, and thermal and mechanical reliability could be arrived at. However, several advantages were perceived. The following are some of the material- and process-related features considered advantageous.

- Lower processing temperature
- Essentially a no-clean process
- No wicking propensity along the component leads
- Less stringent fillet shape and formation concerns
- No tomb-stoning potential
- The interconnection precludes use of any flux
- Environmentally benign process
- Less dependency on solder mask requirements
- Possible low-volatile organics

Despite the above-perceived advantages, several, rather significant, concerns remained. These include the following:

- Joint strength
- Effect of cure temperature and time on the PWB or temperature-sensitive components
- Compatibility of the thixotropic and rheological properties with the prevailing screen printing tools and equipment
- High-volume manufacturing
- The thermomechanical reliability
- Long-term effect of humidity
- Time-variant electrical performance degradation

In addition, the ability to rework and repair conductive adhesive interconnection has always been a concern. Thus, adhesive interconnect technology did not emerge as an interconnection technology of choice for high-volume, low-cost electronic assembly and manufacturing. The technology, owing to some of its inherent advantages, found some niche applications in electronic packaging. Thus, it is discussed only from a contextual perspective in this chapter.

Conductive adhesives are broadly categorized as follows:

- Isotropic conductive
- Anisotropic conductive
- Intrinsically conductive

These are described in some detail. The base polymeric materials used in the first two categories are either thermoset or thermoplastic materials. Thermoset materials, once cured, cannot be softened on subsequent heating, while thermoplastic materials soften on heating after curing.

Isotropic conductive adhesive interconnections

As the name implies, the electrical conductivity of this interconnection is isotropic, i.e., it is same in all the three coordinate axes. These materials find applications in hybrid circuits and coarse-pitch surface-mount assembly. Isotropic conductive adhesives, thermoplastic or thermoset resins, contain a large volume fraction of highly conductive particles such as silver, silver-plated copper, nickel, or gold as filler material, and the filler loading can be as much as 80%. The metal particles are generally in the shape of thin flakes 40–50 μm long and about 5 μm thick. Conductivity is accomplished by the overlapping flakes.

The resin systems can be one- or two-component epoxies, polyimides, modified silicones, etc. (Hvims 1994, 1995). These filled-resin systems can be screen- or stencil-printed, dot-placed, or dispensed on to the carrier. The ability to screen-print the paste depends on its composition and thixotropic properties. Solvent evaporation during application can result in surface skin formation and may require processing in a saturated solvent atmosphere. Component placement pressure, if not optimized, may squeeze the paste and can cause bridging and shorting of adjacent leads. This risk is higher when the lead pitch is finer. Cure temperature is in the range of 100°C –200°C, and the cure time can be a few minutes to 1 hr. These factors influence the assembly process throughputs. Also, silver-filled epoxies are associated with the risk of silver electromigration in high humidity environment, and application of a suitable conformal coating as a moisture barrier may be warranted. Figure 5–17 shows a schematic of an isotropic adhesive interconnection.

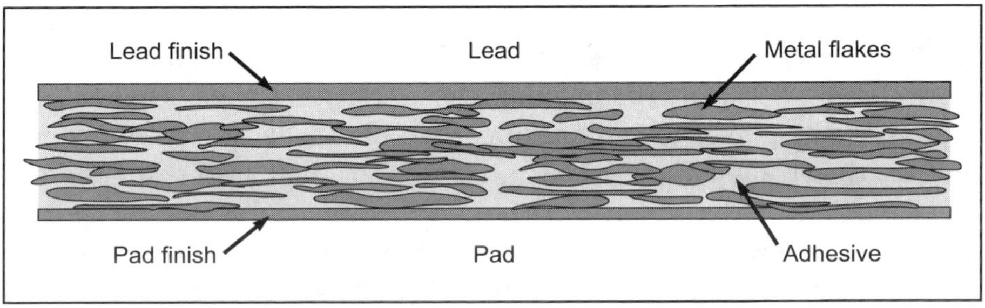

Fig. 5–17. Schematic of an isotropic adhesive interconnection

When used as a die attach adhesive, silver-filled epoxies provide mechanical strength, ability to electrically ground, and also the heat dissipation path to the substrate or carrier. The polymerization or curing process of the adhesive, depending on the T_g and the modulus of the material, can give rise to residual stresses in the interconnection. The higher the T_g of the material, the higher the curing temperature and hence the larger the residual stresses. It is important to note that there can be residual stresses in the interconnection, and considerable variation in the mechanical and electrical properties of the interconnection can be experienced depending on the properties of the resin material. Shear strength can decrease during high-temperature storage. Fully polymerized high T_g materials can show more pronounced degradation than low T_g ones (Whaley et al. 1993).

Reliability testing of conductive adhesive interconnection assemblies generally includes thermal cycling from –55°C to 125°C, temperature and humidity testing at 85°C/85% R.H., and sometimes mechanical testing such as bend and drop tests. Some materials exhibit contact resistance increase with time. Catastrophic fails can also occur. While some may provide acceptable electrical performance, they suffer from mechanical failures. There can be differences in performance between temperature and humidity testing with and without a bias voltage. During bias testing, the component can be hotter with greater moisture egress and also additional curing and may result in improved performance.

Another important factor in the integrity and reliability of the interconnection is the surface finish of the mating surfaces. The component surfaces can be nickel-plated gold, palladium over nickel, nickel, or tin. Printed circuit board surface finishes can be flash gold, organic solder preservative such as polytriazoles, or tin/lead. In one study, it was demonstrated that Pd alloy-plated surface finish exhibited superior performance in both thermal cycling and temperature and humidity tests (Gaynes et al. 1995). In another study it was reported that Ag/Pd-terminated packages exhibited superior performance compared to Sn/Pb finish (Jagt, Beris, and Litjen 1995). It is also recognized that the more compliant the interconnection, the less susceptible it is to failure in a low-cycle fatigue environment since viscoelastic deformations accommodate the strains. The relative movement of the cured resin and the metal particles within can alter the contact points and hence the contact resistance. As a result, with increasing number of thermal cycles increase in contact resistance is a distinct possibility. As most polymeric materials are porous, slow ingress of air and/or moisture can result in surface oxidation of filler material and degrade electrical performance.

Aniosotropic conductive adhesive interconnection

These are also called *z-axis* or *uniaxial* conductive adhesives since their electrical conductivity is only along one axis, namely the vertical or z-axis, and they are nonconductive in the x and y directions. These were developed primarily to address flip chip attach (FCA), TAB, and other fine-pitch surface-mount assemblies. They also find applications in liquid crystal display (LCD) module attachment to flex circuits and LCD display screen attachment (Liu and Rorgren 1995).

The resin matrix can be made up of a host of thermosetting resins. A variety of conductive particles have been explored. These include silver on nickel, nickel on polymer cores, gold on nickel, silver on copper, low-melt metals on high-melt metals, etc. (Chang et al. 1993). The metal loading is far lower than in the case of isotropic conductive adhesives and is generally in the vicinity of 5% by volume and the particles themselves are in the range of 10–25 μm diameter range. The general principle of effecting the interconnection consists of depositing the metal-loaded paste on the component footprint, placing the component, and application of pressure and heat to form the interconnection. During the process, the resin softens or melts and flows around the metal particles; the metal particles come in contact with the two mating surfaces, namely, the pad and the lead; and the resin is cured locking in place the metal balls. Thus, the electrical conductivity is only along the vertical axis since the cured resin separates the conductive particles in the x and y directions. Figure 5–18 depicts a schematic of an anisotropic conductive adhesive interconnection. Alternatively, the conductive adhesive can also be applied in the form of a film to the footprint.

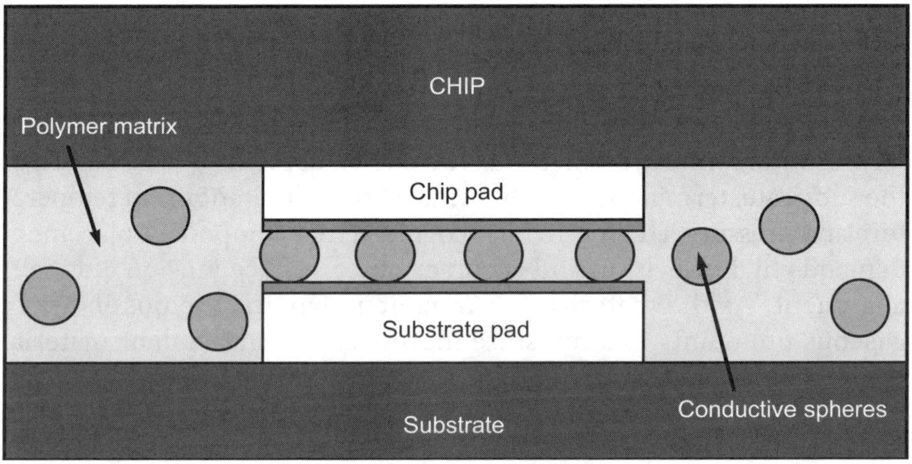

Fig. 5–18. Schematic of an anisotropic conductive adhesive interconnection

In a variation of the approach, the filler spheres are first encased in a dielectric resin to constitute a microcapsule filler (MCF). These particles are then dispersed in an adhesive matrix to form the microcapsule adhesive. This is then applied to the flip-chip footprint; the chip is then placed on the footprint on the PWB; pressure (25 g/bump) is applied to exude the resin, thus facilitating electrical contact under pressure; and the assembly is cured at 170°C for 30 s while maintaining the pressure on the deformed spheres. The interconnection bumps have a resistance of 3mΩ (Date et al. 1994).

As mentioned previously, anisotropic conductive adhesives are also fabricated in customized tape formats. Mylar-backed epoxy adhesive tape preforms that match the component footprint and containing a matrix of silver-loaded adhesive dot pattern are known (Bolzer et al.

1995; Bolzer et al. 1994). The attachment process consists of first tacking the Mylar-backed preform precisely on to the component and heating to about 80°C at 2–3 psi for 1–2 s. The Mylar is then peeled off and the component is located on to the carrier footprint at 80°C, 2–3 psi for 2–4 s followed by final curing at 160°C–175°C for half to one hour.

In still another variation of the anisotropic conductive adhesive attach technique, metal particles such as Ag, Cu, Pd, Pt, or Al are coated with low-melting Bi, Sn, In, Sb, or Zn and dispersed in polymer matrices such as polyesters, polyimides, siloxanes, or polyimide–siloxanes, A perceived advantage of this technique is that the low-melting component of the particle facilitates metallurgical bond formation either through a solid–solid or solid–liquid reaction at the same temperature at which the resin matrix is cured, thus providing a better interconnection. By varying the filler loading, the concept can be used for both isotropic and anisotropic conductive interconnection applications (Kang, Rai, and Purushothaman 1998).

Thus, there exist a number of variations in the materials and formats for this interconnection method. The choice depends on the nature of the substrate, termination pitch, surface finish, limitations on temperature and pressure, etc. It is to be recognized that component placement demands high precision and accuracy since surface tension aids self-alignment. As the materials are organic polymers, the possibility of gaseous pollutants and moisture ingress and the attendant material degradation as a function of time resulting in joint strength degradation cannot be overlooked. However, many niche applications have been found for this interconnection technique in portable electronics and are extensively used including for display applications.

Intrinsically conductive adhesive interconnection

This class of materials, by virtue of their molecular structure, is known to be electrically conductive and have been extensively investigated since 1970s. They owe their electrical conductivity to the conjugated π-electron clouds in their structure that provides mobile electrons to facilitate electrical conductivity. The classes of compounds include acetylenes, pyrroles, thiophenes, anilines, etc. Typical compounds are polyacetylenes doped with AsF_3, polyacetylenes doped with iodine, poly(*p*-phenelyne) doped with AsF_3, polypyrroles doped with iodine, polyanilines, etc. The dopant materials are used to enhance conductivity. Current conductivity range of these materials is $10^2 - 10^4$ $S.cm^{-1}$. Figure 5–19 shows the typical molecular structures of some of these materials.

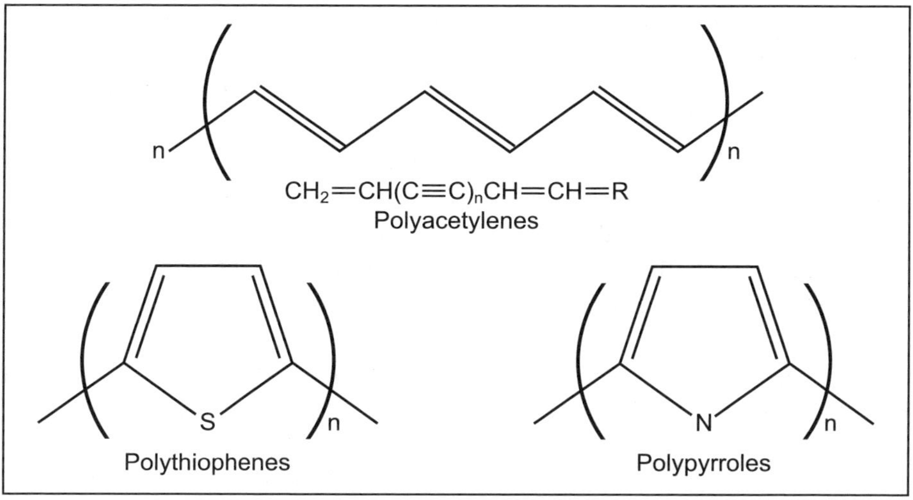

Fig. 5–19. Molecular structures of typical intrinsically conductive materials

An application may involve placing the monomer between two conductive surfaces and allowing polymerizing, and also physically bonding the two surfaces, thus establishing an interconnection. Owing to their relatively low conductivity, the use of these materials is limited to preventing static electricity buildup, which in itself is a significant application.

The materials are not very stable and degrade in air or humidity environments and can pose a severe limitation in their use in electronic packaging. Also, the lowest bandgap attainable in these materials is at best 0.5 eV, which is still not narrow enough for many applications (Scotheim, Elsenbaumer, and Reynolds 2007). It is important to recognize the potential of these materials in future electronic applications as their properties are improved and stability is enhanced.

Optical interconnects

Optical transmission has already become the de facto standard in long-haul telecommunications and data pipes because of the higher bandwidth and lower losses. However, at the printed circuit board level, especially in space-constrained portable electronic applications, optical interconnections are not common. As the interconnection density (number of components per chip, the number of chips per board, etc.), the modulation speed, and the degree of integration evolve, electrical interconnects can run into fundamental bottlenecks, such as speed, packaging, fanout, and power dissipation. In some cases, multichip module (MCM) technology is employed to provide higher data transfer rates and circuit densities. The employment of optical interconnects will be one of the major alternatives for upgrading the interconnection speed within a portable electronic product whenever conventional electrical interconnection fails to provide the required bandwidth.

One embodiment of an optical interconnection scheme, as shown in figure 5–20, incorporates a vertical-cavity surface-emitting laser (VCSEL), surface-normal waveguide couplers, a polyimide-based channel waveguide functioning as the physical layer of optical bus, and a photoreceiver (Chen et al. 2000). The driving electrical signal to modulate the VCSEL and the demodulated signal received at the photoreceiver are all through electrical vias connected to the surface of the PWB. The optical equivalent of a PWB trace is a polymer-based optical waveguide.

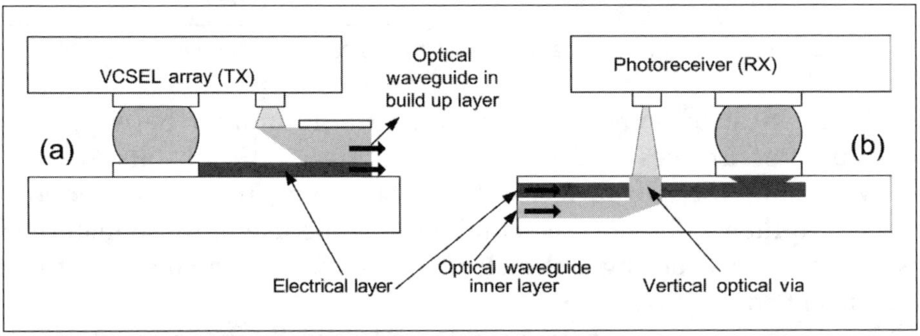

Fig. 5–20. Schematic diagram of optoelectronic PWB showing optical waveguides with integrated micromirrors fabricated (a) as optical buildup layer on the board surface and (b) as optical inner layer inside the board with optical via and vertical waveguide for z-directional light guiding (Immonen, Karppinen, and Kivilahti 2005)

Major hurdles for implementing optical interconnection at the second-level interconnection into high-performance microelectronic systems are packaging incompatibility and manufacturability. A critical requirement is that highest possible compatibility with existing PCB technology is maintained, especially design and manufacturing aspects such as automated placement and mounting. The placement accuracy needs of optical interconnection schemes are higher than for purely electrical interconnection schemes. In addition, the requirements of alignment precision and accuracy are much higher.

A key challenge is to realize efficient optical coupling between waveguides and emitter/detector devices. Positional or alignment inaccuracies in the optical path decrease the coupling efficiency or can even lead to total transmission failure of the device. Although direct coupling enables connections between active devices and the waveguide, to form a 3-D optical signal distribution, for example, in optical multilayer boards, beam turning and directional optical interconnections are required. Achieving alignment accuracy at the optical interfaces requires high-precision bonding tools, and the cost and complexity of the final product are thereby increased. Alternative approaches that incorporate these couplers directly at the board level will need to be developed.

Warpage is another major concern and can give rise to misregistration, delaminations, and excessive cracking in fabricated optical components (Immonen, Karppinen, and Kivilahti 2005). In their study, board warpage was reduced by applying thinner optical buildup layers, using smaller surface coverage or using a more rigid substrate. In addition to increasing the scattering loss of the waveguide, microcracks can degrade the long-term device stability and serve as initiation sites for crack propagation caused by contaminants or humidity. Furthermore, stresses inside the optical film can cause local anisotropy or inhomogeneous distribution of the refractive index resulting in optical birefringence and polarization-dependent loss (Huang 2003).

Polymer-based materials are attractive for making waveguides in guided-wave optical interconnection. They can be spin-coated on a myriad of substrates with a relatively large interconnection distance, and a large variety of organic polymers can serve as microelectronic and optoelectronic materials with potential applications such as interlayer dielectric, protective overcoat, ray shielding, optical interconnects, or even conductive electrical interconnects. In addition to their ease of processing, they possess favorable electrical and mechanical properties such as high resistivity, low dielectric constant, light weight, and flexibility. Since compatibility with conventional microelectronic fabrication processes is essential, the polyimide PMDA/ODA, prepared from pyromellitic dianhydride (PMDA) and 4,4–oxidianiline (ODA), is attractive as it exhibits thermal stability up to approximately 400°C. For example, Kapton is physically stable over a wide range of temperatures (270°C–400°C) and has excellent heat resistance and chemical resistance and therefore provides the required high-temperature process compatibility (Chen et al. 2000).

To efficiently couple optical signals from the signal (light) source to polymer waveguides and then from waveguides to photodetectors, two types of waveguide couplers are in vogue: tilted grating couplers and waveguide coupling mirrors that depend on total internal reflection. The latter couplers are much less sensitive to wavelength and can be easily manufactured using reactive ion etching similar to the procedure for fabricating polyimide-based channel waveguides.

Lastly, there are also several choices for signal (light) sources, and the VCSEL is attractive for optoelectronic interconnection at the board level as it provides a high coupling efficiency using the earlier described waveguide couplers. The control of the emitting aperture and the wavelength make VCSELs a good choice because of the large separation of the adjacent longitudinal modes due to the short cavity length (Chen et al. 2000). The planar configuration of VCSELs allows these devices to be fabricated and wafer-scale-tested with conventional microelectronics manufacturing processes. Because the laser beam is emitted normal to the surface of the device, the same packaging scheme can be used for coupling light from VCSEL into waveguide as that used for coupling light from waveguide into photodetector.

Emerging Trends

With higher and higher levels of integration at first- and second-level packaging, combined with increasing emphasis on miniaturization, interconnections continue to be of finer and finer pitch and smaller and smaller. Over the years, the interconnections have migrated from 100 mil pitch plated-through-hole connections to 16 mil pitch surface-mount interconnections. Solder bumps in FCA on substrates are of even smaller pitch in the 6–10 mil pitch range. Thus, finer and finer pitch interconnections will be the norm. From the second-level interconnects perspective, 0.3 mm pitch interconnects may set the limits for high-volume, high-throughput manufacturing with acceptable yields for portable electronic products. It is envisioned that migration from micro to nano interconnects would be a natural pathway for high-density packaging. In that migratory path of integration and miniaturization, until one achieves total transition to nanoelectronic systems, the industry has to manage a technology combination of microelectronics and nanoelectronics (Aschenbrenner et al. 2006). Thus there is likely to be a micro–nano transition in the interconnection technologies. It is similar to the transition industry experienced migrating from insertion-mount technology to surface-mount technology.

One of the fastest growing technologies is the nanotechnology. Nanomaterials exhibit unusual and unique properties that are advantageously harnessed in areas including first and second-level electronic packaging. Carbon nanotubes have been shown to have superior mechanical, electrical, and thermal properties.

Nantero developed CMOS process-compatible high-density nonvolatile memory technology based on carbon nanotubes that is considered superior to traditional technologies such as SRAM, DRAM, Flash, etc. The processes involve conventional spin-coating, etching, and patterning common to CMOS manufacturing. According to Nantero, specially fabricated carbon nanotubes are suspended over two electrodes of a silicon substrate. As a voltage is applied, the fiber bundle is drawn towards the underlying electrode making an electrical contact and thus giving a bias of one. The suspended noncontact position represents the zero bias state. The fibers are held in place by van der Waals forces. NRAMs are considered to be faster than traditional DRAMs. Failure-free operation up to 10^{15} times has been demonstrated owing to the superior mechanical strength of the carbon fibers. Future portable electronic gadgets such as MP3, players, PDAs, etc., that require gigabyte memories are considered the beneficiaries of this new noteworthy technology. Working modules have been demonstrated with many of the manufacturing issues successfully resolved.

Reactive nano technology (RNT) is an interconnection technology that utilizes a multilayer nanofoil to join dissimilar materials. The process is considered predictable, controllable, and affordable. The foils are a new class of nano-engineered materials which consists of thousands of nanoscale layers that alternate between elements with large negative heats of mixing, such as Ni and Al. Combinations of similar or dissimilar materials like ceramic-to-ceramic, metal-to-ceramic, and metal-to-metal joints can be made by these techniques by local heating (Wang et al, 2004). Joining of LEDs with tin-plated copper terminations on to a PWB with a matte tin surface finish has been demonstrated. Other finishes that are suitable include gold, silver, etc. The bonding pressure is 7 MPa and requires an electrical pulse of 1 V at 100 A, and interconnection is effected in a few milliseconds. The process eliminates the need for any fluxing. NanoBond eliminates the need for conventional reflow processing or for the use of fluxes. The resulting bond has a shear strength of 30 MPa and seems to survive the standard thermal cycling, high-temperature storage, mechanical shock, temperature, and humidity, and vibration tests (JESD 22-108, A 105 C, A 103 C, IEC 60068 2-27 and 2-6 tests (Duckham 2006).

Attachment of a Au-plated stainless steel connector to a Au-plated PWB using a preformed 60-μm nanofoil and a free-standing 25 μm Au-Sn solder perform has been demonstrated (Rude et al. 2003).

RNT is also used to provide hermetic sealing of a Au-plated stainless steel lid to a Au-plated stainless steel base, which has shown helium leak rates of less than 1×10^{-10} atm. cc/s. This is considered more attractive and effective than the traditional laser welding process (Rude et al. 2005). Another concept of interconnection, termed *nano Velcro connection* involves the formation of multitudes of conductive nanotubes on both the carrier and component terminations so that they can be enmeshed to provide a compliant electrical and mechanical connection. Thus, this type of interconnection eliminates the need for the traditional soldering process in the second-level assembly (Levin et al. 2005). It is important to recognize, however, that concepts like this are in a formative stage and might take extensive research and development effort to realize on a manufacturing scale as well as to integrate with conventional interconnection technologies (Turlik 2004; Ashenbrenner et al. 2006)

References

Amagai, M., Y. Toyoda, T. Ohmishi, and S. Akita. 2004. High drop test reliability: lead free solders. Paper presented at the Proc. Tech. Prog. 54th ECTC, Las Vegas, NV: 1304–1309.

Amelinckz, S., W. Bontnick, Dekeyser, and F. Seitz. 1957. On the formation and properties of helical dislocations. *Philosophical Magazine*, 8 th ser., 2 (15): 355–377.

Aschenbrenner, R., S. Fiedler, T. Loher, K. F. Becker, H. Reichl. 2006. Development of nanostructures based on interconnection technology. Paper presented at the NaoTech Northern Europe, Helsinki, Finland.

Bolzer, J. C., M. Reynolds, and J. Popylarczyk. 1995. Area bonding conductive adhesive (ABC) for flex circuit connection to LTCC/MCM substrates. Paper presented at the Proc. IEEE ECTC: 529–533.

Bolzer, J. C., M. Reynolds, and J. Popylarczyk, and K. Gileo. 1994. Area bonding conductive epoxy adhesive preforms for grid array and MCM substrate attach. Paper presented at the Proc. IEEE MCM Conf., Santa Cruz, CA: 15–17.

Buckley, D. 1994. Replacing solder with conductive adhesives. *Electronic Production*: 25–28.

Chen, R. T., L. Lynn, C. Choi, Y. J. Liu, B. Behari, L. Wu, S. Tang, R. Wickman, B. Picori, M.K. Hibbs-Brenner, J. Bristow, and Y.S. Lee. 2000. Fully embedded board-level guided-wave optoelectronic interconnects. *Proceedings of the IEEE* 88 (6): 780–793.

Chang, D. D., J. A. Fulton, H. C. Ling, M. B. Schmidt, R. E. Sinitski, and C. P. Wong. 1993. Accelerated life test of z-xis conductive adhesives. *IEEE Trans. Components Hybrids Manuf.* Technol. CHMT 16 (8): 836–842.

Cobb, H. L. 1946. Cadmium whiskers. *Monthly Rev. Am. Electroplaters Soc.*, 33 (28): 28–30.

Date, H. Y., Hozumi, H., Tokuhira, H., Usui, M., Horikoshi, E., and M. Sato. 1994. Anisotropic conductive adhesive for fine pitch interconnections. *SPIE* 2369: 570–575.

Duckham, A. 2006. Nano-foil bonding. *Advanced Materials and Processes*: 37–38.

Dunford, S., S. Canumalla, and Viswanadham P., Intermetallic morphology and damage evolution under thermomechanical fatigue of lead free solder interconnections. Proc. of 54th ECTC (2004), Las Vegas, NV: 726–736.

Eshelby, J. D. 1953. A tentative theory of metallic whisker growth, Phy. Rev., 91: 755–756.

Findlay, A., A. N. Campbell, and O. N. Smith, 1951. *The Phase Rule and Its Applications*. New York: Dover Publications.

Fisher, R. M., L. S. Darken, and K. G. Carroll. 1954. Accelerated growth of tin whiskers. *Acta Metallurgica*, 2: 368–373.

Frank, F. C. 1953. On tin whiskers. *Phil. Mag.* XLIV(7): 854–860.

Galyon, G. T., and L. Palmer. 2005. An integrated theory whisker formation: The physical metallurgy of whisker formation and the role of internal stresses. *IEEE Trans. Electron. Packaging Manuf.* 28(1): 17–30.

Gaynes, M. A., R. H. Lewis, R. F. Saraf, and J. M. Roldan. 1995. Evaluation of contact resistance of isotropic conductive adhesives. *IEEE Trans. Comp. Packaging, Manuf. Technol. B. Advanced Packaging* 18 (2): 299–304.

Glazer, J., P.A. Morris, and J. W. Morris. 1991. The effect of Au on the reliability of fine pitch surface mount solder joints. *J. Suface Mount Technol.* 4 (10): 15–27.

Glazer, J. J. 1994. Microstructure and mechanical properties of lead free alloys for low cost electronic assembly—A review. *J. Electron. Mater.* 23 (8): 693–700.

Glazunova, V. K., and N.T. Kudryavtsev. 1963. An investigation of the conditions of spontaneous growth of filiform crystals on electrolytic coatings. *J. Appl. Chem. USSR* (translated from Zhurnal Prikladnoi Khimii). 36 (3):, 519–525.

Hua, F., M. Zequn, , and J. Glazer. 1998. Eutectic Sn–Bi as an alternative to Pb-free solders. Paper presented at the Proc. 48th IEEE Electronic Components and Technology Conf.: 277–283.

Huang, B., H. S. Hwang, and N. C. Lee. 2007. A compliant and creep resistant SAC-Al(Ni) alloy. Paper presented at the Proc. Electronics Packaging Technology Conference, Singapore: 693–702.

Hvims, H. L. 1994. Solder replacement. *Soldering Surface Mount Technol.* (17): 12–19.

Hvims, H. L. 1994. Testing of solder replacement technology. *Qual. Reliab. Eng.*, 10 (5): 423–434.

Hvims, H. L. 1995. Conductive adhesives for SMT and potential applications. *IEEE Trans. Comp. Packaging. Manuf. Technol. B. CPMT-18*: 284–291.

Immonen, M., M. Karppinen, and J. K. Kivilahti, 2005. Investigation of environmental reliability of optical polymer waveguides embedded on printed circuit boards, *Proc. 5th Int. Conf. on Polymers and Adhesives in Microelectronics and Photonics*: 23–26.

Huang, M. 2003. Stress effects on the performance of optical waveguides. *Int. J. Solids Structures* 40: 1615–32.

Jagt, J. C., P. J. M. Beris, and G. F. C.M. Litjen 1995. Electrically conductive adhesives: A perspective. *IEEE Trans. Components, Packaging Manuf. Technol. B. CPMT*. 19 (2): 292–98.

Kabade, R. 2003. Metallurgical investigations of Sn-Ag-Cu solder with the addition of gold. Master's Degree Thesis, University of Texas at Arlington.

Kang, S. K., P. A.Lauro, D. Y. Shih, D. W. Henderson, and K. J. Puttlitz. 2005. Microstructure and mechanical properties of lead-free solders and solder joints used in microelectronic applications. *IBM J. Res. Dev.* 49 (4–5): 607.

Kang, S. K., R. S. Rai, and S. Purushothaman, 1998. Development of high conductivity lead (Pb)-free conducting adhesives. *IEEE Trans. CPMT A* 21: 18–22.

Koonce, S. E., and S. M. Arnold. 1953. Growth of metal whiskers. *J. Appl. Phys.* 24 (3): 365–66.

Koonce, S. E., and S. M. Arnold. 1954. Metal whiskers. *J. Appl. Phys.* 25 (1):134–35.

Lee, B. Z., and D. N. Lee, 1998. Spontaneous growth mechanism of whiskers. *Acta. Mater.* 46: 3710–714..

Liu, W., and N. C. Lee. 2006. Novel SACX solders with superior drop test performance. Paper presented at the Proc. SMTA Int. Conf., Rosemont, IL, Sept. 25–28: 134–147.

Liu, J., and R. Rorgren. 1993. Joining displays using thermosetting anisotropically conductive adhesives. *J. Electron. Manuf. Technol.*, 3 (4): 205–14.

Loomans, M. E., and M. E. Fine, 2000. Tin-silver-copper eutectic temperature and composition. *Metall. Mater. Trans.* A. 31A: 1155–162.

McCormick, M. et al. 1995. The design of new, lead free solder alloys with improved properties, Proc. *IEEE Int. Symp. Electronics and Environment*: 171–176.

Mei, Z., F. Hua, and J. Glazer. 1999. Sn-Bi-X solders, *Proc. SMTA-I*: 399-403.

Miller, L. F. 1969. Controlled collapse chip joining, *IBM J. Res. Dev.* Vol.13: 239–250.

Park, J. Y., R. Kabade, C. U. Kim, T. Carper, S. Dunford, and P. Viswanadham. (2003). Influence of Au addition on the phase equilibria of near eutectic Sn3.8Ag0.7Cu Pb-free solder alloy, *J. Electron. Mater.*, Vol. 32: 1474–1482.

Park, J. Y., Kim, C. U., Carper, T., and Viswanadham, P. (2003). Phase equilibrium studies of Sn Ag-Cu eutectic solder using differential cooling of 3.8 Ag 0.7Cu-Sn alloys. *J. Electron. Mater.* Vol. 32: 1297–1302.

Romm, D., and D. A. Abbott. 2003. Whisker evaluation of tin plated logic component leads, TI Application Report, SZZA037A.

Rorgren, R. S., and J. Liu. 1995. Reliability assessment of isotropically conductive adhesive joints in surface mount applications. *Trans. IEEE CPMT Advanced Packaging* 18:305–12.

Rude, T., J. Subramanian, J. Levin, Van D. Heerden, and O. Knio. 2005. Hermetic sealing of microelectronics packages using a room temperature soldering process. Paper presented at the Proc. Tech. Prog. IMAPS, Philadelphia, PA: 286–290.

Rude, T., et al. 2003. Mounting of microelectronic components at room temperature with solders and local heat sources. Paper presented at the Proc. Tech. Prog. IMAPS, Boston, MA.

Levin, J., J. Subramanian, T. Rude, and T. O. Knio. 2005. Room temperature soldering of connectors to PCB using reactive multilayer foils, Paper presented at the Proc. IMAPS Tech. Conf. on Connectors, Philadelphia, PA.

Scotheim, T. A., R. L.Elsenbaumer, and J. R. Reynolds, 2007. *Handbook of Conducting Polymers*. Vols.1 and 2, New York: Marcel Dekker Publishing Co.

Singh, P. 1977. Reliability evaluation of an electrical connector. Paper presented at the Proc. NEPCON: 464–482.

Smigelskus, A. D., and E.O. Kirkendall 1947. Zinc diffusion in alpha brass. *Trans. AIME* 171: 130–142.

Snugovsky, L., D. D. Perovic, and J. W. Rutter, 1977. Solidification of ternary and near eutectic alloys in the Ag-Cu-Sn System. *Journ. Mat. Sci. & Technol.*, Vol 20, part II: 1403–1413.

Tu, K. N. 1994. Irreversible processes of spontaneous whisker growth in intermetallic Cu-Sn thin reactions. *Phys. Rev. B* 49 (3):2030–034.

Turlik, I. 2004. The next technology revolution—Nanotechnology, A presentation. American Physical Society March meeting, Montreal, Canada.

Wang, J., Besnoin, E., Duckham, S. J. Spey, M. E. Reiss, D. M. Kuio, and T. P. Weihs. Joining of stainless steel specimens with nano-structured Al/Ni foils. *Journal of Applied Physics*, vol. 95, No. 1: 248–256.

Whaley, D. C., D. J. Williams, Y. Ogunjimi, O. A. Boyle, and J. M. Goward. 1993. A comparison of the behavior of isotropic and anisotropic adhesives, in *Manufacturing Aspects in Electronic Packaging, ASME, Production Engineering Division*, Vol. 65: 81–91.

Wild, R. N. 1973, 1975. Some fatigue properties of solders and solder joints. IBM No. 73Z000421, 74Z000481, Owego, NY: IBM Publications.

Wild, R. N. 1971. Properties of some low melt fusible solder alloys with improved properties. *Inter NEPCON*, Brighton, England:

Wolters, J. 1977. *Zur Geschichte der Lottechnik*, Degussa, Hanau, West Germany.

6

Printed Wiring Board Assembly

Introduction

The focus of this chapter is on populating the printed circuit boards with surface mount components as very few portable electronic assemblies, if any, have insertion mount components. Assembling the various single chip modules, multichip modules, resistor packs, capacitors, connectors, and such parts as switches, sockets, etc., is termed second-level packaging. In portable electronic products, the components are mostly leaded surface mount packages such as quad flat packs or J-leaded packages, area-array packages like ball grid array, chip scale, and/or land grid array packages.

For high volume portable consumer electronic products, manufacturing processes have to be highly automated with high throughputs and yields. In a competitive, high-volume manufacturing environment, even small changes in materials and/or processes that save a few cents or a few seconds will have a significant impact on the profitability of the business. Breakdowns and interruptions in manufacturing can be very costly. Often, if line defects in manufacturing are not discovered and rectified in a timely manner, it could result in considerable number of defective assemblies, warranting either rework or discarding the product altogether. In either case, the cost impact can be considerable and devastating. In this chapter the reader is assumed to be familiar with general electronic assembly processes and hence only aspects pertinent to portable electronic assembly are addressed.

Assembly process

The assembly process steps in portable products are essentially similar to those for conventional printed wiring board (PWB) assembly. Both single- and double-sided assemblies are in vogue depending on the functionality and complexity of the product. The process includes (a) stencil printing of solder paste, (b) component placement, (c) reflow, and (d) visual solder joint inspection, as shown in figure 6–1. The standard assembly processes are well documented in literature (Prasad 1997; Behun et al. 1997; Matteson and Kandelid 2001). The complexities in PWB assemblies arise out of demands imposed by the high levels of miniaturization and integration necessary to package more functions in a given area of the board. Packages consisting of very fine-pitch components and high packaging density (namely, number of packages per unit area of the PWB) are common. The boards are much thinner than those used in traditional desktop and business products. PWBs for desktop products are usually 8–10 layers in structure and about 0.062" thick. On the other hand, for most portable products the boards have 0.020–0.030" thickness generally containing 4–8 layers with high-density wiring and interconnect structure involving micro technology. Owing to the small size of the product, the assembly is carried out in panel format, with each panel containing four to six product cards, as shown in figure 6–2.

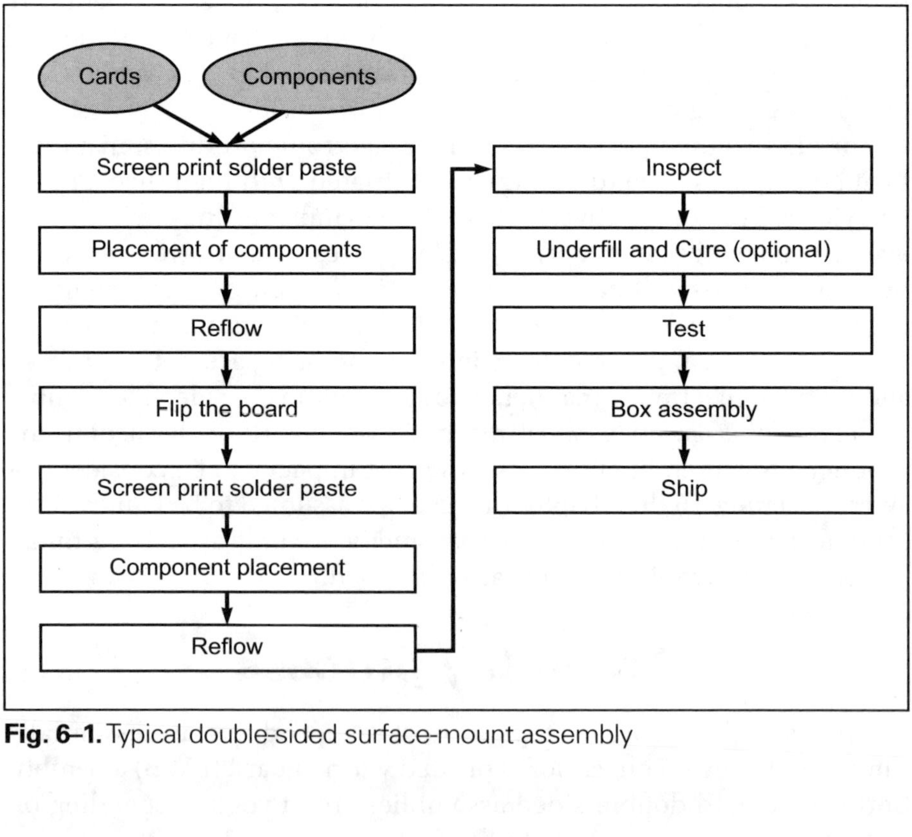

Fig. 6–1. Typical double-sided surface-mount assembly

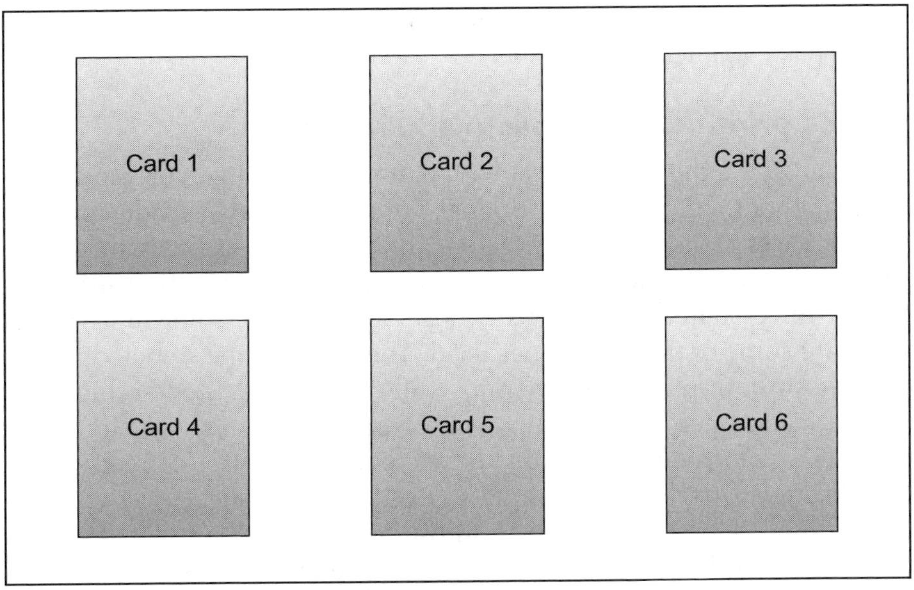

Fig. 6–2. Schematic of printed wiring board panel and cards

In the ensuing paragraphs, the salient features of the assembly process as pertinent to fine pitch and high density are discussed.

The process consists of inspecting the incoming boards for any defects such as contamination, plating defects, fiducials, etc., and ensuring that the solderable surfaces are protected with suitable surface finishes. In portable products, organic solderability preservative (OSP), electroless nickel-immersion gold (ENIG), immersion tin, or immersion silver is used. Of these, ENIG was more prevalent owing to the long shelf life, superior solderability, and good electrical contact for battery contacts, etc. However, owing to the occasional sporadic occurrence of the "black pad" phenomenon and associated solderability and reliability concerns, OSP is preferred. But OSPs are subject to mechanical abrasion and hence exposed corrosion-prone Cu pads do not constitute good electrical contact surfaces. A compromise is selective OSP, where the pads to be soldered are coated with the OSP and the other areas are coated with ENIG. Choice of surface finish has a strong impact on both the assembly parameters and reliability. A surface finish that affords acceptable reliability under thermal loading conditions may not provide the same reliability under mechanical loading conditions. The reliability can depend on, among other factors, the interfacial intermetallics, their morphology, and the strain rates involved. There can also be differences among studies due to variations in the card structures, component menu, etc. (Hossain et al. 2008). After inspection, the boards are moved to the solder paste application station.

Stencil printing/solder paste application

A stencil is a thin metal sheet with patterned openings corresponding to component footprints on the panel. For most portable product builds, the thickness of the stencil is in the 4–6 mils range. Stencils are made in a variety of ways depending on the sophistication. Laser cutting and electropolishing (stainless steel for the most part) and electroforming (nickel only) are some of the techniques used. The shape of the stencil opening and the smoothness of the opening walls are critical. Stencil thickness variations are generally minimal in laser-cut stencils. On the other hand, electroformed stencils are known to provide better defined stencil opening owing to the nature of the fabrication technique. However, there can be considerable variation from the nominal thickness. Unacceptable variations by as much as 30 μm on a stencil thickness of 165 μm, almost 19%, have been noted (Rahn 2006). The stencils are stretched and mounted in a frame. The stencil is aligned with the board so that its openings match with the footprints on the PWB panel. Also, thinner stencils have a tendency to stretch over a period of usage, thus resulting in mismatch between the stencil openings and the footprints on the PWB panels and should be watched carefully.

The solder paste is an intimate mixture of fine solder particles, rosin flux, and other chemical substances that include rheology modifiers, surfactants, and organic solvents. Solder pastes are classified into different types according to the diameters of the solder particles: the finer the particles the higher the solder type. Table 6–1 shows the solder paste types and their relation to the particle size in micrometers.

Table 6–1. Solder paste type and its relation to solder alloy particle diameter

Solder Paste Type	Particle Size/μ
1	75–150
2	45–75
3	20–45
4	20–36
5	15–25
6	5–15

It is important to recognize that, as the particle size decreases, for a given solder amount there will be an increase in surface area and hence more surface oxide content, which influences solderability.

The solder paste usually contains 80%–90% metal content, and the remainder is flux. However, the volume percentage of the metal to flux ratio is around 50% owing to the lower density of the liquid flux. With the implementation of ROHS and WEEE guidelines and elimination of lead (Pb) in electronic assemblies, Pb-free interconnection alloys have been developed and are increasingly being used in recent years. In most portable electronic product applications, solder pastes used are of the no-clean type. That means that after the interconnection reflow process, any residues left on the panel are not cleaned. The activators used in these fluxes are relatively mild and non-chloride-bearing. Thus, the residues are mostly benign and do not cause any deleterious effects like corrosion or electromigration. However, attention must be paid to their impact on underfilling the package–board interspace.

An amount of solder paste is dispensed onto the stencil and is kneaded to obtain the appropriate thixotropy. The paste is forced into the stencil openings across the board with a squeegee. It is an interactive process involving the board surface, paste, the squeegee, the stencil, and the operating environment. It is important to understand the implications of these interactions when dealing with fine-pitch printing. An ideal print is one in which the paste deposit on the component footprint is flat on the top and has straight angles and vertical sides that represent the stencil opening. Also, the printed paste should not slump prior to the placement of the component. The viscosity of the paste is controlled by the metal to flux ratio, the solvents, the thixotropic compounds, and the nature of the flux itself. Some thixotropic agents are derivatives of castor oil. Solder paste is generally stored between 5 and 10°C. Too low a storage temperature can cause crystallization of the organic rosin/resin, while storage at warmer temperature can cause loss of solvents and may initiate activation reactions. It is preferable to store the paste in the dark in order to avoid any light-induced polymerization reactions. Thus it is important to observe several precautions to preserve the uniform composition of the paste and obtain the best performance from the paste. The higher the storage temperature the shorter can be its shelf life. Pb-free solder pastes, because of their inherent poorer solderability, are more sensitive in this regard.

In most paste printing operations, metal squeegees are used. Squeegee speed, the angle it subtends with the board, and the pressure are critical parameters for printing solder paste. Temperature and humidity of the manufacturing floor also play a role in successful operations. Higher temperature can cause evaporation of solvents, formation of skin on the printed paste, etc. Higher humidity can cause moisture absorption by the paste and affect its viscosity as well as thixotropy and thus affect the printing process as well as post-printing behavior of the paste during placement and interconnection. It is not uncommon that the solder paste materials, the handling of the paste, and the printing process can contribute to almost 40%–70% of the assembly defects, and, hence, adequate care should be taken to minimize the defects due to this process step. The solder paste print is then inspected for defects. These defects include bridges due to excessive paste, pads with missing paste, pads with insufficient paste, misaligned paste print, slumped paste, stringed conical prints, etc. When the print quality and definition is not acceptable, the panel and the stencil are cleaned, the cause established, and remedy identified, and then the process is repeated.

Components and their placement

Subsequent to satisfactory solder printing, the PWB panel is transported to the placement station where the different component packages are placed on their respective pads with high precision and accuracy with high-speed placement machines also called, descriptively, *pick and place machines*. These include actives, passives, connectors, and all other special components. These can have a variety of lead forms and terminations: J-leaded, gull wing leaded, leadless, area array of solder balls, etc. The proper storage of these plastic encapsulated microelectronic packages is of utmost importance, as many are prone to different degrees of moisture uptake during storage.

When a moisture-laden package is used, depending on the amount and package geometry, the moisture exudes with explosive violence during the high-temperature reflow assembly causing package damage. This is termed the *popcorn effect*. The extent of the damage depends, in addition to the moisture content, on the temperature and time conditions of reflow, the package design, and properties of the packaging materials set. They are categorized into six different levels based on their propensity to absorb moisture. Packages of a given level have a defined use-time limit once they are opened from their vacuum-packaged containers. Each container also has a humidity indicator card (HIC) to indicate humidity level inside the package. While a blue color on the HIC indicates acceptable level, a pink color suggests that the packages have been exposed to unacceptable levels of moisture and may not be usable as such. Moisture resistance classification and the floor life and use conditions are shown in table 6–2. A majority of packages are at level 3 with a floor life of one week. More moisture resistance packages are generally more expensive. Between uses in the manufacturing, it is important to store them in a dry atmosphere and ensure that the total actual exposure time does not exceed the designated floor life. When the packages exceed their designated floor life, they need to be baked at 125°C for 24 h prior to assembly use. The possibility of some solderability degradation due to lead/termination oxidation during baking cannot be ignored, and baking in an inert atmosphere such as nitrogen may be considered. The components are then presented to the placement station.

Table 6–2. Moisture sensitivity levels of plastic packages and floor-life conditions

Level	Floor Life	Use Conditions
1	Unlimited	≤30°C/85% r.h.
2	1 year	≤30°C/60% r.h.
2A	4 weeks	≤30°C/60% r.h.
3	168 hours	≤30°C/60% r.h.
4	72 hours	≤30°C/60% r.h.
5	48 hours	≤30°C/60% r.h
5A	24 hours	≤30°C/60% r.h
6	Time on the label	≤30°C/60% r.h

A camera on the placement machine moves over to the site fiducial, which is a reference point in proximity to the designated footprint and on to the package footprint. The x–y position and angle (Θ) offset is calculated. Once the coordinates are determined, the machine turret head moves the component presentation stage to the pick-up location such as a feeder, tray, magazine, tubes, etc. The head picks up the component, presents it to the camera, and the center of the component is aligned with the center of the nozzle and then places the part on the component footprint with solder paste on it. This entire process is highly automated, programmed, and extremely rapid. There are several variations in the placement machines in terms of their operational sequences, turret head designs, nozzles, transport, etc. The important aspects are placement accuracy, placement pressure, repeatability, tolerances, and, ultimately, the speed, namely, the number of parts that can be placed per hour.

As indicated earlier, components are shipped to the user in different types of containers, and the placement machines need to have the flexibility to handle the different container system. Table 6–3 shows typical component packaging options. Sometimes a plurality of placement machines are employed to accommodate the diversity in package styles, sizes, and packaging formats.

Table 6–3. Typical component carriers for different packages

Component	Typical Container
Resistors, capacitors, SOT, & SOICs	Tape and reel
PLCC, QFP, TSOP	Tray feeders
Flip chip	Gel pack
Small resistors and capacitors	Bulk feeders
Small volume parts	Sticks/tubes
Fine pitch devices, BGAs, & large PLCC	Waffle packs

When dealing with fine-pitch leaded packages such as 0.5 mm pitch and finer, depending on the lead material, lead fragility can be an important factor in handling them. Alloy-42 leads are stiffer than copper alloy leads and easier to handle without damage. Sometimes, package designs with bumpers or special container designs are incorporated to minimize the handling and transport damage.

Lead and ball coplanarity is an important parameter in the assembly process. All the leads or the termination balls should be in the same seating plane. The height of the lead or termination from the seating plane is a measure of the nonplanarity. If some leads are bent up, they will not touch the solder paste and as a result may not get soldered. The nonplanarity is generally 4 mils for most surface-mount assembly. This may not be acceptable to fine-pitch SMT components. An alternative requirement could be that the nonplanarity be less than the solder paste height for all the terminations to be soldered.

The smallest passive components used in the portable electronic product used are 0201 and 01005. A 0201 device is 20 mils long and 10 mils wide and a 01005 device is 10 mils long and 5 mils wide. These are extremely small dimensions. Solder paste screening and the control of solder paste volume are very critical to their assembly, since a defect called tombstoning can occur. Board pad designs and stencil openings have to be optimized in order to obtain satisfactory results. Wider pad designs and larger stencil openings, lowest pad separation, and higher overlap between component termination and the printed solder separation are aspects that can help achieve high yields. For the 01005 resistors, a 9-mil-wide, 10-mil-long stencil was reported to provide good yields (Schalke 2007). Precautions against electrostatic discharge are also helpful. Placement pressure can be an important parameter in placing these components without causing latent damage to sensitive components.

The electronics manufacturing industry routinely performs 0.5-mm-pitch leaded- and area-array chip-scale package (CSP) surface-mount assembly with high yields. Many can assemble 0.4 mm CSPs too. However, the packaging industry is migrating to even finer pitch CSPs, namely, 0.3 mm pitch CSPs, to meet the demands of higher density assemblies. It will be very difficult to fabricate stencils that would permit high-yield printing at that fine pitch. Also, extremely fine particle solder pastes with appropriate printable characteristics may be difficult to make at an acceptable cost. Two alternatives are available: (a) Instead of printing paste on the site, the package is dipped in a suitable flux to a predetermined height and is placed on the site and reflowed along with other packages. This is a practice that is generally employed for assembling flip chips. Of course, flip chip assemblies are invariably underfilled with suitable epoxy and cured to achieve the requisite reliability. This procedure assumes that adequate reliability is obtainable without the additional paste. (b) The other choice is to dip the package in a specially formulated paste to a predetermined height, place the component on the footprint, and reflow as usual. It is important to ensure that this does not result in excess solder or bridging of adjacent solder joints.

Reflow

The panel with components placed appropriately in their designated locations is then transported on conveyor to a reflow oven where the package-to-board solder interconnections are made. While the two most important reflow methods are vapor-phase reflow and multizone infrared-convection reflow, most portable products utilize only the latter technique. In vapor-phase reflow, the boards enter a chamber where a boiling, inert fluorocarbon fluid whose boiling temperature is close to and slightly above the melting point of the soldering alloys transfers the latent heat of vaporization to the board, thus melting the solder alloy and making the interconnection. As the board leaves the chamber, the solder solidifies. This process is beset with thermal shock to the board and its components, and also the undesirable ozone-depleting fluorocarbons. Multizone infrared convection is considered superior in terms of avoiding any thermal shock and better temperature control. An oven may have as few as three to as many as seven zones depending on their sophistication and customer demand. Some contain as many as five heating zones and two cooling zones.

As the board and components enter the reflow oven, they go through four different zones. They are sequentially (1) preheat, (2) soak, (3) reflow, and (4) cool down. Each of these zones has specific functions as described below. Majority of interconnection defects occur due to paste application and the reflow process. Identification of the optimal oven recipe is critical (Ramkumar et al. 2008a, b, c).

Preheat Zone. In the preheat zone, the board and the components are slowly brought from room temperature to approximately 150°C at a heating rate of 2–4°C/s . The exact temperature depends on the nature of the flux, its activation temperature, and the solder alloy. The temperature is usually slightly below the flux activation temperature. The slope of the heating ramp is fairly linear. High preheat rate and temperature can (a) induce thermomechanical stresses (the component surface temperature being higher than the interior temperature), (b) cause component damage, (c) result in solder balling, (d) cause solvent volatilization and paste dry out, and (e) result in solder particle oxidation. On the other hand, low-temperature slow rate can cause poor flux activation, cold solder joints, etc. The optimized preheat step thus avoids any thermal shock to either the board or the components, and facilitates the solvent evaporation below its boiling point avoiding any splattering of the solder balls.

Soak Zone. In this zone, the temperature is raised at an even slower rate from 150°C to 180°C. During this phase, the flux wets all the surfaces and starts reacting with the surface oxides on the component leads or balls and the copper pads on the board, reducing them to pure metal. Too high a soak temperature can result in paste dry-out and solder splattering. Too low a soak temperature results in inadequate flux activation. Also, too long a soak time may result in excessive flux volatilization and paste dry-out. Thus the temperature and length of soak need to be optimized. The residence in the soak zone may be 60–90 s duration.

Reflow zone. The board and the components are rapidly brought to a temperature that is 20°C–30°C above the liquidus of the solder alloy. For tin–lead solders, it is in the range of 210°C–225°C and for lead-free solders it can be 230°C–245°C. The dwell time of the board and components, namely time above liquidus (TAL), is generally kept in the range of 30–60 s depending on the complexity of the bill of materials (BOM). This fast heating rate and TAL are important to ensure that all the components reach the requisite temperature, the lead and pad surfaces are completely wetted, the metallurgical bond is formed at the solder/lead interface and the solder/pad interface, and any damage to the board in the form of solder mask charring or blistering is prevented, etc. The TAL is a critical parameter. Too short a time can result in poor bonding, and too long a time can result in overheating of any heat-sensitive components and formation of excessive intermetallics, which can embrittle the solder joints.

Cooling zone. The board along with its components is then subjected to cooling before exiting the oven. In this zone, the molten solder is solidified. It is desirable to adjust the cooling rate to less than 4°C/s so that the assembly does not experience any thermal shock, excessive growth of intermetallics is inhibited, and a fine-grain solder joint microstructure is obtained.

Thus every stage of the reflow process has to be carefully optimized for a given product.

It is important to note that some metals and alloys undergo supercooling. A supercooled state is a metastable equilibrium state in which the liquid, instead of solidifying at the freezing point, continues to remain in liquid state longer, and then cools suddenly as the temperature is being lowered continuously. The difference between the solidification temperature and the melting point is the extent of supercooling. In the case of solder alloys, intermetallic phases continue to grow as plates, rods, needles, etc., within the solder joint. Hence, in situations where supercooling occurs, the resulting solder microstructure can be quite different from when supercooling is absent. It is not unlikely that two adjacent solder joints can exhibit different microstructures. Since solder microstructure has significant impact on solder joint behavior under thermal and mechanical loading, it is imperative to minimize or eliminate supercooling by suitable preventive measures. It may involve alloy composition modifications. Figure 6–3 shows a typical reflow profile indicating the different reflow process zones.

If the assembly is a double-sided one, the entire process starting from the solder paste stenciling operation is repeated. Special board fixtures are used in order to protect the already assembled parts. As the board with the second-side components go through the reflow process, the first-side interconnections undergo solder reflow a second time. This is termed *secondary reflow*. During this step, the backside components are held in place by the surface tension of the molten solder. A slight elongation in the solder joint can occur, which will only help the reliability aspect.

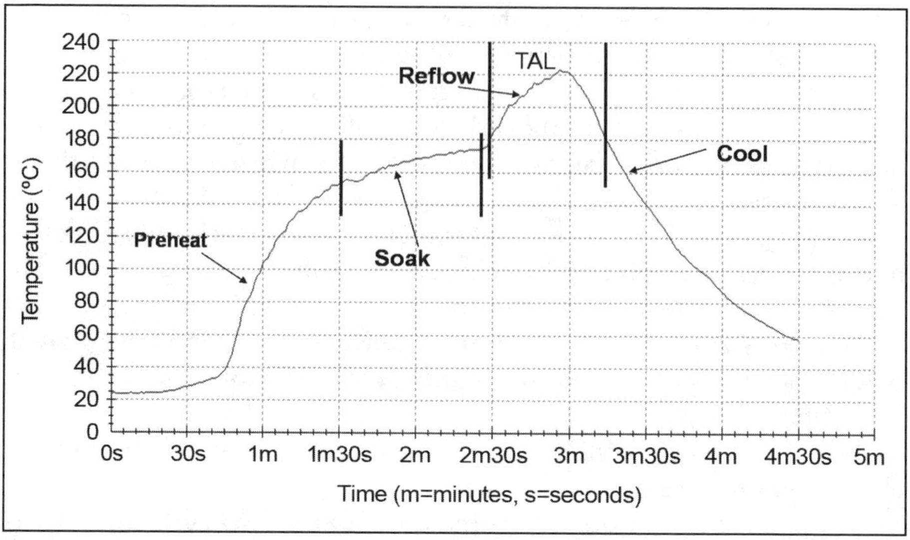

Fig. 6–3. A typical reflow profile for eutectic Sn–Pb reflow

Generally, this does not impact the quality of the final assembly. Only in extremely rare cases, a component may fall off when the component in question is a very heavy ceramic component that cannot be supported by the molten solder. This is not the case with any portable electronic product.

Pb-free interconnect alloys, in general, are not as wetting as the tin–lead solders in ambient air processing. Use of inert or nitrogen atmosphere alleviates this concern, albeit at some additional expense. Given the product design life and the cost pressures portable products are under, many do not perceive the need of nitrogen atmosphere in these assemblies.

Inspection and test

At the egress of the board from the reflow oven, the assembly is inspected for solder joint integrity and board damage. Visual, optical, and, sometimes, X-ray inspection are performed to examine a number of assembly features. These include: (1) adequate fillet size and shape, (2) excess solder, (3) insufficient solder, (4) dewetting, (5) nonwetting, (6) opens, (7) shorts, and (8) size and location of any solder voids. The assembly is also tested for functionality. Appropriate rework and repair operations are carried out to ensure assembly integrity and functionality. The assembly is then separated into individual product cards for the next level of assembly.

Underfilling

In the vast majority of portable electronic products, the package-to-board interconnections on the circuit card assembly meet or exceed the product quality and reliability requirements. Many components are rigid or flexible substrate CSPs, gull-wing leaded packages, fairly small land grid array (LGA) packages, and small resistor and capacitor packs, etc. Also, the product design life is shorter than many desktop machines. Thermal cycling requirements for many types of portable electronic equipment are well below the 1000 cycle –40°C–125°C standard. CSPs with flexible interposers and organic laminate interposers generally do not require any underfilling for thermal cycling reliability. However, occasionally one encounters situations where, due to the component choice, design life stipulations, or product operating requirements, reliability enhancements are warranted. An example of this is use of thin small outline packages (TSOP I and IIs) where the package is extremely thin, with a low package CTE, has a large chip-to-package ratio, has low standoff with 50 µm to zero, and consist of stiff alloy-42 leads. Low coefficient of thermal expansion (CTE) and low standoff with stiff leads make the assembly less compliant with poor solder joint reliability (Viswanadham et al. 1993). Thus this assembly step, namely, underfilling was indicated as an optional process step to enhance solder joint reliability. CSPs with inorganic/ceramic interposers, land grid array (LGA) packages, QFNs etc., might also require some reliability enhancements.

Major factors in the package-to-board interconnection reliability exposure are thermal stresses and/or mechanical stresses. The displacements due to mismatch of CTE are the main strains under thermal loading, while the displacements under mechanical loading are due to bend, drop, or shock. Cracks originate at the highest stress point in the interconnection and propagate, eventually leading to opens. Various polymeric materials are used to couple the entire bottom area or the perimeter of the package and the carrier to distribute the stresses. The process of filling the interspace between the package and the board is called *underfilling* and the material is called the *underfill*. Of the many choices available, epoxies are the most common class of compounds that are used for this application. In order to obtain maximum stress relief, it is desirable to choose a material that matches the mechanical properties of the interconnection material, namely, the solder. Epoxy materials in general have CTEs much higher than solder alloys and hence they are filled with appropriate amounts of low-CTE filler materials such as silica or alumina powders to reduce the effective CTE.

Several important material and process aspects are to be taken into consideration for the choice of the underfill material. These are listed in table 6–4. The properties listed are only a broad guideline. The actual values can be dependent on package–board combination and application. For example, surface tension can be dependent on the package and board surface characteristics. Flow speed can depend on the package size and whether the package is full, perimeter-, or custom-depopulated array. Speed also depends on the gap between the package and the board. Cure time and temperature may be product dependent. When thermal curing is not an option, one may choose UV cure. Thus, selection of material with appropriate properties is critical to achieve the desired results.

Table 6–4. Typical properties of underfill for CSPs and flip-chip attach

Category	Property	Typical value/range
Material	Coefficient of thermal expansion	18–30 ppm/ °C
	Viscosity	40–50cps
	Surface tension	20-25 dynes.cm
	Moisture absorption	<1%
	Glass Transition temperature	>125°C
	Ionic impurities	<10 ppm
	Modulus	5–10 GPa
	Elongation	1% or greater
	Thermal stability (<1% weight loss)	250 °C or greater
Process	Flow	0.5 mm/s
	Adhesion	>50 MPa shear
	Cure temperature	150°C
	Cure time	1 minute
	Shelf life	6 months or better
	Ambient pot life	8 hrs or greater
	Outgassing/volatiles	< 1% mass loss

While the flow characteristics of an underfill in the interspace between the package and board can be quite complex, a simple parallel-plate model provides an idea of the aspects involved. The time t required for an underfill to flow between the package and the board is given by

$$t = (3\eta L^2)/(h\gamma \text{Cos}\Theta) \tag{6–1}$$

where η is the viscosity, γ is the surface tension, Θ is the wetting angle, L is the length of the flow, and h is the package–substrate gap. It can be seen that smallest wetting angle and highest surface tension provide the shortest fill time.

The underfill is dispensed using a nozzle in a preselected pattern such as L, double L, or a U pattern, as desired. Sometimes dispensing the underfills at the four corners or at the four edges of the package alone may provide the desired reliability. Other underfilling techniques include dispensing through opening in the radio frequency (RF) shields or through an opening in the bottom side of the board. Several automatic dispensing tools are available. In the underfill process, assembly temperature is important. The assembly is heated and maintained at ~80–120°C. Since viscosity decreases with temperature, it aids the capillary flow of the materials. Preheating and maintaining the hardware at elevated temperatures exudes any absorbed or occluded moisture from the PWB and package, thus reducing the void propensity due to entrapped moisture.

The assemblies are cured in an oven after the dispensing operation at a predetermined cure time–temperature schedule.

Underfills certainly provide package-to-board interconnection reliability enhancements of several orders of magnitude in many instances such as flip-chip, TSOPs, ceramic BGA assemblies, and the like (Viswanadham et al. 1993). Dispensing underfill materials on the perimeter leads of TSOPS I and II assemblies has been shown to increase the second-level assembly reliability significantly (Emerick, A. et al. 1993).

However, caution needs to be exercised in regard to CSP assemblies. CSP packages have interposers that alleviate thermal stresses due to CTE mismatch between the package and the board. Underfilling the organic and flexible interposers like tape automated bonding (TAB)-based CSPs makes the assembly more rigid and reduces the compliancy, thus negatively impacting thermal cycling reliability. For the same package-to-board set, double-sided assemblies exhibit less reliability enhancements owing to localized stiffness caused by the double-sided assembly. Thus, while some packages exhibit reliability enhancements, others may not, and it is important to understand and evaluate the package and assembly structures (Ghaffarian and Kim 2000). In addition, an underfill that provides good thermal cycling reliability enhancement may not necessarily provide the same degree of enhancement under mechanical loading such as drop. While high-modulus materials are required for thermal cycling reliability improvements, a low modulus material may perform better under drop loading. Thermal cycling failures resemble ductile fractures, while drop failures are typically brittle in nature with little deformation because the strain rates involved in drop loading are several orders of magnitude higher. A typical example of ductile and brittle fractures is shown in figure 6–4.

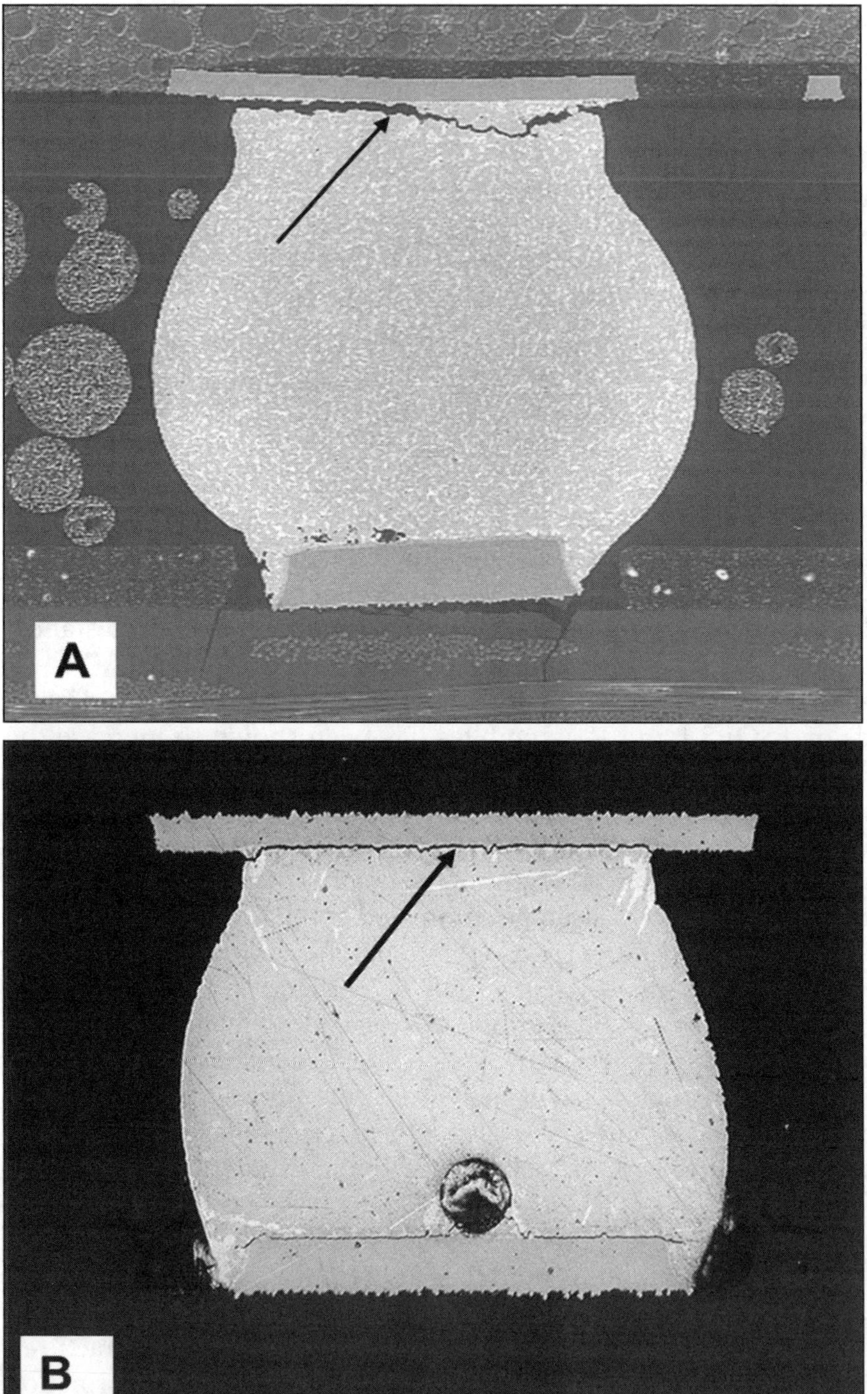

Fig. 6–4. Typical ductile and brittle fracture example (Courtesy of Steve Dunford, Nokia, Inc.)

In a study using three different underfills with modulus in the range of 2 GPa–2 MPa, a small but noticeable improvement was observed in drop reliability with decreasing modulus of the underfill and is shown in figure 6–5. During drop, the assembly experiences a combination of competing bending and inertial dynamic stresses. Improved reliability is attributed to the better energy dissipation and strain redistribution characteristics of the low- modulus rubbery underfill materials (Canumalla et al. 2002).

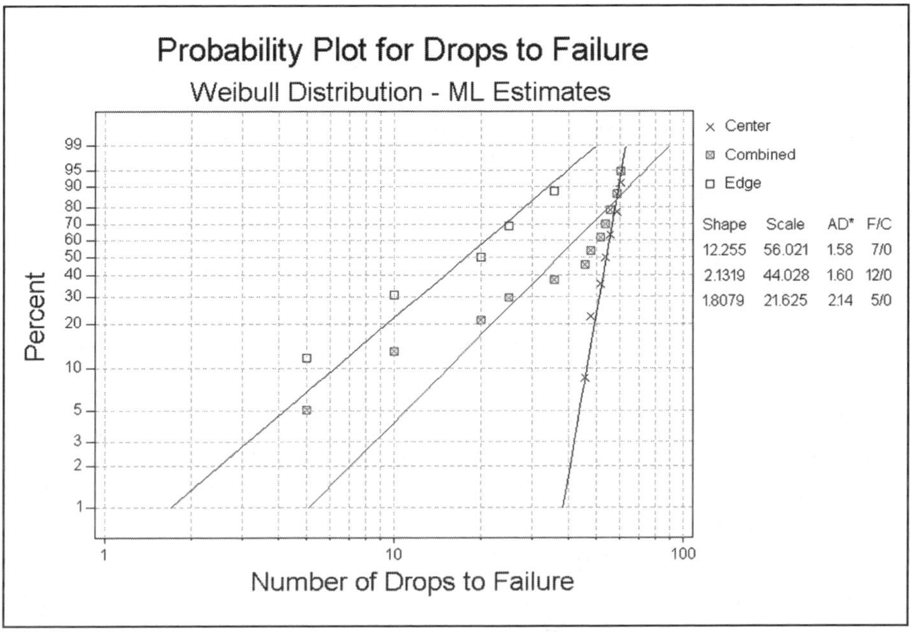

Fig. 6–5. Two-parameter Weibull plots of cumulative failures vs. number of drops for assemblies with no underfill, and with 2 MPa, 1 GPa, and 2 GPa modulus underfill materials

Actual product operational stresses involve both thermal and mechanical loads, not necessarily concurrent or simultaneous. Thus, portable electronic products pose a challenge in the judicious choice of the underfill materials that meet the requirements of thermal loads as well as mechanical loads. CTE, modulus, and glass transition temperature of the underfill material are to be carefully optimized in order to meet a specific application. Finite-element modeling and simulation should be considered an essential and integral part of the design. Ultimately, product testing under accelerated testing provides the validity and verification of the material choice and its efficacy.

Once encapsulated, the assembly is almost impossible to repair or rework. Thus, any defect discovered in the encapsulated region after encapsulation renders the assembly unusable and the entire assembly is discarded, which implies that adequate testing be performed prior to encapsulation. It is generally recognized that encapsulation involves additional materials, process, equipment, labor, and other costs, and is generally deferred until it is deemed absolutely necessary.

No-flow underfills

Underfilling process for any packaging application whether perimeter- or area-array package or flip-chip assembly is a standalone operation since it involves preparation of the assembly, dispensing the material, curing, and testing. It requires additional tooling, operator training, etc. Also, it is a batch operation and affects throughput and yield, and adds to the cost of the product. In an effort to alleviate some of these limitations of the underfilling process and enable cost-effective manufacturing, several new formulations have been investigated to eliminate some of the process steps. These efforts were originally aimed at flip-chip assembly, but high levels of miniaturization and integration in portable electronic application of these materials are extended to chip-scale packages also. The materials are variously called *compression flow underfills*, *reflow underfills*, or *no-flow underfills* (Shih and Wong 1999; Condos and Borgeson 2000).

The process consists of first depositing a precise amount of no-flow underfill liquid onto the component or die footprint on the substrate or PWB. This is accomplished by screen printing, jetting, or needle dispense. The die/CSP is then placed on the underfill-dispensed footprint using a high-speed placement tool. The placement process involves, in addition to locating the die/CSP footprint with precision, appropriate downward pressure and dwell. Placement force for flip chips is in the range of 400 –1600 g depending on the chip. This step provides the spreading of the underfill to form fillets and also ensures physical contact of the solder bumps or balls with the contact pads. The assembly is then passed through a multistage reflow oven where simultaneous solder reflow and partial or complete curing of the underfill is attained. Thus, in principle, use of reflow, or no-flow underfills not only reduces the number of manufacturing operations but makes underfilling integral to SMT assembly (Houston et al. 2005).

While the process is simple conceptually, it is important to be cognizant of several challenges involved. The following aspects deserve careful consideration in the choice and implementation of no-flow underfill materials (Kim and Baldwin 2002; Pascarella and Baldwin 1998):

- Solder ball/bump pitch and size dependency: decrease of bump size and pitch may require greater placement force.
- Voiding may depend on the bump size and pitch.
- Placement pressure may depend on the underfill print height.
- Bond pad configuration: square versus circular; circular may be preferred.
- Placement dwell in the formation of the fillets: can give rise to chip float defects.
- Reflow profile optimization may be very critical.

With a material of appropriate properties and optimized process parameters, assemblies with acceptable yields and reliability have been reported in the literature. Accelerated thermal shock and thermal cycling tests indicate failures due to delamination at the underfill/package, or underfill/substrate interfaces, solder fatigue cracks, solder extrusion and associated bridging between adjacent bumps, underfill bulk cracks, etc.

Some of these new materials can be completely cured during the reflow process, while others require some post-reflow curing. Some may work utilizing the traditional reflow profile, while others require significant modifications, even to the extent of reversing the profile, some times called an *inverted profile* where the soak follows the reflow segment (Condos and Borgeson 2000).

Considerable research work is perhaps needed to make this technique user friendly and universally applicable to encompass diverse package menus and package–substrate combinations.

Reworkable underfills

One of the reasons 'flip chip on organic laminate' technology has not become pervasive in consumer and portable electronics is the CTE mismatch between the board and the silicon die and the associated second-level assembly reliability. In order to achieve the requisite reliability, the assemblies have to be underfilled with a suitable encapsulant/underfill to distribute the stresses. Once underfilled and cured, further repair and rework of the assembly is almost impossible. Thus the inability of repair and rework of underfilled assemblies has dampened the adaptation of underfills. In cases where the assembly had to be reworked, methods such as mechanically grinding off the chip were attempted, of course, with limited success. The proliferation of portable consumer electronics with emphasis on miniaturization and integration and high-volume manufacturing, reworkability of circuit card assemblies, and hence development of a repairable underfill materials and processes, is a highly desirable feature (Hannan et al. 2000). The critical properties of such a material include viscosity, CTE, modulus, glass transition temperature (Tg), adhesion to the various surfaces involved such as solder mask, solder, and passivation layer, cleanability, etc. Important process parameters include easy removal, cycle time, acceptable reflow time–temperature profile, etc. Finally, the underfill is capable of providing reliability equivalent or better than the conventional underfills (Hannan and Viswanadham 2001; Hannan et al. 2001a).

For an underfill to be reworkable, the essential features are the ability to remove the defective package and site cleanability. In general, thermoset plastics are not amenable for rework. Thermoplastics are more amenable for rework as they soften at elevated temperatures.

Recently, several attempts have been made to formulate reworkable underfills and have met with considerable success.

The initial underfilling procedure is conventional as described earlier. The repair process consists of (a) removal of the defective component, (b) removal of residual underfill material from the package footprint and site dressing, (c) localized site cleaning, (d) new component attach, and, finally, (e) underfilling the new component.

a) *Removal of defective underfilled package*: The board assembly is preheated to about 100°C and then the package is heated to a temperature slightly above the solder melting temperature. This temperature is well above the glass transition temperature of the underfill material and the material is softened. The package is either lifted or twisted off the board. This removal operation takes approximately 3 min. The success of the removal operation can be challenging. The package-to-package spacing on the board can limit the ability to remove the defective component. When the packages are too close to each other, adjacent good packages may also get heated, resulting in secondary reflow of their solder joints and also latent internal unquantifiable damage to the devices. If the packages are shielded, the shield has to be removed prior to package removal. Thus the assembly is subject to additional thermal exposures. If the assembly is double-sided with components on the backside also, these components are also subject to additional thermal stresses, especially in the case of thin boards.

b) *Residual underfill removal and site dressing*: Removal of the defective component invariably leaves some of the underfill material as well as solder spikes on the site. Rework station solder removal tool is generally used to remove the residual underfill as well as the solder. It is important to remove any solder bridges of adjacent pads. It is also important not to expose the intermetallics on the pads since it will be difficult to solder to an intermetallic layer. A thin layer of solder on the pad is essential for interconnection. The residual underfill removal can be material-specific and needs to be optimized.

c) *Localized site cleaning*: A Dremmel brush is sometimes used to remove traces of underfill as final cleaning step followed by an isopropanol wipe. In steps (c) and (b) it is important to be cognizant of possible damage to the board in the form of solder mask damage, pad liftoff, etc., and proper care and diligence is important.

Steps d) and e) are the normal component attach and underfill processes, and hence are not addressed here. The total rework process time is generally under 10 min. In a rework operation, solder on the pads is not replenished and the new component is attached with no additional solder. It is assumed that this has little or no reliability impact of significance, given the short design life of the product.

Thermomechanical and mechanical reliability tests, namely, thermal cycling and drop test results, indicate reworkable underfills provide reliability on par with the traditional underfills (Hannan et al. 2001b).

Another discouraging aspect of underfill operations is one of occupational safety. Some of these materials are perceived to cause allergies, skin and eye irritations, etc., and hence exercise of appropriate caution is always recommended.

It is important to emphasize that in high-volume manufacturing of portable electronic products such as cell phones, digital cameras, and the like, the assembly yields have to be very high. The defects need to be in a few parts per million (ppm) ranges when millions of gadgets are being manufactured. The processes should be under control at all times, manufacturing errors and omissions need to be caught in time, assembly lines are to be interrupted expeditiously to avoid waste, and corrective steps and measures incorporated efficiently in order to maintain yields, throughputs, quality, reliability, and profitability.

As new technologies become available, they need to be introduced and seamlessly transitioned into the practicing technologies to be the first to the market place.

Flexible Electronic Assemblies

These can be generically defined as a patterned array of conductors supported by a flexible dielectric film on to which active and passive components are assembled. The film materials are generally polyimide, polyester, and the like. They provide higher conductor density capability than the rigid FR-4 PWBs. The flexibility required may be a one-time flexing as in three-dimensional circuit structures, which is termed *static flexing*, or repetitive flexibility during use called *dynamic flexing* as in digital cameras, camcorders, cell phones, disk drives, etc. They provide ability and versatility in rotating, folding, etc. The choice of use of flex circuit technology is dictated by the end use. Owing to the flexibility and thermal and mechanical reliability, these assemblies are perceived to be superior to rigid assemblies.

The assembly of the flex circuits is similar to the rigid board assemblies except that special work board holders are required to hold the flex flat in order to mount the surface-mount components and reflow them. These work board holders are made of either epoxy-glass or aluminum and are held flat either mechanically and/or by vacuum during solder paste printing or component mounting. Sometimes, component lead angles are opened up to 65°–75° from the traditional 55°–65° to provide additional strain relief (Gileo 1992; Khandpur 2006; Feljstad 2006). A typical flex circuit assembly is shown in figure 6–6. It shows populated as well as unpopulated flex circuit assemblies. Increasing use of flexible circuit assemblies will be seen in portable electronic products owing to their significant advantages in weight, volume, versatility, and cost.

As portable electronic products encompass to include wearable electronic products, many of the conventional assembly designs and techniques are not aptly suited to their fabrication and manufacture since stretchability becomes a desirable product attribute. These products need to conform to the shapes of the wearer, enclosures, etc. A concept called *stretchable electronics for large area applications* (STELLA) has recently been proposed. The structure consists of rigid sections where components are mounted, and areas in between the rigid sections are made flexible to permit certain amount of stretchability. Stretchability is accomplished by rendering the circuitry meandering rather than linear. In one example, 100-μm-thick thermoplastic urethane film in a roll format 120 cm wide and ~100 m length is used as the flexible substrate. Circuit-grade 35-μm-thick copper is then laminated to the film at a temperature around 200°C. The copper is structured and circuitized by conventional PWB fabrication techniques such as imaging, developing, etching, and stripping steps. However, the circuit features in some sections are meandering, sinusoidal-shaped structures with predefined wavelengths and amplitudes to permit stretchability. Smaller wavelengths obviously provide better elongation. Components are mounted and encapsulated on the rigid sections. Specially designed Cu structures are incorporated to protect components from stretching stresses. The interconnection structures are then embedded into a stretchable matrix by laminating to a second polyurethane film. This second film also serves as a solder mask. Bond pads are then opened by UV laser drilling. Component interconnection is effected either with low-melt Sn/Bi alloy or a suitable conductive adhesive. The low-profile components are further encapsulated with polyurethane glob top (Ostman et al. 2008).

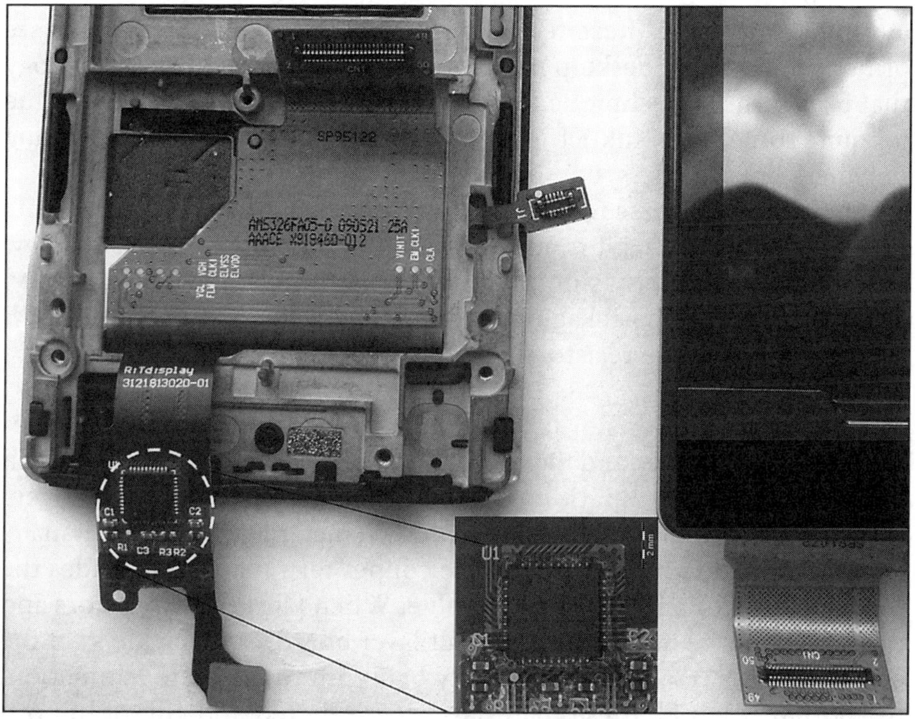

Fig. 6–6. Typical examples of flex circuit samples and assembly

In another variation of stretchable electronics manufacturing, called the *stretchable molded interconnect* (SMI) technology, components are first mounted on a copper foil using a Pb-free solder and reflow process. The components and the Cu foil are then laminated into a polyurethane encapsulation at 200°C and 20 bar pressure. A specially designed and milled aluminum block with appropriate component cavities was used to laminate the encapsulation. The copper foil with the encapsulated components was then turned upside down and the copper foil structured and circuitized, thus creating the desired circuit patterns. The circuitized copper was then laminated with another film of polyurethane to protect the circuitry. The assembled system is impermeable to moisture and air. These stretchable systems have been shown to withstand at least 100 stretch cycles with 10% elongation and are known to be capable of integrating into wearable electronics (Lacour et al. 2005).

Printable Electronics

Printable electronics is an emerging technology which is based on screen printing. Stencil printing techniques have been in practice for a number of years. These techniques have been used in solder-paste printing and used in desktop printers, etc. Multilayer ceramic technology that uses printing technique for generating circuitry has been in vogue for quite some years. Ink-jetting technique has also been in use in jetting polymers, inks, solders, etc.

Application of these techniques to fabricate PWBs and components is relatively new. Thus it is an extension of known techniques for new applications for increasing manufacturing efficiencies, adapting to fine-pitch electronic packaging, and also saving real estate on the PWB by incorporating the resistors and capacitors in the inner layers of the multilayer board.

One method of resistor fabrication utilizes loading carbon particles into an organic matrix and then screen-printing the resulting ink between the copper pad terminations on the PWB and curing it in an oven. Capacitors are similarly fabricated with such materials as barium titanate. Predetermined spacing between the copper terminations provides the requisite resistor and capacitance value. When fabricating resistors and capacitors on the inner layers of multilayer boards, the thickness of the cured passives is kept in the vicinity of 50 μm to facilitate lamination. In another method, these materials are ink-jetted and cured. In both cases, the viscosity, surface tension, temperature, and other properties are carefully controlled for successful fabrication. The jetting process involves a piezoelectric pump that is programmed to dispense the material on demand, and either the board or the dispensing head moves in the x–y directions. The dispensing head is also kept at a well-controlled temperature.

An important aspect of these types of passives is, as they are in a polymer matrix, they are prone to absorb moisture and the values may drift as a function of temperature and humidity. It is important to choose the right material set to obtain robust packages. Another significant aspect is the tolerance. They can easily be fabricated to 10% tolerances. Fabrication yield drops significantly when tighter tolerances of 5% or below are desired (Lee et al. 2005).

Embedding resistors and capacitors in the inner layers of the multilayer PWB releases significant real estate on the top and bottom layers, thus facilitating high packaging density of the active devices.

Inkjet technology has generated considerable interest in electronic packaging community in recent years to explore and extend this technology to create low-cost fine-feature packaging and production technologies. This technology is a fully additive, noncontact, direct material deposition process. It also has the advantage of eliminating several masking and etching process steps common to conventional printed wiring substrate fabrication, thus significantly reducing waste.

Inkjet technology has been demonstrated for eutectic and Pb-free solder bumping of semiconductor wafers. Jetting of polymers has also been known for quite some time. In any application, the performance requirements have to be established. These include such aspects as feature sizes aimed at linewidth and line spacing, resistivity, line thickness, line definition, uniformity of deposit, etc. The basic process for creating fine-line circuits on substrates involves surface treatment of the substrate surface, printing the conductor material, sintering the pattern, printing the dielectric on the top, and curing the dielectric. This process is repeated as many times as required for the creation of a multilayer structure. A schematic of the print process is shown in figure 6–7.

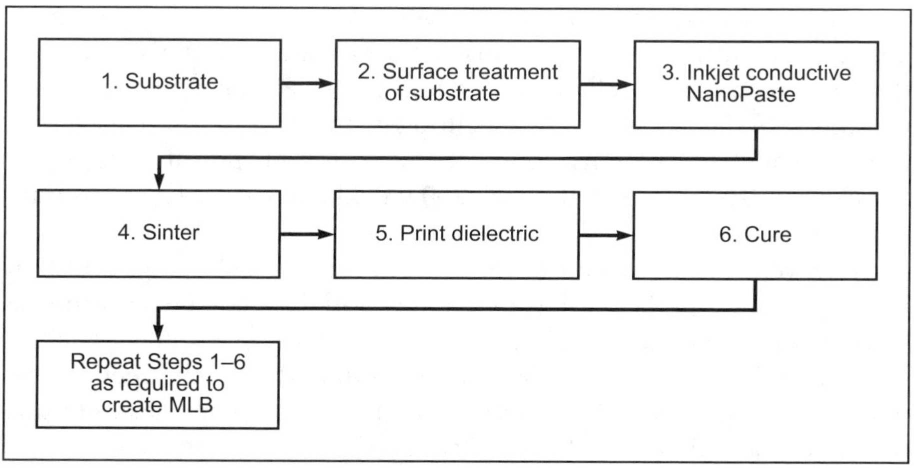

Fig. 6–7. Schematic of a typical inkjet process

The substrate can be either rigid or flexible. In case of a flexible substrate, it is held flat either by a mechanical or a vacuum fixture. The surface is treated with a material that enhances the wetting characteristics of the substrate. The choice of the treatment depends on the nature of the substrate and the ink that is being deposited. One example is a dilute solution of 3M EGC170 in 3M HFE-7100. The next step, as indicated in the schematic, is dispensing the conductor ink. The surface energy of the substrate needs to be greater than the surface energy of the ink that is being deposited in order to achieve good wetting. The planarity of the surface, format, and the material interface are critical.

While several conductive materials such as copper, silver, gold, etc., are possibilities, silver appears to be the preferred choice. An intimate mixture of silver nanoparticles, a dispersant, and solvents is used. The nature of the carrier liquids and their viscosity, their surface energy, the conductor particle size, the cure temperature, all have to be compatible with the carrier and performance requirements. Nanoparticle ink is then deposited dropwise by the ink-jetting system. Instrument and jetting parameters that are taken into consideration and optimized are the print head and nozzle size and geometry, as well as printing parameters like firing wave form, temperature, etc. The drop size can be as small as 35 μm. Cured linewidths and spacings of 75 μm each have been reported. After the printing step, the material is sintered at a temperature around 230°C for about 1 hr. This step is followed by the dielectric printing step and curing the dielectric with ultraviolet radiation and heat. This sequence of steps completes constructing one layer of circuitry, and these steps are repeated to create a multilayer circuitry. Line resistivities of 3 μΩ cm, compared to that of pure silver (1.6 μΩ cm), can be achieved. A module fabrication with 136 μm fine pitch device with 65 μm pad interconnections with inkjet process was demonstrated. Thus the feasibility of the inkjet technology to first- and second-level electronic packaging was demonstrated. Reliability of this technology in terms of thermal cycling, thermal shock, corrosion and electromigration propensity, etc., is yet to be demonstrated (Gamato et al. 2004; Pekkanen et al. 2007; Matisalo et al. 2007).

However, it is important to recognize that this technology is still in its infancy and significant developmental work is needed to accomplish high-volume manufacturing in terms of materials, process, and reliability. The technology, as it matures, can be combined with the reel-to-reel manufacturing of flexible circuits. One of the first beneficiaries of this technology could be the RF identification (RFID) tags industry.

Roll-to-roll Assembly

The versatility of polymeric materials is well established in the electronic industry through their use as photoresists, conductive and nonconductive adhesives, solder masks, board materials, sealants, underfills, etc.

With ever-increasing time-to-market and cost pressures, alternative advanced PWB and assembly technologies are constantly being explored. Integration of technology platform polymers and electronics is evolving to meet the growing demands of the portable electronics industry. They afford design freedom and flexibility. Flexible laminates are highly amenable to fast and low-cost reel-to-reel (R2R) manufacturing. Combined with this, advanced printing techniques such as inkjet printing are used to generate extremely fine circuit lines. Passive components such as resistors and capacitors are directly printed as opposed to traditional surface mounting. This not only simplifies the assembly process but also eliminates separate part procurement and stocking, and the associated expensive inventory management.

Emerging Trends

PWB assembly techniques are undergoing rapid evolutionary and revolutionary changes. Fine-pitch assembly is already at 0.4 mm pitch components attach. As one migrates to 0.3 mm pitch, traditional solder paste screen printing is likely to encounter materials and process limitations, affecting yields and throughputs. Special paste formulations and component dipping in the paste and assembling them instead of screen printing, as is done in flip chip attach with flux, could be an option in the short run. Increased use of multilayer flex with embedded actives and passives is being actively explored. Innovative assembling techniques will be forthcoming to accommodate the many configurations of wearable portable electronics. Thus, embedded electronics is on a migratory path for portable electronics appliances. As the industry migrates towards increasing use of nanomaterials and molecular electronics, it will experience migration to micro-nanoelectronic transition and finally to nanoelectronics. The industry envisions innovative assembly techniques in the horizon.

References

Behun, J. R., T. Caulfield, M. Cole, T. C. Riley, P. Singh, and P. Viswanadham. 1997. Package to board interconnection, in R. R. Tummala, E. J. Rymaszewski, and A. G. Klopfenstein, ed., *Microelectronics Packaging Handbook, Part III*. 2nd ed., New York: Chapman and Hall: 129–244.

Canumalla, S., S. Shetty,, and N. Hannan, 2002. Effect of corner underfill voids on chip scale package (CSP) performance under mechanical loading. Paper presented at the Proc. 28th ISTFA: 361–370.

Condos, P. A., and P. Borgeson, 2000. Flip chip assembly with reflow encapsulants. Paper presented at the Proc. SMTA-I: xx–yy.

Emerick, A., J. Ellerson, J. McCreary, R. Noreika, and P. Viswanadham. 1993. Enhancement of TSOP solder joints reliability using encapsulation. Paper presented at the Proc. 43rd Electronic Component and Technology Conf., Orlando, FL.187–192.

Feljstad, J. (2006). Flex Circuit Technology, 6th Ed. BR Publishing, Seaside, OR.

Gamato, D. R., P. Brazis, K. Kalyanasundaram, J. Zang, 2004. *Printed Organic and Organic and Molecular Electronics*. Kluwer Academic Publishers. 695.

Ghaffarian, R., and N. P. Kim, 2000. Does underfill affect CSP reliability? *Electronic Packaging and Production*: 28–35.

Gileo, K. ed. 1992. *Handbook of Flexible Circuits*. New York: Van Nostrand Reinhold.

Hannan, N., P. Viswanadham, K. Kulojarvi, and J. Ahlsted. 2000. Investigation of repairable underfill materials for reliability enhancements. Paper presented at the Proc. Tech. Prog. SMTA-Int. Conf., Chicago: 858–870.

Hannan, N., and P. Viswanadham. 2001. Critical aspects of reworkable underfills for portable consumer products. Paper presented at the Proc. 51st Electronic Components Technology Conf., Orlando, FL: 181–187.

Hannan, N., P. Viswanadham, L. Crane, E. Yaeger, A. Torres, R. W. Johnson, and H. Lasto. 2001a. Reworkable underfill materials for improved manufacturability and reliability of CSP assemblies, Proc. APEX Technical Conf., San Diego: AT8–3–10. 1–8.

Hannan, N., P. Viswanadham, and H. Quinones. 2001. Underfilling chip scale packages with reworkable underfills for consumer product applications. Paper presented at the Proc. 2001 Pan Pacific Microelectronics Symp., Kauai, Hawaii: 76–88.

Hannan, N., P. Viswanadham, G. Carson, B. Chan, and G. Lohrentz. 2001b. Reworkable underfill materials for high volume surface mount assembly. Paper presented at the Proc. APEX Technical Conf., San Diego: MT2-3 1–8.

Hossain, M., N. Lakhhar, P. Viswanadham, and D. Agonafer. 2000. The influence of final finish on lead-free assembly reliability. *Printed Circuit Design and Fab*. 25 (11): 36–39.

Houston, P. N., B. A. Smith, and D. F. Baldwin. 2005. Implementation of no-flow underfill material for low cost flip chip assembly. Paper presented at the Proc. SMTAI.

Khandpur, R. S. 2006. *Printed Circuit Design, Fabrication and Assembly*. New York: McGraw Hill.

Kim, C., and D. F. Baldwin, 2002. No-flow underfill process modeling and analysis for low cost, high throughput flip chip assembly. Paper presented at the Proc. 52nd ECTC: 1402–1410.

Lacour, S. P., J. Jones, S. Wagner, T. Li, and Z. Suo. 2005. Stretchable interconnects for elastic electronic surfaces. *Proc. IEEE Flexible Electronics Technol.* 93 (8): 1459–1467.

Lee, K. J., S. Bhattacharya, M. Varadarajan, L.Wan, I. R. Abotha, V. Sundaram, , P. Muthana et al. 2005. Design, fabrication, and reliability assessment of resistors and capacitors on multilayered organic substrates. Paper presented at the Proc Advanced Packaging Conf.,

Mantysalo, M., P. Manikkamaki, J. Meittinen, K. Kaija, S. Pienimaa, R. Ronkka, K. Hashizumme et al. 2007. Paper presented at the Proc.Tech. Prog. 2007 ECTC Conf.: 89–94.

Matteson, F., and S. Kandelid, 2001. Fundamentals of board assembly, in *Fundamentals of Microsystems Packaging*, R. R. Tummala ed., New York: McGraw Hill Books: 659–693.

Ostman, A., T. Loher, M. Seckel, L. Bottcher, and H. Recht. 2008. Manufacturing concepts of stretchable electronic systems. Paper presented at the Proc. Microsystems, Packaging, Assembly and Circuits Technol. Conf., IMPACT: 24–27.

Pascarella, N. W., and D. F. Baldwin. 1998.Compression flow modeling of underfill encapsulants for low cost flip chip assembly, *Proc. 48th ECTC*: 463–470.

Pekkanen, V., M. Mantysalo, and P. Manikkamaki. 2007. Design considerations for inkjet printed interconnection and packaging. Paper presented at the Proc.Tech. Conf. IMAPS, San Jose, CA: 1076–1083.

Prasad, R. P. 1997. *Surface Mount Technology, Principles and Practice*. 2nd ed., New York: Chapman and Hall.

Rahn, A. 2006. Pastes, stencils and process controls. *SMT Magazine* 20 (2): 28–34.

Ramkumar, S. M., A. Kannabiran, A. Baskaran, P. Viswanadham, and D. Dahle, 2008a. Understanding hidden reactions and the importance of profile in reflow soldering, Part 1. *Global SMT and Packaging* 8 (7): 10.

Ramkumar, S. M., A. Kannabiran, A. Baskaran, P. Viswanadham, and D. Dahle. 2008b. Understanding hidden reactions and the importance of profile in reflow soldering, Part 2. *Global SMT and Packaging* 8 (8): 22.

Ramkumar, S. M., A. Kannabiran, A. Baskaran, P. Viswanadham, and D. Dahle. 2008c. Understanding hidden reactions and the importance of profile in reflow soldering, Part 3. *Global SMT and Packaging* 8 (10): 34.

Schlake, J. 2007. Foundations for successful 010005 resistor array process. Paper presented at the Proc. Tech. Prog., SMTA-International: 878–886.

Shi, S. H., and C. P. Wong. 1999. Recent advances in the development of no-flow underfill encapsulants—A practical approach towards actual manufacturing application. Paper presented at the Proc. 49th ECTC: 770–776.

Viswanadham, P., M. Stennett, and R. Haggett. 1993. Second level assembly and reliability aspects of thin small outline packages advances in electronic packaging. Paper presented at the Proc. ASME Int. Electronic Packaging Conf., Vol. 2: 1127–1134.

Viswanadham, P., D. Coplin, and H. Roger. 1993. Solder joint reliability of TSOPs–An overview. Paper presented at the Proc. 43rd Electronic Component and Technology Conf., Orlando, FL: 863–870.

7

Essentials of Reliability Statistics

Introduction

A brief review of reliability statistics is pertinent to understand the reliability of portable electronic products. There are several excellent monographs and texts on the subject, and since an exhaustive treatment is outside the scope of this book, only the salient aspects of reliability statistics relevant to portable electronic products are discussed along with illustrative examples from reliability testing of portable electronic products (Nelson, 1982; Tobias and Trinidade, 1995).

Concepts of reliability statistics

The measurements of interest to reliability testing contain small variations from part to part and from time to time. These variations, whether arising from the measurement technique or inherent in the part, can be represented by a random variable. For example, the paint thickness on a plastic part can be expected to vary from 0.5 to 0.7 mm, with the actual paint thickness assuming an infinite number of values between these two extremes. Often, mathematical functions can be used to describe the variations, and these are commonly referred to as *parametric distributions* or simply as *models*. Some distributions relevant to portable electronic product design and reliability are discussed, with particular reference to the following useful forms of the functions.

Probability density function

The probability density function (pdf) describes the probability distribution of a continuous, random variable. This includes such quantities as failure time, length, width, etc. The pdf can be considered as a continuous representation of a frequency (or probability) histogram of random variables such as cycles to failure, paint thickness, length of the antenna in a mobile phone, etc. If numerous enough measurements are made, the histogram would resemble a pdf. For a continuous random variable x, a pdf is the function $f(x)$ with the following properties:

- $f(x) \geq 0$ (it takes only values greater than or equal to zero)
- $\int_{-\infty}^{+\infty} f(x)dx = 1$ (the total area under the curve equals 1)
- $P(a \leq X \leq b) = \int_{a}^{b} f(x)dx$ (probability that x takes values between a and b is equal to the area under the curve from a to b)

Cumulative distribution function (cdf)

An alternative way to represent the continuous distribution of a random variable is the cumulative distribution function $F(x)$, where the cdf represents the probability that the variable takes a value less than or equal to a.

$$F(x) = P(X \leq x) = \int_{-\infty}^{a} f(u)\, du \text{ for } -\infty < a < \infty \tag{7–1}$$

The cdf is related to the pdf by the relationship

$$f(x) = \frac{dF(x)}{dx} \tag{7–2}$$

So, if the failure time is the random variable being modeled, the cdf represents the probability that the unit will fail before that time, or, alternatively, it is equal to the proportion of units failing prior to time a.

Reliability function

Just as the cdf represents the probability of failure before time a, the reliability function represents the probability of survival or mission success up to time a. Then, the sum of the cdf and the reliability function is always unity. In other words,

$$R(x) = 1 - F(x) \tag{7–3}$$

Hazard function

The hazard function is defined as the ratio of the pdf to the reliability function, and is given by

$$h(x) = \frac{f(x)}{R(x)} = \frac{f(x)}{1 - F(x)} \tag{7–4}$$

Some Commonly Used Distributions

In this section, some distributions commonly encountered in the design and reliability of portable electronic products are described briefly. The reader is referred to several excellent treatises on the subject (e.g., Nelson 1982; Tobias and Trinidade 1995; and Montgomery and Runger 2000).

Exponential distribution

The exponential distribution is one of the most commonly used distributions and is characterized by a constant failure rate, such as would be experienced due to random failure events. The pdf, cdf, reliability function, and the hazard function for the exponential are given, respectively, as

$$f(x) = \lambda e^{-\lambda x} \tag{7–5}$$

$$F(x) = 1 - e^{-\lambda x} \tag{7–6}$$

$$R(x) = 1 - F(x) = e^{-\lambda x} \tag{7–7}$$

$$h(x) = \frac{f(x)}{1 - F(x)} = \lambda \tag{7–8}$$

The exponential distribution is used if there are no significant wear-out mechanisms over the intended life of the application and the effect of early life failures is discounted. Then, the failure time distribution can be described by a nearly constant failure rate (λ). The estimate for the *mean time to failure* (MTTF) is simply the inverse of the failure rate λ.

Normal and lognormal distributions

The normal distribution has particular applicability in modeling measurement errors encountered in everyday industrial applications where the histogram has a bell shape. This distribution is used widely in quality control and monitoring of processes. Random variables such as strength, dimensions, weights, etc. can be described by the normal pdf, which can be defined in terms of the mean (μ) and standard deviation (σ) with the relation

$$f(x) = \frac{1}{\sigma\sqrt{2\pi}} \exp\left[\frac{-(x-\mu)^2}{2\sigma^2}\right] \qquad (7\text{–}9)$$

where $-\infty < \mu < \infty$ and $\sigma > 0$. The variance is given by σ^2. The normal cdf for the normal distribution is given by the relation

$$F(x) = P\{X \le x\} = \int_{-\infty}^{x} \frac{1}{\sigma\sqrt{2\pi}} \exp\left[\frac{-(x-\mu)^2}{2\sigma^2}\right] dx \qquad (7\text{–}10)$$

$$R(x) = P\{X > x\} = 1 - F \qquad (7\text{–}11)$$

A related model of particular importance in reliability engineering is the lognormal model. The lognormal distribution has been successfully used for describing applications where the multiplicative degradation processes cause failures, such as corrosion, metal migration, fatigue crack propagation, electromigration, and diffusion, where the processes typically involve a speeding up of the rate of damage. Also, the lognormal distribution is useful in cases where the data range over several orders of magnitude, and for repair times of equipment. The pdf of the lognormal distribution is given by the relation

$$f(x) = \frac{1}{\sigma x\sqrt{2\pi}} \exp\left[-\frac{(\ln x - \ln T_{50})^2}{2\sigma^2}\right] \qquad (7\text{–}12)$$

where $0 < \sigma < \infty$ is the shape parameter and $\ln T_{50}$ is the median lifetime. The cdf, reliability function, failure rate, mean, and variance are given as, respectively,

$$F(x) = \int_{0}^{t} f(x)dx \qquad (7\text{–}13)$$

$$R(x) = 1 - F(x) \tag{7–14}$$

$$h(x) = \frac{f(x)}{1 - F(x)} \tag{7–15}$$

$$\text{Mean} = T_{50} \exp\left[\frac{\sigma^2}{2}\right] \tag{7–16}$$

$$\text{Variance} = (T_{50})^2 \exp(\sigma^2)[\exp(\sigma^2) - 1] \tag{7–17}$$

The shape parameter of the lognormal distribution is different from the standard deviation of the normal distribution even though both are represented by σ. The lognormal pdf can assume a variety of shapes based on different values of the shape parameter. This flexibility plays an important role in fitting skewed distributions. If population of random failure times t_f can be modeled by a lognormal distribution with median parameter T_{50} and shape parameter σ, then the natural logarithm of failure times is simply a normal distribution with mean μ (= ln T_{50}) and standard deviation σ. In other words, taking the logarithm of lognormal distribution yields a normal distribution. The flexibility of the lognormal and Weibull distributions sometimes makes it difficult to choose between them, especially when limited data are available. Although both models may fit the experimental data well, predicted failure rates at the tails of the distribution can be highly inaccurate if the wrong distribution is chosen. Guidelines on choosing an appropriate distribution are discussed in a later section.

Weibull distribution

The Weibull distribution has gained widespread applicability in reliability engineering because of its flexibility in describing different kinds of data with increasing, decreasing, or constant failure rates. In general, if there are many identical and independent competing damage processes leading to failure and the final failure occurs when the damage reaches a critical level, the Weibull distribution finds good applicability. Since its initial introduction for predicting ball bearing and fatigue failures, the Weibull distribution has been applied across a wide range of applications including material strengths, dielectric breakdown strengths, and fatigue failure. The Weibull pdf is given by the equation

$$f(x) = \frac{\beta}{\eta}\left(\frac{x-\gamma}{\eta}\right)^{\beta\text{-}1}\exp\left[-\left(\frac{x-\gamma}{\eta}\right)\right]^{\beta} \tag{7–18}$$

where β is the shape parameter, η is the scale parameter, and γ is the location parameter and

$f(x) \geq 0$
$x \geq \gamma$
$\beta > 0$
$\eta > 0$
$-\infty < \gamma < \infty$

The above form of the Weibull distribution is the most general and is generally known as the *three-parameter Weibull distribution* or *shifted Weibull distribution*. Such a form of the Weibull distribution is useful for cases where a "failure free" life can be rationalized. The shift parameter is also known as the *failure free life* or *location parameter*, and it is important that there is a physical basis for incorporation of a failure free life in the modeling. For example, the Weibull distribution with three parameters including the location parameter should be considered as the distribution function at a stress level near the fatigue limit of a material (Tanaka and Sakai 1979). It is well known that some materials, when subjected to stresses lower than the fatigue limit, do not exhibit fatigue failures. Since structures are often designed on the basis of the fatigue limit of the material, it is important for the reliability analysis to include a failure free strength. However, the justification for a location parameter to describe reliability data is uncommon in electronic packaging and portable electronic product development. Therefore, the two-parameter Weibull distribution pdf is given by the equation

$$f(x) = \frac{\beta}{\eta}\left(\frac{x}{\eta}\right)^{\beta-1}\exp\left[-\left(\frac{x}{\eta}\right)\right]^{\beta} \tag{7–19}$$

The quantities of interest in reliability engineering for a population that can be characterized by the Weibull distribution are

Mean $$\bar{x} = \gamma + \eta\Gamma\left(\frac{1}{\beta} + 1\right) \tag{7–20}$$

Median $$\breve{x} = \gamma + \eta(\ln 2)^{1/\beta} \tag{7–21}$$

Standard deviation $$\sigma_x = \eta \left\{ \Gamma\left[\frac{2}{\beta} + 1 \right] - [\Gamma(1/\beta + 1)]^2 \right\}^{1/2} \quad (7\text{–}22)$$

Reliability $$R(x) = \exp\left[-\left(\frac{x - \gamma}{\eta} \right) \right]^{\beta} \quad (7\text{–}23)$$

Failure rate $$h(t) = \frac{\beta}{\eta} \left(\frac{x - \gamma}{\eta} \right)^{\beta - 1} \quad (7\text{–}24)$$

Cumulative failure rate $$F(x) = 1 - \exp\left[-\left(\frac{x}{\eta} \right)^{\beta} \right] \quad (7\text{–}25)$$

It is straightforward to obtain the corresponding equations for the two-parameter Weibull distribution by substituting $\gamma = 0$ in Eq. (7–20, 7–21, 7–23, and 7–24).

The Weibull distribution also has special implications in accelerated testing. In most cases, the use conditions in the field cannot be employed for reliability testing in product development because the failure times would be excessively long and failure rates may be too low. Both these factors render reliability testing prohibitively expensive and time consuming. Therefore, it is common practice to force samples to fail earlier and at higher failure rates while retaining the failure mechanisms observed in the field. The impact of acceleration on the probability plot is discussed in a later section of this chapter.

Classification of data

Data can be classified into life and non-life data. For example, strength, dimensions such as length and width, mass, etc. are considered non-life data, while number of cycles to failure, number of drops to failure, time to failure, etc. are considered life data. One significant difference between the two is that life data, whether from reliability tests or from the field, can be incomplete in the sense that the exact failure time of each unit is not known, whereas non-life data are almost always complete. For example, while experimental data characterizing the paint thickness of phone covers will almost always be complete and each measurement will have a known value, reliability data may consist of partial number of units with exact failure times while the rest of the units may not have failed prior to test completion. Most commercial reliability analysis software have built-in schemes to accommodate different types of data, but the burden is placed on the data analyst to classify the data correctly and input the data in the correct form.

When performing reliability tests, complete data with exact failure times for each unit are preferred most, because the parameters of the life distribution can be determined with a minimum of uncertainty. However, for some kinds of highly reliable electronic devices, obtaining such data may be prohibitively expensive because of the required testing times or sample sizes for obtaining failure data. For system-level reliability of portable electronic products in environments such as mechanical shock, complete failure data is generally achievable in a reasonable amount of time. For semiconductor reliability, however, two different types of truncated (or censored) reliability evaluations are normally encountered. In the first kind, called Type I censoring, the reliability evaluation is terminated after a predetermined duration. In this case, the number of units failing is a random quantity. The advantages of Type I (time based) censoring is that it is easier to schedule the evaluation, while the disadvantage is that if the sample size or test conditions are nonoptimal, the test may end in zero failures. In the second kind of censoring, called Type II, the test is terminated after a predetermined number of units have failed, say 63.2%, and, in this case, the duration of the test is a random quantity. The advantage of Type II (failure based) censoring, is that adequate reliability information is guaranteed, while the disadvantage is that this type of test is not easy to schedule. Whether Type I or II censoring is adopted, failure times need to be measured during a reliability evaluation. Often, in the case of portable electronic products, the failure is defined in terms of electrical failure: open, short, or parametric shift.

If exact failure times are desired, continuous in situ monitoring of the product or device is often necessary via daisy chain (DC) testing, where the test unit resembles a functional unit in all physical aspects but without electrical functionality. The term "daisy chain" refers to a continuous loop spanning several solder joints or wire bonds, sometimes extending across first- and second-level interconnections. When the device is subjected to reliability testing, if any single interconnection breaks open, the resistance across the entire loop approaches infinity, and the in-situ continuity monitoring will detect a "failure." In this way, a daisy-chained chip scale package (CSP) can be used to test the reliability of interconnection without actually having the functionality of the real device. Sometimes, the entire portable electronic product can be "daisy chained" wherein a number of critical CSPs are also employed in daisy chain form. One downside to this approach is the cost of the DC devices or products. A second, more intangible downside is that, although the DC device or product resembles the functional unit, because of process differences in fabricating these DC units subtle differences in materials and strength are unavoidable. Therefore, questions remain about extrapolating the results of the DC devices to the functional devices. Sometimes, a practical compromise in reliability testing, especially for semiconductor devices with high reliability, involves the use of interval or grouped or readout data collection. In this approach, the reliability test is interrupted at predetermined readout times so that each unit may be evaluated for electrical failure, and the surviving units are inserted back into test while the failed units are removed for failure analysis. The advantage of this approach is that even functional units may be evaluated for reliability. The disadvantages include imprecise information about failure times and, if the readout times are non-optimal, the data may be inadequate if failures occur over intervals that are too close or if the failures are too few.

A further classification can be made on the basis of the kinds of censoring: right, left, or interval. The most commonly encountered censoring is right censoring, where it is only known that the unit has not failed before a certain time. In the case of interval censoring, the failure is only known to have occurred between time t_1 and time t_2. In the case of left censoring, the failure is known to have occurred before a certain time. Left censoring may be thought of as a special case of interval censoring where time t_1 is zero.

Parameter estimation methods

Once a distribution has been selected as appropriate for the life data at hand, the next step in the reliability analysis is to estimate the parameters of the underlying distribution. Only a brief review of commonly used parameter estimation methods is given here, recognizing that most readers will probably use commercially available data analysis software, and the goal is to cover enough detail to make a recommendation about the appropriateness of different options found in the software. The least mathematically intensive method of parameter estimation is the method of manual probability plotting, wherein the median ranks of the life data are physically plotted on a special graph paper and the parameters are estimated visually. Obviously, this subjective method is labor intensive and requires the use of special graph papers. In addition, the determination of confidence intervals is not easy to accomplish using this method. In terms of complexity, the next step is the linear regression method of fitting a line through the data points using a least-squares minimization of the error between the data and the estimate. This method can be quite successful in cases where the cdf can be linearized and the data is complete with no censored units. But when censored data are present, a more powerful mathematical method called the *maximum likelihood method* is preferable, and the benefits are discussed in brief in a later section (Nelson, 1982; Tobias and Trinidade, 2008).

Sample size calculations

The sample sizes needed to estimate various parameters with a desired accuracy are important for test planning purposes. Several useful formulae for calculating sample sizes for estimating proportions, mean, and standard deviation of a normal distribution and the shape parameter of a Weibull distribution are listed below as an aid to manual calculations (Nelson, 1982). In general, as the need for accuracy in determining the population parameters increases, the sample sizes required also increase.

Proportions: To estimate the proportion of defective samples (p) in a population to within ±w% with approximately α% confidence, the number of samples needed is given by

$$n \cong p(1-p)\left[\frac{K_{\alpha}}{w}\right]^2 \tag{7–26}$$

where K_α is the standard normal percentile (z_P) corresponding to $100(1+\alpha)/2$th from table 7–1. The unknown proportion p refers to the population, while the proportion $\hat{p}$ refers to the proportion of defective units in the test subpopulation, and α is the confidence level.

Sometimes, solving this equation is not straightforward because the right-hand side includes the proportion of defective units p, which is an unknown. In that case, the value of $\hat{p}$ can be used as an approximation instead of p. Alternatively, a conservative (overly large) sample size estimate may be obtained by using a value of $p = 0.5$, which assumes that 50% of the units in the population are defective. These formulae use the normal approximation of the binomial distribution for $\hat{p}$, and are reported to be adequate for np or $n(1-p)>10$. If sample size n is large, an approximate α% confidence on the proportion p of the population is given as

$$\hat{p} - K_\alpha\sqrt{\frac{\hat{p}(1-\hat{p})}{n}} \le p \le \hat{p} + K_\alpha\sqrt{\frac{\hat{p}(1-\hat{p})}{n}} \tag{7–27}$$

where K_α is the standard normal percentile (z_P) corresponding to $100(1 + \alpha)/2$ from table 7–1, and $\hat{p} = x/n$ is the point estimate with x failures in a subpopulation of size n.

Example 1. A certain number of painted mobile phone cases are defective and we wish to determine this proportion by examining n samples out of a total population of 10,000. Find the sample size to determine the proportion of defective painted covers with an accuracy of 5% of the actual value with 95% confidence, and without prior knowledge of the unknown p.

Solution. Comparing with Eq. (7–26), $w = 0.05$, and assume $p = 0.5$. K_α is estimated to be 1.96 corresponding to $100(1+\alpha)/2\%$ value of 97.5 since α is 0.95.

$$n \cong 0.5(1 - 0.5)\left(\frac{1.96}{0.05}\right)^2 = 384 \qquad (7\text{–}28)$$

Normal distribution: Two quantities are of interest when dealing with a normal distribution: the standard deviation and mean. Procedures for calculating sample sizes differ for these estimates, and both are discussed separately. To estimate standard deviation to within a factor f such that the real standard lies between the limits σ/f and σf with a confidence level of $100\alpha\%$, the required sample size is given by

$$n \cong 1 + 0.5\left[\frac{K_\alpha}{\ln f}\right]^2 \qquad (7\text{–}29)$$

where K_α is the standard normal percentile (z_P) corresponding to $100(1+\alpha)/2$ from table 7–1. The accuracy of the estimate increases with the sample size.

Example 2. The standard deviation is to be estimated to within a factor $f = 1.2$ with 90% confidence, so calculate the required sample size.

Solution. The value of z_p corresponding to $100(1 + \alpha)/2$ is estimated from table 7–1 to be 1.645. Substituting $K_\alpha = 1.645$ and $f = 1.2$ into Eq. (7–29),

$$n \cong 1 + 0.5\left[\frac{1.645}{\ln(1.2)}\right]^2 = 42 \qquad (7\text{–}30)$$

Table 7–1. Standard normal percentiles z_p from Nelson (1982)

100(1+α)/2 %	z_p	100(1+α)/2 %	z_p
10^{-4}	-4.753	40	-0.253
10^{-3}	-4.265	50	0
10^{-2}	-3.719	60	0.253
0.1	-3.090	70	0.524
0.5	-2.576	75	0.675
1.0	-2.326	80	0.842
2.0	-2.054	90	1.282
2.5	-1.960	95	1.645
5	-1.645	97.5	1.960
10	-0.282	98	2.054
20	-0.842	99	2.326
25	-0.675	99.5	2.576
30	-0.524	99.9	3.090

To estimate the mean of a normal distribution to within $\pm 100w\%$ of the true value m with $100\alpha\%$ confidence, the sample size n is given by

$$n \cong \sigma^2 \left(\frac{K_\alpha}{w} \right)^2 \tag{7–31}$$

Example 3. What is the sample size required to estimate the mean to within ±2% with 90% confidence, if the standard deviation $\sigma = 0.0624891$ and variance $\sigma^2 = 0.003905$?

Solution. The value of w is 0.02 and $K_\alpha = 1.645$ corresponding to $100(1 + \alpha)/2 = 95$ in table 7–1. Substituting into Eq. (7–31),

$$n \cong 0.003905 \left(\frac{1.645}{0.02} \right)^2 = 27 \tag{7–32}$$

Weibull distribution: The sample size required to estimate β of a Weibull distribution to within a factor f such that the real β lies between $f\beta$ and β/f, with about $100\alpha\%$ confidence is given by

$$n \cong 1.1 \left[\frac{K_\alpha}{\ln(f)} \right]^2 \tag{7–33}$$

Example 4. Estimate the sample size required to estimate β to within 30% (or $f = 1.30$) with 90% confidence.

Solution.

$$n \cong 1.1\left[\frac{1.645}{\ln(1.30)}\right]^2 = 44 \qquad (7\text{–}34)$$

Sample size can also be calculated on the basis of a need to demonstrate that a product meets a required average failure rate (AFR), and the general procedure is illustrated for the case of a Weibull distribution (Moura 1991). The inputs required for this calculation are as follows:

1. λ_R is the specified average failure rate
2. t_R is the field usage time related to the average failure rate
3. A_F is the acceleration factor for the laboratory test
   β is the shape parameter of the failure distribution
4. t_S is the duration of the life test being planned
5. t_{EQ} is the life at use-equivalent condition
   α is the confidence with which the AFR is to be determined
6. f_D is the number of failures expected in the test (demonstration number)

The cumulative probability of failure for any distribution is given as

$$F(t_R) = 1 - \exp\,(-\lambda_R t_R) \qquad (7\text{–}35)$$

The characteristic life parameter η that will just satisfy the AFR is given by

$$\eta = \frac{t_R}{\{-1\text{n}[1 - F(t_R)]\}^{1/\beta}} \qquad (7\text{–}36)$$

The cdf at use-equivalent conditions is calculated using $t_{EQ} = t_s \cdot A_F$ as

$$F(t_{EQ}) = 1 - \exp\left[-\left(\frac{t_{EQ}}{\eta}\right)^{\beta}\right] \qquad (7\text{–}37)$$

Then, the minimum number of samples is given by using the chi-square percentile (obtained from standard tables) corresponding to degrees of freedom defined as $2(f_D + 1)$, and is given by

$$n = \frac{x^2\,(2(f_D{+}1),\alpha)}{2F(t_{EQ})} \tag{7–38}$$

A closer examination of the above formula reveals that prior knowledge of the failure distribution (β) and the acceleration factor for the test are required as inputs into the calculations. If these are not known, trial experiments may be needed or initial assumptions need to be made. Furthermore, the test duration needs to be known or determined on the basis of several practical constraints including time. Often, when a new technology is being evaluated, these quantities are not known *a priori* and a few iterations are needed to understand the reliability of the product.

Example 5. The life of a glass display under mechanical loading follows a Weibull distribution with shape parameter $\beta = 2.5$. The application in a portable product requires that the AFR does not exceed 0.1%/1,000 h. Calculate the minimum number of samples required to demonstrate compliance to the AFR requirement with 90% confidence for a 300-h test where an acceleration factor of 20 is applicable. The demonstration number or number of allowed failures is 10.

$\lambda_R = 0.1\%/1{,}000$ hour $= 0.1 \times 10^{-2} \times 10^{-3}$ fails/hour
$t_R = 1{,}000$ hours
$\beta = 2.5$
$A^F = 20$
$\alpha = 0.90$
$f_D = 10$

$$F(t_R) = F(1{,}000) = 1 - \exp[-1 \times 10^{-6} \cdot 1{,}000] = 1 - 0.999 = 0.001 \tag{7–39}$$

$$\eta = \frac{1{,}000}{\{-\ln[1 - 0.001]\}^{1/2.5}} = \frac{1{,}000}{0.0631} = 15{,}848.9 \tag{7–40}$$

$$t_{EQ} = 300 \times 20 = 6{,}000 \tag{7–41}$$

$$F(t_{EQ}) = 1 - \exp\left[-\left(\frac{6{,}000}{15{,}848.9}\right)^{2.5}\right] = 0.084406 \tag{7–42}$$

The degrees of freedom for the chi-square percentile calculation is 2(10 + 1)=22, and using $\gamma = 0.90$, the $\mathcal{X}^2_{(22,0.90)}$ is estimated to be 30.813 from standard tables. Or, the *chiinv* function in Microsoft Excel can be used with an argument of 1 – 0.9 = 0.1 for the probability and 22 for degrees of freedom. Thus, the minimum sample size is calculated to be

$$n = \frac{30.813}{2 \times 0.084406} \cong 183 \tag{7–43}$$

Probability plots and their interpretation

A probability plot is special kind of graph in which either a cdf or cumulative hazard function can be depicted as a straight line. Probability graph paper was initially developed to allow for drawing straight line fits to experimental data but nowadays computer software is used almost exclusively to make probability plots. A probability plot offers the following advantages:

- Simplicity and speed
- Data can be presented in a format that is easier to understand, visualize, and draw conclusions from.
- Several simple estimates such as nominal life, characteristic life, and cumulative failure probability can be deduced from probability plots alone.
- Comparisons of two data sets to answer questions such as "Is there an improvement?" or "Which materials set offers better reliability?" without resorting to complex hypothesis or likelihood ratio tests are possible when probability plots are constructed with confidence intervals.
- One can quickly determine how well an assumed distribution fits the data.
- Probability plots can help in identifying anomalies data such as competing or multiple failure modes and decide whether more complex models are needed to describe the data.
- From a practical perspective, the reliability engineer can use the probability plot to identify samples for failure analysis and explore deeper into failure mode identification.

Thus, probability plots reveal information and provide insight to the reliability engineer that might otherwise be hidden. It is recommended that engineering decisions be taken after a critical examination of the probability plot even if purely numerical methods such as (maximum likelihood estimation (MLE) methods are used for parameter estimation. On the other hand, probability plots do not lend themselves easily to objective decision making and different people may obtain slightly different estimates for the parameters. A thorough analysis should combine both graphical and analytical methods in conjunction with careful failure analysis to form a complete picture of the reliability evaluation.

The importance of failure analysis cannot be stressed enough, and it is not an overstatement to assert that reliability data analysis and physical/chemical failure analysis are but different sides of the same coin.

Linearization of the cdf

On a linear scale, the cdf of even the simpler distributions such as the exponential distributions are not linear functions of time (life). The benefit of linearizing these functions is the computational simplification in that a wide variety of mathematical schemes have been developed to fit "straight lines" to data points.

Exponential distribution: The cdf of the exponential distribution is given by

$$F(t) = 1 - \exp(-\lambda t) \tag{7–44}$$

Rearranging, after taking natural logarithms on both sides, we get

$$\ln\left[\frac{1}{1-F(t)}\right] = \lambda t \tag{7–45}$$

Therefore, if $1/(1 - F(t))$ is plotted on a log scale against time on a linear scale to yield a semilog plot, and the data falls approximately on a straight line, the slope of which will yield the parameter λ.

Weibull distribution: The equation of the 2-P Weibull distribution is given by

$$F(t) = 1 - \exp(-(t/\eta)^{\beta}) \tag{7–46}$$

Rearranging this equation and taking natural logarithms twice on both sides gives,

$$-\ln(1 - F(t)) = (t/\eta)^{\beta} \tag{7–47}$$

$$\ln[-\ln(1 - F(t))] = \beta\ln(t) - \beta\ln(\eta) \tag{7–48}$$

Typically, the $-\ln(1 - F(t))$ quantity is plotted versus the time to failure on log–log scales. If the data can be described by the Weibull distribution, the data will fall on a straight line on such a plot. In general, in reliability evaluations there are a limited number of data points (failures) available to estimate the form of the hypothesized underlying population distribution function. The linearized form of the cumulative distribution function $F(t)$ is relatively more data-efficient in comparison to histograms representing the pdf or other sigmoidal functions such as the reliability or hazard functions.

The ordinate (y-axis) in a probability plot is the cumulative percent of failures. Since the n samples tested represent a subset of the whole population, the failure time is a random variable and can be expected to vary each time the reliability experiment is carried out. For example, if a reliability evaluation is repeated over and over, the distribution of failure times for the second failure can be determined. For calculations with limited sample size, it is customary to evaluate the population cdf at the median of this sampling distribution. The median cdf value for the second failure out of 30 units is also called the median rank for the second failure with a sample size of 30, and can be calculated by the approximation proposed by Bernard and Bos-Levenbach (1953):

$$r_i = \frac{i - 0.3}{n + 0.4} \tag{7–49}$$

where i is the progressive order of the failed specimen and n is the sample size.

So the median rank for our example is calculated as $r_2 = \frac{2 - 0.3}{30 + 0.4} =$ 0.05592. It has been shown that for calculating the cumulative probability of failure, the median rank approximation is by far the most accurate (Fothergill 1990). There are several methods for estimating the shape parameter (β) and the scale parameter (η). In general, for uncensored life data, the fractional bias in determining η decreases for increasing sample sizes and increasing β. For $\beta < 4$, the least-squares regression technique can be as inaccurate as the MLE method (Montanari, Mazzanti, Cacciari, and Fothergill 1997). However, it has been suggested that by using a weighted version of the least-squares technique or the correction proposed by Ross, much more accurate estimates for η and β can be obtained, particularly for small sample sizes (Ross 1996). Other researchers have shown that it is difficult to outperform MLE methods in any systematic way for complete failure data, but that they show promise when used with small sample sizes, depending on the value of β (Watkins

1996). On the other hand, for censored failure data, MLE methods are generally preferred, especially for the case with no failures. For the case of small number of samples with heavy censoring, the estimated values for β and η have to be treated with caution because of the propensity for bias irrespective of the estimation technique used.

Appearance of accelerated test data on probability plots

Almost always, the use conditions in the field cannot be employed in product development because the failure times would be too long and failure rates may be too low, and reliability testing using these conditions will be prohibitively expensive and time consuming. Therefore, it is common practice to force samples to fail at shorter life times and at higher failure rates while maintaining the same field failure mechanism by performing accelerated tests. This section discusses the appearance of accelerated test data on the probability plot.

If the acceleration is "linear", the acceleration factor A_F is given as the ratio of the life time under field-use conditions and the life time under more severe laboratory conditions.

$$A_F = \frac{t_U}{t_L} \tag{7–50}$$

Exponential distribution: If failure times follow an exponential distribution, the cdf at use conditions is

$$F(t_U) = 1 - \exp[-\lambda_U t_U] \tag{7–51}$$

From the definition of acceleration, substituting $t_U = t_L \cdot A_F$, we get

$$\begin{aligned} F(t_U) &= \mathrm{F}(t_L \cdot A_F) \\ &= 1 - \exp[-\lambda_U A_F t_L] \end{aligned} \tag{7–52}$$

If $\lambda_L = \lambda_U \cdot A_F$, the above equation reduces to the expression $F(t_U) = 1 - \exp[-\lambda_L \cdot t_L]$. Thus, failure rate under accelerated conditions is simply the failure rate under use conditions multiplied by the acceleration factor AF.

Weibull distribution: If the failures follow a Weibull distribution at use conditions, the cdf at use conditions is given by

$$F(t_U) = 1 - \exp\left[-\frac{t_U}{\eta_U}\right]^{\beta_U} \tag{7–53}$$

One can calculate the cdf under accelerated conditions by substituting $t_U = t_L \cdot A_F$ into the above equation to yield

$$F(t_U) = 1 - \exp\left[-\frac{t_L \cdot A_F}{\eta_U}\right]^{\beta_U} = 1 - \exp\left[-\frac{t_L}{\eta_U/A_F}\right]^{\beta_U} \quad (7\text{–}54)$$

On the other hand, under accelerated conditions the cdf is given by

$$F(t_L) = 1 - \exp\left[-\frac{t_L}{\eta_L}\right]^{\beta_L} \quad (7\text{–}55)$$

Comparing the above two equations, it can be concluded that only the characteristic life (η) is scaled by the acceleration factor A_F, and the shape factor is not affected. That the shape factor remains unchanged upon acceleration of the failure mechanism follows as a consequence of assuming linear acceleration and Weibull distribution. On a probability plot, life data at different acceleration factors should appear as nearly parallel lines.

Conversely, if upon analyzing accelerated life test data the shape parameters are not similarly unchanged, it can be concluded that either the acceleration was not linear or the Weibull distribution is inappropriate for failure data.

Normal distribution: If failure times follow a normal distribution, the cdf at use conditions is given by

$$F(t_U) = \Phi\left[\frac{t_U - \mu_U}{\sigma_U}\right] \quad (7\text{–}56)$$

One can calculate the cdf under accelerated conditions by substituting $t_U = t_L \cdot A_F$ to yield $F(t_U) = F(A_F \cdot t_L)$

$$= \Phi\left[\frac{A_F \cdot t_L - \mu_U}{\sigma_U}\right]$$

$$= \Phi\left[\frac{t_L - \mu_U/A_f}{\sigma_U/A_F}\right] \text{ where } \mu_L = \frac{\mu_U}{A_F} \text{ and } \sigma_L = \frac{\sigma_U}{A_F} \quad (7\text{–}57)$$

$$= \Phi\left[\frac{t_L - \mu_U}{\sigma_L}\right]$$

Thus, for a normal failure distribution, both mean and standard deviation are scaled by the acceleration factor. Thus, the standard deviation at laboratory (accelerated conditions) also must be multiplied by the acceleration factor to obtain the standard deviation at use conditions.

Lognormal distribution: For phenomena such as electrochemical migration, when the life times follow lognormal distribution, the cdf at use conditions is given by

$$F(t_U) = \Phi\left[\frac{\ln\left(t_U/t_{50,U}\right)}{\sigma_U}\right] \tag{7–58}$$

One can calculate the cdf under accelerated conditions by substituting $t_U = t_L \cdot A_F$ into the above equation to yield

$$\begin{aligned} F(t_U) &= F(A_F \cdot t_L) \\ &= \Phi\left[\frac{\ln\left(t_U/t_{50,U}\right)}{\sigma_U}\right] \\ &= \Phi\left[\frac{\ln\left(A_F \cdot t_L/t_{50,U}\right)}{\sigma_U}\right] \\ &= \Phi\left[\frac{\ln\{t_L/(t_{50,U}/A_F)\}}{\sigma_U}\right] = \Phi\left[\frac{\ln\left(t_L/t_{50,L}\right)}{\sigma_L}\right] \end{aligned} \tag{7–59}$$

where $t_{50,L} = \dfrac{t_{50,U}}{A_F}$ and $\sigma_L = \sigma_U$.

Thus for an accelerated test in which the failures follow the lognormal distribution, the shape parameter (β) remains unchanged and only the median life is scaled by the acceleration factor, in a fashion similar to the case of the Weibull distribution. Again, on a probability plot, life data at different acceleration levels appears as parallel lines, and this is a consequence of the model rather than an assumption. If such data do not appear to be approximately parallel, then either the assumption of a lognormal distribution is inappropriate or the assumption of linear acceleration is inaccurate.

Choosing an Appropriate Distribution

One of the first tasks in parametric analysis of reliability or other measurement data (weight, dimension, etc.) is the choice of a parametric distribution model that is most appropriate for the data at hand. The choice must often be made on the basis of a small number of available data from samples taken from a larger population. The following considerations are pertinent to the selection of the model:

- The failure mechanism and failure rate behavior agree with the statistical argument of the model. For example, an approximately constant failure rate due to random causes can lead one to consider the exponential distribution as the statistical model for describing the failure lifetimes.
- Historical precedence: Numerous analyses of thermal fatigue reliability test data have used Weibull 2P distribution to model the failure lifetimes. Therefore, that is the first choice for new test data where the same failure mechanism is operative.
- The model should fit the data to an appropriate degree, and this is often evaluated by caculating a goodness of fit (GOF) or similar measure.
- The model must be selected with the view that it will be used to estimate failure rates via acceleration models to a use condition far removed from the test data, because selecting the wrong model can have severe repercussions in terms of predicting a failure rate that is inaccurate by orders of magnitude.

The adoption of a particular distribution is not a trivial task, and the presence of irregularities such as multiple failure modes can make it very difficult to identify the correct underlying distribution, especially if the sample size is small. Therefore, engineering judgment, failure analysis data, and robust statistical analysis are essential.

It is relatively easy to evaluate the applicability of distributions such as the exponential distribution because it only fits data with a fairly constant failure rate. However, distributions such as Weibull and lognormal distributions are both very flexible, and it becomes difficult to decide which of these two distributions should be selected to fit the available data, especially when the sample size is not very large. In such cases, choosing an inappropriate distribution can inject significant errors in the predictions for failure rates, especially when the test conditions are highly accelerated. The method suggested by Croes et al. (1998) can help in selecting the correct distribution. The method is based on a calculation of Pearson's correlation coefficient (ρ) and the ratio $\rho_{Weibull}/\rho_{lognormal}$ is computed. If the ratio is greater than 1, the Weibull distribution is recommended, and if the ratio is less than 1, the lognormal distribution is more appropriate. It has also been shown that the accuracy of this approach increases with increasing sample size, and especially for substantial fraction of failed units (Cain 2000). In other words, it is easier to pick the correct distribution in a test where a greater portion of units fail.

Illustrative Examples

Example 6. How to interpret a Weibull Probability Plot of Life Data

A package under development is required to pass 500 cycles of board-level thermal cycling –40/125°C with no failures at 500 cycles. For test case A, which was a pass/fail kind of test, 20 units were tested to 500 cyles but electrical continuity tests did not reveal any failures. To increase the margin, the test was extended to 1,000 cycles, and finding no failures after electrical probing, the test was stopped. For test case B, the board level reliability (BLR) test was carried out with continuous electrical monitoring of the daisy-chained units on 20 units. However, in this case, the test was carried out until all the units failed, and the failure times were recorded as 1229, 1282, 1328, 1418, 1434, 1531, 1613, 1701, 1703, 1749, 1763, 1781, 1809, 1894, 1954, 1996, 2032, 2048, 2125, and 2173 cycles. Analyze the data for the two test cases and draw conclusions from the probability plot.

Solution. For test case A, the units PASS the requirement of no failure until 500 cycles. While the point estimate of the reliability is 100%, the lower 95% estimate for the reliability can be calculated to be 83.2% using confidence intervals on a binomial proportion. It is relatively straightforward to argue that this kind of analysis does not give a complete picture of the anticipated reliability for number of cycles greater than 1,000.

Consider test case B, where representative units from the same population are subjected to thermal cycling to obtain actual failure times. From a testing cost perspective, this test can be expected to take more time and cost more. The data was analyzed using a Weibull probability plot using maximum likelihood method to estimate the parameters as shown in figure 7–1. The parameter estimates are given as η = 1,845.9828 and β = 7.3031303. The 95% lower and upper confidence bounds on η and β are estimated to be 1,723.0924 and 1,969.5436, and 4.9890751 and 10.117151, respectively. On the probability plot, the Weibull fit is shown as a solid line drawn through the data and the lower and upper

95% confidence intervals (CIs) are shown as dashed lines bounding the estimates of the Weibull fit. It is a good practice to plot these CIs to understand the spread in the estimates. A visual examination of the goodness of fit shows that a straight line appears to be a reasonably good fit of the data points, and there is no pronounced curvature that would indicate the need for a three-parameter Weibull fit. The horizontal line drawn on the plot at 0.632 cumulative unreliability (y-axis) represents the characteristic life of 1,846 cycles. The lower and upper confidence bounds on the characteristic life can be estimated from the plot to be approximately 1,700 and 2,000. Of course, a more accurate estimate was obtained by calculating it with the software. The reliability and failure probabilities were also estimated using the analysis software, along with the 95% confidence bounds on the estimates, and are shown in table 7–2 for 500, 1,000, 2,000, and 5,000 cycles. So, even though there were no failures at 500 cycles, it can be seen that the point estimate for the reliability is 99.993%. This is much higher than the reliability estimate obtained in the pass/fail test (83.2%), and illustrates the benefit of using a parametric model fit rather than assuming just a binomial distribution. A further advantage is that the reliability at life times beyond the last fail can also be estimated. For example, the reliability at 2,000 cycles can be estimated either graphically or using the maximum likelihood method. From the plot, a vertical line is drawn at 2,000 cycles, and the intersection with the lower, fit, and

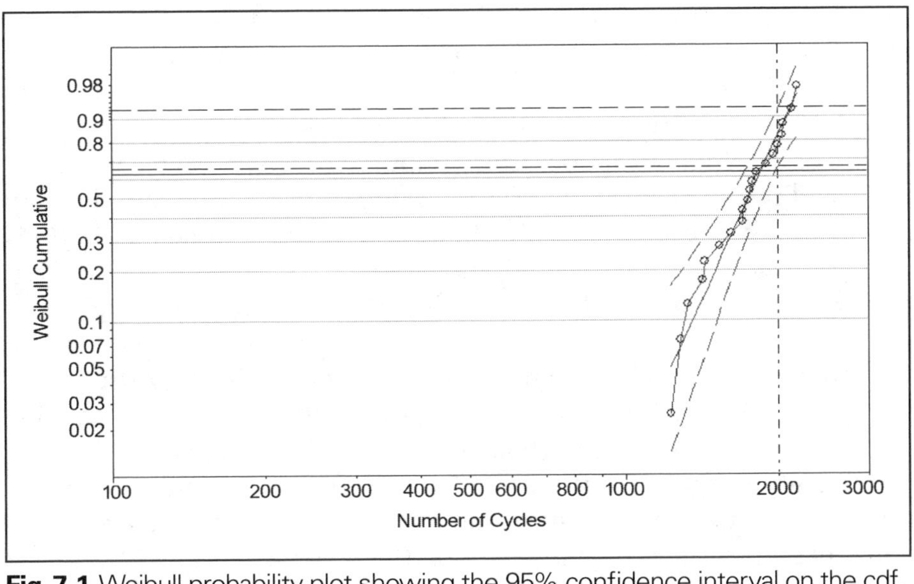

Fig. 7–1. Weibull probability plot showing the 95% confidence interval on the cdf.

estimate lines yields the cumulative unreliability estimates as approximately 65%, 85%, and 94%, respectively. The values from table 7–2 at 2,000 cycles are 65.97%, 83.39% and 92.86%, respectively. If the investigation were conducted as a pass/fail test up to 2,000 cycles, the lower 95%, point estimate, and upper 95% estimate of the unreliability using a binomial distribution would have been 62.47%, 0.8%, and 0.98%. It can be seen that the Weibull estimates are narrower because more information is used to generate these estimates.

Example 7. Reliability requirements based on probability plots

Instead of specifying reliability in terms of *x* number of failures when *y* number of units are tested until *z* number of cycles or hours are completed, an alternative is to specify the cumulative unreliability that can be tolerated for a certain number of cycles. This kind of requirement makes sense for units and reliability test environments for which it is relatively easy to encounter test failures in a reasonable amount of time. Board-level solder joint reliability tests such as thermal cycling or mechanical drop are typically cases where it is customary to have a significant portion of the units under test fail. However, as long as there are at least a few failures (5 to 8), such reliability requirements ensure that testing is carried out until package weaknesses are better understood. The downside to this approach is that the cost of qualification can easily become prohibitively expensive for large number of qualifications, and a balance between the two may be found by employing such reliability requirements for developmental efforts which are often fewer than final products. One example of a reliability requirement for the data in the previous example is that 95% reliability (or 5% cumulative failure rate) at 1,000 cycles must be demonstrated with 95% confidence. The data are plotted in figure 7–2, where it is seen that the upper confidence bound of the Weibull analysis is very close to the point representing 5% unreliability and 1,000 cycles. Typically, the software used for reliability analysis can also estimate the reliability and the 95% confidence bounds at any particular life. In table 7–2, the estimates for 1,000 cycles are given as 1.13%, 0.19%, and 6.35% for the point estimate, lower, and upper 95% confidence bounds, respectively. Therefore, even though the first failure occurred at 1,229 cycles, the reliability requirement is not met because the upper 95% confidence bound on the cumulative unreliability at 1,000 cycles is greater than 5% (reliability is less than 95%). Since there is some uncertainty in the estimation of Weibull parameters, and this uncertainty depends not only on the sample size but also on the failure times, constructing reliability requirements on the Weibull (or other parametric model) analysis accounts for uncertainty.

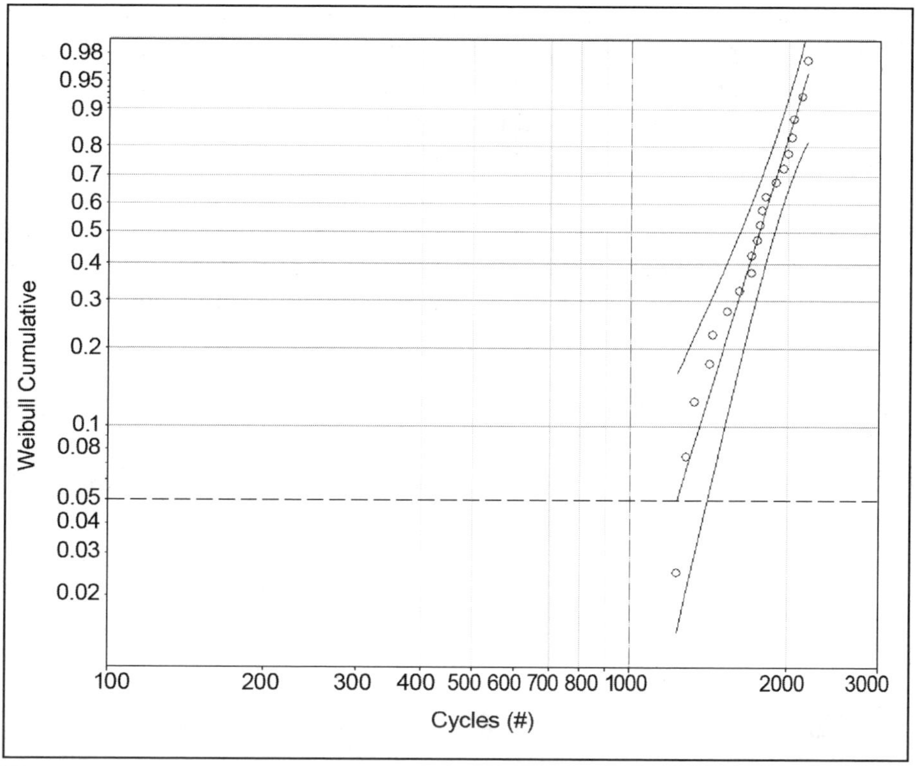

Fig. 7–2. Parametric analysis-based reliability requirement: 5% cumulative failure rate (95% reliability) at 1,000 cycles must be exceeded with 95% confidence

Table 7–2. Estimates of the reliability and failure probability at different cycles

Time	Prob Failure	Std Error	Lower 95%	Upper 95%	Prob Survival
500	0.00007	0.00013	0.00000	0.00242	0.99993
1000	0.01130	0.01016	0.00192	0.06356	0.98870
2000	0.83394	0.06726	0.65968	0.92863	0.16606
5000	1.00000	0.00000			0.00000

Example 8. Comparison of reliability life data

During package development, two different pad surface finish options (A & B) are being evaluated and the criterion for selection is the solder joint reliability under a BLR drop test. Which material set is better from a reliability perspective if the reliability specification requires that there be no more than 1% failures after 50 drops with 95% confidence?

Solution. The probability plot of the data is shown in figure 7–3 using Weibull parametric analysis. The estimates of the Weibull parameters are given in table 7–3 and the reliability estimates and 95% confidence intervals are given in table 7–4. From the reliability estimates, it is seen that the upper 95% confidence bound on the reliability at 50 drops for cases A and B are, respectively, 3.9% and 9%. This means that both surface finish options do not satisfy the reliability requirement, but option A seems to be somewhat better with h estimate of 353 drops as compared to the h estimate for B of 183 drops. But, are the differences between two different results statistically significant?

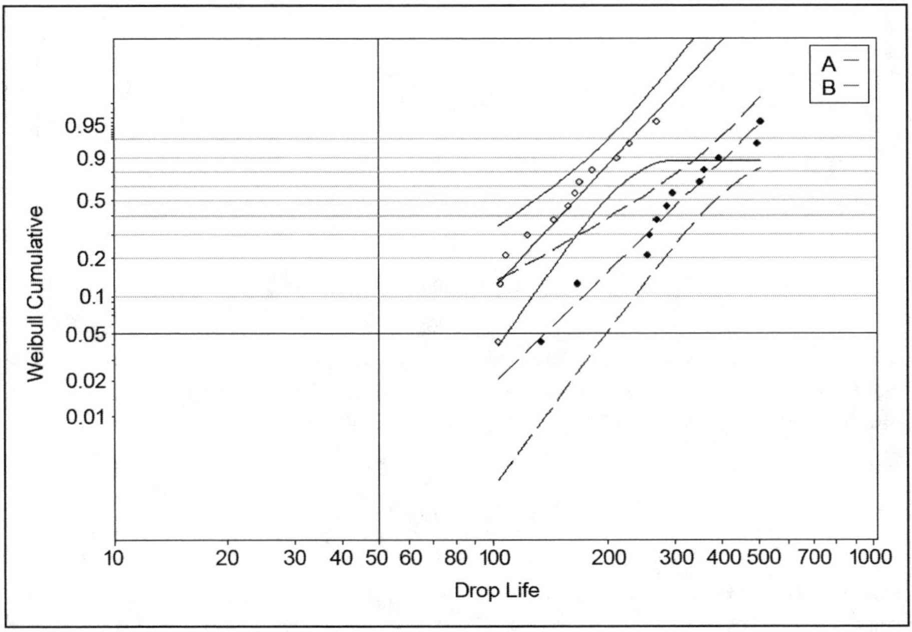

Fig. 7–3. Weibull probability plot of bend fatigue reliability performance comparing two different pad surface finishes A (dashed) and B (solid)

Table 7–3. Weibull parameter estimates for pad finishes A and B in BLR drop test

ID	Parameter	Estimate	Lower 95%	Upper 95%
A	η	352.72835	285.12965	430.63252
A	β	3.1608039	1.928139	4.7087536
B	η	183.23176	151.26675	219.27553
B	β	3.5010784	2.1634725	5.1543421

Table 7–4. Survival (reliability) estimates for pad finishes A and B

ID	Time	Prob Failure	Std Error	Lower 95%	Upper 95%	Prob Survival
A	50	0.00208	0.00314	0.00011	0.03876	0.99792
B	50	0.01054	0.01184	0.00115	0.08977	0.98946

Confidence interval method: The β Estimate for B (3.5) is contained by the 95% confidence bounds for A (1.92–4.7). Therefore, it can be concluded that there is NO statistical difference in β and the assumption of similar failure mechanism is supported by the statistical analysis. The η estimate for set B (183) lies outside of 95% confidence bounds for option A (285–430). Therefore, it can be concluded that there is a statistical difference in characteristic lives, and option A is indeed better than option B.

Likelihood ratio test method: There is another more general method that applies to most distributions and statistical models, called the *likelihood ratio test method*. This method, while requiring software capable of calculating log likelihood, is applicable to almost any kind of complex data. Only a brief application of this method to the problem at hand is presented next, and a more detailed description can be found elsewhere (Tobias and Trinidade 1995; Nelson 1982). In this method of hypothesis testing, first, the null hypothesis and alternate hypothesis are formulated. In our case, the null hypothesis is that the two distributions are identical and can be described by the same parameters ($\eta_A = \eta_B$ and $\beta_A = \beta_B$), and the alternate hypothesis is that the two distributions are different ($\eta_A \neq \eta_B$ and $\beta_A \neq \beta_B$). Therefore, for the case of the null hypothesis, we need only two parameters (η and β) to fit the entire data, while for the case of the alternate hypothesis, we need four parameters to fit the entire data (η_A, η_B, β_A, and β_B). Using a statistical software, such as JMP, the log likelihood (L_{reduced}) can be calculated to be 17.059 for the case of the null hypothesis or the reduced model with only two parameters to describe the entire data from both surface finishes. For the full model, with four parameters, the sum of the loglikelihoods (L_{full}) turns out to be 4.9475 + 3.3504 = 8.2979. Now, following the procedure described by Nelson or Tobias and Trinidade, the loglikelihood test statistic is calculated to be $2 \times (L_{\text{reduced}} - L_{\text{full}}) = 2 \times 8.7611 = 17.522$. The difference in the number of parameters used for the two different hypotheses is 4 – 2 = 2. The one-tailed probability of the chi-squared distribution is calculated to be 0.000157, which implies that the null hypothesis can be rejected. Therefore, the two distributions A and B are different and cannot be modeled using only two parameters.

Now, consider the null hypothesis that both distributions have the same shape parameter β, which implies that a total of three parameters are to be considered (η_A, η, and β). The loglikelihood using a single shape factor and two different scale parameters to model the entire data is calculated to be 8.35. The test statistic is 2 × (8.35 – 8.2979) with one degree of freedom. The one-tailed probability of the chi-square distribution is calculated to be 0.74687. Therefore, the null hypothesis that the shape parameters for A and B surface finishes are identical cannot be rejected.

The conclusion from the likelihood ratio test is the same as that from an examination of the confidence intervals for the parameters.

References

Bernard, A., and E. D. Bos-Levenbach. 1953. The plotting of observations on probability paper. *Statistica Neerlandica* 7: 163–173.

Cain, S. 2000. Distinguishing between lognormal and Weibull distribution. *IEEE Trans. Reliability* 51(1): 33–38.

Croes, K., J. V. Manca, W. De Ceunick, L. De Schepper, and G. Molenberghs. 1998. The time of guessing your failure distribution is over. *Microelectronics Reliability* 30: 1187–1191.

Fothergill, R. C. 1990. Estimating the cumulative probability of failure data points to be plotted on Weibull and other probability. *IEEE Trans. Elec. Insulation* 25 (3): 489–492.

Montanari, G. C., G. Mazzanti, M. Cacciari, and J. C. Fothergill. 1997. In search of convenient techniques for reducing bias in the estimation of Weibull parameters for uncensored tests. *IEEE Trans. Dielectrics and Elec. Insulation* 4 (3): 306–313.

Montgomery, D. C., and G. C. Runger. 2000. *Applied Statistics and Probability for Engineers*. New York: John Wiley.

Moura, E. C. 1991. *How to determine sample size and estimate failure rate in life testing.* The ASQ Basic References in Quality Control: Statistical Techniques, eds. S. S. Shapiro and E. F. Mykytka. Vol. 15, Milwaukee: ASQ Quality Press.

Nelson, W. 1982. *Applied Life Data Analysis*: Ch. 12, 522–557, New York: Wiley.

Ross, R. 1996. Bias and standard deviation due to the Weibull parameter estimation for small test sets. *IEEE Trans. Dielectrics* 3 (1): 28–42.

Tanaka, T., and T. Sakai, 1979. Estimation of three parameters of Weibull distribution: Relating to parameter estimation of fatigue life distribution. *J. Soc. Mater. Sci. Japan*, 28 (304): 13–19.

Tobias, P. A., and D. C. Trinidade. 1995. *Applied Reliability*. 2nd ed., New York: Chapman & Hall, 174

Watkins, A. J. 1996. On maximum likelihood estimation for the two parameter Weibull distribution. *Microelectronic Reliability*. 36 (5): 595–603.

8

Reliability of Electronic Assemblies

Introduction

Most hand-held/portable consumer electronic products can be characterized as high-volume, low-cost devices. Original equipment manufacturers (OEMs) face pressure to develop new, more advanced technology products in record time, while at the same time improving productivity, product field reliability, and overall quality. Reliability is defined as the probability that a product will perform its intended function under encountered operating conditions for a specified period, whereas quality, in narrow terms of reliability alone, can be defined as the reliability at time zero.

The highly personal use profile and mobility for these products imply that consumers will take these products with them wherever they go, and expect the same dependable performance irrespective of the exposure of the product to the elements, for example, rain, snow, or accidental drop. Under such conditions, meeting the reliability expectations for portable products can be a challenge, especially since reliability expectations need to be met without compromising profitability. One driver for product reliability is perceived quality and customer satisfaction. Another reason for ensuring reliability is that product field-failure rate, which plays a key role in controlling warranty and repair costs, tends to be higher for an unreliable product. In other words, all other factors remaining the same, a more reliable product will be more profitable. However, in reality, reliability always has associated costs, and there is a level of optimum reliability beyond which additional reliability improvements have a negative impact on the business profits. These optimum reliability levels need to be derived from knowledge of specific end-use environments and customer requirements. Still another reason to strive for product reliability is that reliability (and quality) could be employed as product differentiators in product marketing and/or advertising, which will only increase the business value of product reliability.

The operating environment for mobile electronic equipment differs considerably from that for desktop computers or telecommunication applications, and is often more varied in terms of thermal excursions and exposure to humidity and corrosive environments. The products are more prone to high humidity exposures in both noncondensing and condensing atmospheres. Additionally, portability makes the product more likely to experience mechanical loads such as drop, bend, twist, etc. Mechanical drop impacts from heights of 1.5 m on hard surfaces are not uncommon. The number of power ON cycles, the operational voltages, and other conditions are also different. Table 8–1 shows the typical operating environments for portable hand-held telecommunication devices compared with conventional desktop and automotive under-the-hood electronics.

The data in table 8–1, however, do not completely capture the diverse and relatively "uncontrolled" environments that portable products experience in real life. There have been several attempts at studying the environmental use conditions for portable products by the different standards organizations (ETS 1994; IEC 60721-3-7 2002; JESD94.01 2007).

Table 8–1. Comparison of typical application conditions of desktop, mobile, and automotive electronics

Desktop	Product Life (years)	Power-on Cycles/ day	Power-on-Hours	Relative Humidity/ %	Environment Temperature Range/°C	Operational Temperature Range/°C	Voltage/ V
Desktop	5	1 to 17	13,000	10 to 80	10 to 30	20 to 30	12
Mobile terminal	5	20	43,800	10 to 100	-40 to 40	32 to 70	1.8 to 3.3
Automotive under the hood	15	5	8200	0 to 100	-40 to 125	-40 to 125	12

*Source JEDEC

Hardware

To develop a deeper understanding of the end-use environment, some OEMs have conducted direct measurements of the conditions in the use environment (Vakevainen et al. 2001). A miniature data logger was incorporated into several mobile phones along with sensors for temperature, humidity, etc., and these phones were used by people in different occupations in Texas (USA), Finland, and Bangkok (Thailand). The results of this study suggest that, over a two-week period, there were approximately two temperature excursions every day (one high and one low) upon which were superimposed several mini-cycles. The temperatures ranged from 10 to 35°C, while the humidity varied from 10% R.H. to about 45% R.H. for a representative phone. For the entire population of test phones, however, the temperatures ranged from 3 to 45°C, while the relative humidity ranged from 5% to 65%. The means and standard deviations of the temperatures and relative humidity measured in that study are shown in table 8–2.

Table 8–2. Temperature and humidity from direct measurements

Environment Classification	Temperature/°C		Relative Humidity/%	
	Mean	Std. Dev.	Mean	Std. Dev
Low Temp	22	6	28	9
High Temp	27	4	46	8
Temp Cycling	23	6	29	10
High Humidity	30	3	51	10

Based on such studies, it is estimated that portable electronic products need to be functional between –10 and 50°C for normal applications and an extended range of –25 to 70°C for harsh usage applications such as law enforcement, exploration, etc. The corresponding range of humidity over which the portable product can be anticipated to be functional is 0%–95%.

The mechanical environment, however, is not documented in open literature. Based on the range of experimental conditions in published literature for mechanical loading, it is speculated that the mobile phones can be subjected to drop impacts on to concrete and other hard surfaces from heights of 0.5–1.5 m. It is conceivable that the actual use environment involves a combination of drop impacts from different heights and product orientations. The current understanding of the reliability environment for portable products such as mobile phones remains incomplete, and the reliability engineer will likely have to assume and/or estimate cycle frequency, duration, or even the total exposure for the product. It appears that such knowledge, especially about the mechanical environment, will remain in the competitive and proprietary domain at least for the near future.

Essentials of reliability engineering

Irrespective of the operating environment, programs for producing reliable products require quantitative methods for assessing and predicting various aspects of product reliability. This involves the collection of reliability data from the following (Meeker and Hamada 1995):

- Laboratory life tests to assess product reliability
- Degradation tests of materials, devices, and components
- Design of experiments for reliability improvement
- Tests on early prototype units to learn about possible failure modes and mechanisms
- Monitoring of early production units in the field
- Analysis of warranty data and samples from warranty population
- Systematic longer term tracking of product in the field

The need for shorter design cycle time is a driver for reducing the time and resources spent in reliability testing. Since, non-accelerated tests can take an excessively long time to yield valuable data for reliability improvement, different kinds of accelerated tests have been developed to estimate relatively quickly the failure-time distribution or long-term performance of the product in the field, based on a careful study of the operating environment. However, in and of themselves, accelerated life tests may not always yield actionable data to *improve* reliability. Analysis of the failures to uncover failure mechanisms and the root causes of failure are crucial for formulating corrective actions that can improve reliability. Sometimes, a second round of reliability tests may be required to assess the reliability of the improved products.

The focus of this chapter is on the physics behind accelerated laboratory life tests to assess reliability in thermal, mechanical, and electrochemical environments. Since failure analysis is ideally an integral part of any reliability assessment and improvement exercise, some representative failure mechanisms commonly observed in portable electronic products will be discussed in an ensuing chapter.

The study of interconnection reliability, until a few years ago, was driven primarily by the computer industry. Therefore, a vast majority of literature on electronic packaging reliability comprised primarily of thermal cycling reliability, and to a lesser extent, corrosion and electrochemical migration phenomena. In fact, there was a tacit assumption that reliability at the second level of packaging always implied thermal cycling reliability. As such, the titles of some reliability publications did not even indicate that the investigation pertained only to thermal cycling. Until recently, this did not cause any serious consternation among the packaging community in the days where much of the information processing hardware was confined to environments with controlled temperature and humidity.

The material discussed in this chapter is intended to introduce interconnection reliability issues for thermal, mechanical, and electrochemical environments for portable, consumer electronic products to readers who are primarily familiar with similar issues in business, office, and telecommunication applications. The scope of the chapter is limited to interconnection reliability and excludes topics such as electromechanics or liquid crystal display issues.

Considerations in accelerated life testing

The pitfalls of accelerated life tests (ALTs) need to be considered to avoid seriously incorrect inferences about the product reliability in the field, based solely on laboratory tests (Meeker and Escobar 1998). The following aspects need to be recognized when interpreting the results of ALTs:

Multiple or unrecognized failure mechanisms. High levels of accelerating variables can induce failure mechanisms that would not normally be observed at operating conditions. For example, instead of just accelerating corrosion or electrochemical migration, higher temperatures may cause melting or material deformation or degradation. Higher humidity may cause swelling, warpage, and delamination. In less extreme cases, high levels of accelerating variables will change the relationship between life and the variable. If different failure mechanisms are operative at high levels of the accelerating variables, and this is recognized, failure times for that mechanism can be censored out. Sometimes, such censoring can result in inadequate data. If the presence of undesirable failure mechanisms is not recognized, it is possible that seriously incorrect inferences are drawn.

Failure to properly quantify uncertainty. It is important to recognize that all statistical estimates have some uncertainty associated with them. Using point estimates alone can be misleading in many cases. Uncertainty can result either from the experiment or from the model. In general, statistical confidence intervals do not account for model uncertainties. Extrapolations, fundamentally, are fraught with errors, especially when based on inadequate sample sizes and point estimates. Performing a sensitivity analysis to assess model uncertainty or testing adequate number of samples is one solution.

Multiple time scales and degradation affected by more than one accelerating variable. In ALT, particularly when there is more than one failure mechanism, it should be recognized that all mechanisms may not be accelerated in the same manner or to the same degree. For example, when performing ALT of solder interconnections under accelerated conditions, creep and fatigue are accelerated differently depending on ramp rates and hold times at the different temperatures.

Masked failure mechanism. If there is more than a single failure mechanism, it is possible that one mechanism is accelerated more than the others depending on the parameters of the test. In such cases, the masked failure mechanism may not show up in laboratory testing but can dominate field failures, or vice versa. It is not only prudent but also cost effective to verify that the failure mechanisms seen in the field are the same as the failure mechanisms observed in accelerated testing. For example, during the development of a portable product, chip-scale package (CSP) interconnection failure was noticed in drop reliability testing. The contributing factor was the poor strength of the buildup layer on the printed wiring board (PWB), which led to under-the-pad cracking and eventually a trace fracture. However, in the field, the failure rate was dominated by a display connector rather than by the CSP interconnection. The failure mode involved the shearing off of the connector on the PWB, thereby leading to a display failure. The underlying root cause of weakness in the buildup layer caused both failure modes, but the laboratory test did not reveal the failure mode that eventually occurred in the field because it was masked by the CSP interconnection failure.

Faulty comparison. A popular use of ALT is in the comparison of alternative designs or materials from vendors, in addition to its use in predicting field reliability. The rationale is that if material from one vendor or one design performs better in laboratory tests, relative field reliability would follow a similar relationship. However, in cases where the reliability in the field is governed by a different failure mechanism than that observed in the laboratory test, ALT results can mask the actual field performance and serve as the basis for an inaccurate prediction of field reliability.

Accelerating variables as cause for deceleration. The most common examples involve failure mechanisms that require specific combinations of humidity, stress, and temperature. For example, when the usage rate is accelerated for a connector undergoing wear, the accelerated test can inhibit a secondary corrosion failure mechanism by continuously removing corrosion products and not giving enough time for the reaction to occur. Another example is failure due to tin whisker formation, which has a high propensity at a certain temperature and humidity for certain substrate and coating compositions and thicknesses. Optimum temperatures for tin whisker growth have been reported to be between 50 and 70°C by several researchers (for example, Wassink 1994). Unfortunately, since the working temperature of most electronic equipment is relatively close to the optimum temperature for tin whisker growth, an injudicious selection of temperature acceleration can yield incorrect results.

Differences between prototype and production samples. It is important to test units manufactured under actual production conditions, using materials and parts that will be employed in actual production samples. Sometimes, test methods capable of handling functional products may need to be developed. For example, ball-shear tests are widely used to assess the quality of the ball attachment process for area-array packages such as ball grid array packages (BGAs) or CSPs (Erich et al. 1999). However, the ball-shear test method cannot be applied to assess the interconnection quality or strength in a functional product after surface-mount assembly because an individual ball is no longer accessible for test. To accommodate interconnection strength data requirements on functional products, tests such as the package-to-board interconnection strength test method (PBISS) can be used (Canumalla et al. 2004; Hanabe and Canumalla 2004).

Thermomechanical reliability

Historically, for office and business machines, accelerated thermal cycling tests are carried out in the 0 to 100°C range with 10–15 min dwell times at ramp rates in the 10–15°C/min range. A life requirement of 1,000 cycles translates into a product life of about 7–10 years. These machines hardly experience other mechanical stresses in the operational environment. In contrast, hand-held electronic hardware can experience extreme ambient temperature fluctuations in the range of –30 to 45°C depending on the geographic location. When the appliance is left in an automobile, it can experience even more severe temperature conditions depending on the climate and diurnal variations. Thus, accelerated thermal cycling tests applicable to business machines may not be severe enough to assess the performance of hand-held electronic appliances. Another difference in regard to the portable hardware is the shorter product design life. The average product design life is in the range of 2–5 years instead of the 7–10 years in other consumer products such as desktop machines.

Owing to the aforementioned considerations, portable electronic PWB assemblies are generally subjected to accelerated thermal cycling of –40 to 125°C for 200–800 cycles in order to assess the product performance. The basis for this requirement can be understood in terms of the Norris–Landsberg modification to the Coffin–Manson equation (JESD94.01 2007).

The acceleration factor AF can be calculated using the equation

$$AF = \left(\frac{\Delta T_{\text{lab}}}{\Delta T_{\text{use}}}\right)^{1.9} \left(\frac{f_{\text{use}}}{f_{\text{lab}}}\right)^{1/3} \exp\left[T_{\text{use}}^{\max} - T_{\text{lab}}^{\max}\right] \tag{8–1}$$

where ΔT_{lab} = temperature range for accelerated lab test (–40–125°C),
ΔT_{use} = temperature range for use conditions,
$f_{\text{lab}}, f_{\text{use}}$ = frequency of thermal cycles per day in the lab and in field use, 48 and 1, respectively, and
$T_{\text{use}}^{\max}, T_{\text{lab}}^{\max}$ = maximum temperature of thermal cycles in the lab and in the field use, 125 and 50°C, respectively.

Based on the results of the use conditions measurement study discussed earlier, the actual use condition for a particular mobile phone can be interpreted in at least two different ways. The conservative deterministic estimate is that each user will subject the phone to temperature extremes of –10 and 50°C, everyday. The acceleration factor is determined to be approximately 3.98 using the above equation. However, such point estimates can be overly simplistic, since it is unlikely that every user of a mobile phone will subject the portable product to the same temperature range. If the portable product life is assumed to be five years with 365 thermal cycles per year, the number of cycles that would signify end of life is approximately 459 cycles.

A more realistic view of the requirements can be estimated by using Monte Carlo simulations. For example, consider that the lows in the end-use conditions can be described by a uniformly random sampling of temperatures ranging from –10 to 22°C, and the highs can be described by a uniform distribution of temperatures from 27 to 50°C (Vakevainen et al. 2001). A Monte Carlo simulation is then carried out using the modified Coffin–Manson equation. Then, for 1,000 simulations the mean AF is estimated to be about 24. Similarly, mean number of cycles corresponding to end of product life is 147 cycles. These data are shown as a 3-D plot in figure 8–1(a) and (b). Analysis of the data (percentiles) indicates that the deterministic AF calculation would be sufficient for approximately 75% of the population, but the test cycles to equivalent end-of-life estimates are good for only 50% of the population. In this case, the reliability engineer interested in estimating the test duration most representative of product life should rely on the more conservative Monte Carlo estimates that indicate 490 laboratory test cycles being equivalent to a five-year product life as the worst case. Therefore, board-level reliability requirements could be specified to be 500 cycles with low cumulative failure rate of, say, 0.1%.

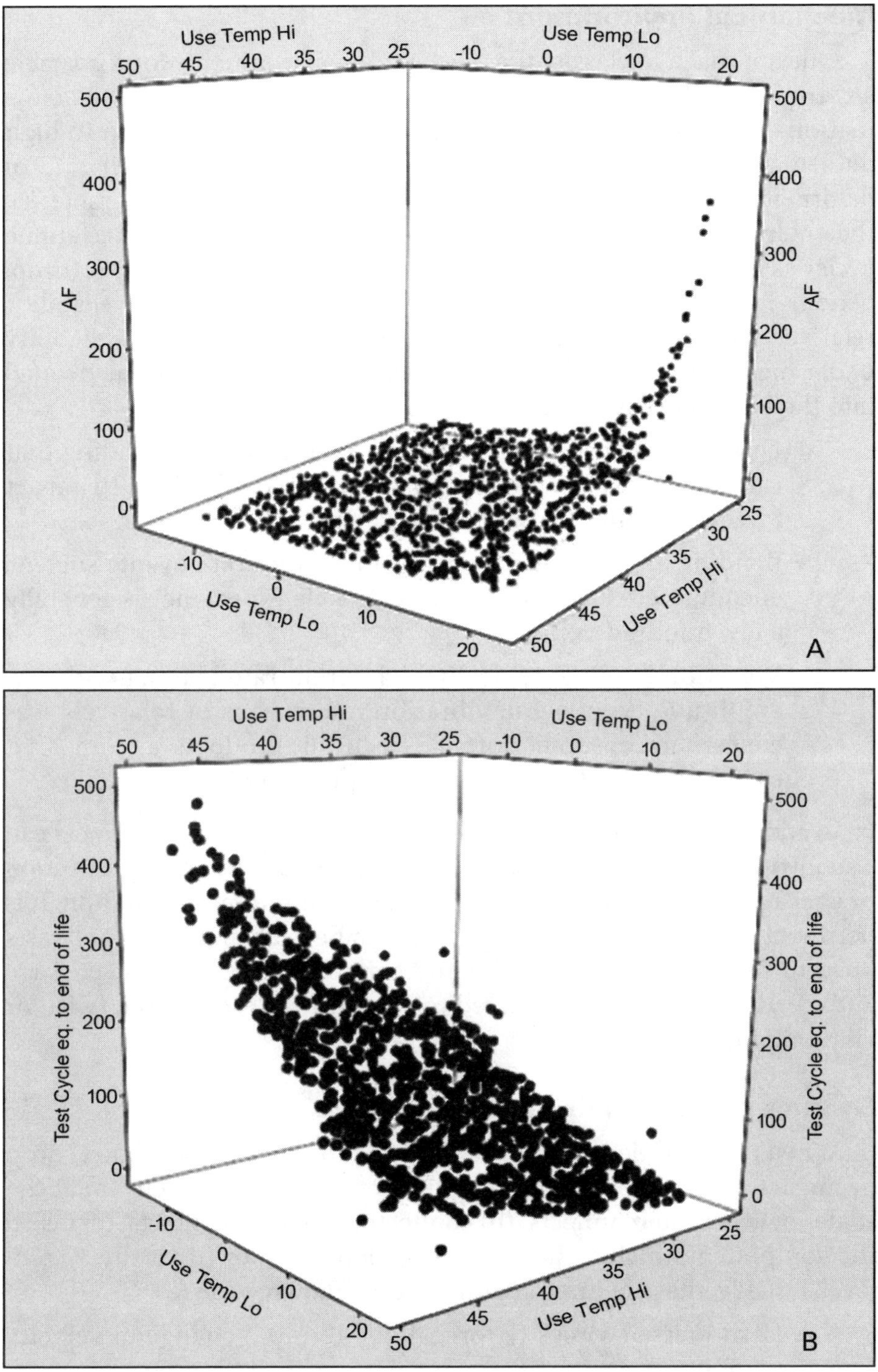

Fig. 8–1. Monte Carlo simulation of (a) AF and (b) number of cycles corresponding to end of product life

Mechanical environment

One approach to classify the mechanical environments for a portable electronic product is based on the rate of deformation: (a) low deformation—as experienced in bending and twisting, (b) medium to high deformation rate—as experienced in vibration, or (c) high rate of deformation—as in case of drop or shock. Another way to characterize the environment is based on the life expectancy in number of fatigue cycles as being high-cycle fatigue (vibration) or low-cycle fatigue (drop, bending, and twisting). In comparison to thermomechanical reliability, relatively little has been published in the public domain on reliability under mechanical loading. Broadly, mechanical loading can be divided into the following categories:

- Drop or impact loading: typically high strain rate loading that can also cause bending and twisting of the product due to impact forces. The number of drops to failure is generally low.
- Bending and twisting: typically low strain rate events such as encountered during key presses. The life expectancy is generally a few hundred cycles.
- Vibration loading: typically high strain rate loading with low amplitude. In general, vibration failures are of relatively less concern in consumer portable electronic products, as compared to industrial and military portable products.

In addition, reliability evaluations of portable electronic products can also involve either shear or pull testing performed at the interconnection or package level for purposes of determining the strength variation. It is pertinent to include them in the discussion because shear and pull tests serve to define the strength of the interconnection between the package and PWB, which is closely related to reliability in drop, bend, twist, or vibration loading.

Drop or impact environment

When portable electronic products are subjected to mechanical drop or impact, it is important to recognize that failure can occur (a) at the solder or other interconnects, (b) connector or spring contacts, (c) inside the components such as LCD, housing, lens, etc., or (d) at the system level. Usually, these failures are due to the following causes:

- High inertial forces (*g*-forces) due to rapid change in velocity upon impact

- Large strains in the solder interconnects between the PWB and package due to excessive dynamic buckling, flexure, or twisting of the PWB, and/or
- Shock waves that travel through the product assembly upon impact

It is reasonable to assume that all three effects can coexist during any single event and that the interactions among them can be relatively complex.

The drop tests carried out can be at the product level or at the board assembly level. Product-level drop tests involve tests on the entire product including the housing, while the board-level drop tests are performed on just the PWB assembly with components mounted on it, as described below.

Product-level drop or impact testing. Product-level drop testing can be classified as constrained or free. In constrained drop testing, which is by far the most common, the product is clamped rigidly to a heavy table that is guided along vertical rails to have a single impact against a target surface.

Clatter is probably best understood in terms of drop impact of an elongated or flat object onto a surface. Invariably, one corner touches down first, the object begins to rotate, and clattering occurs as the various corners encounter the impact surface before the object finally comes to rest. In that sense clatter refers to the condition where the second or third impact of the object likely occurs before the deformation from the first impact has returned to zero. On an oscilloscope time readout, strain or acceleration data due to clatter will resemble an extended but single impact sequence. In contrast, *multiple impacts* refer to the condition when the object bounces up and lands at a different location and orientation. In the case of multiple impacts, it is probable that the deformation in the assembly has had a chance to return to zero after the first impact, and the second impact occurs a short time later. On a time scale, they appear as two distinct events rather than as a single event. A third type of secondary impact, *chatter*, refers to the condition when subsystems or components impact each other within the product, for example, a battery impacting the case or a component. On an oscilloscope display monitoring the deformation–time response, chatter will appear as two events superimposed on each other and not distinctly separated in time. One of the main effects of the resultant secondary impacts in real-life situations is that, depending on the moment and the coefficient of restitution, the ends of the object can strike at much higher velocities than during the first impact. The increased amplitude of velocity shocks, the possibility of exciting resonant conditions, and repetitive shocks are reasons why the damage in a "real-life" drop can be significantly higher (Goyal and Buratynski 2000).

On the other hand, free-fall testing replicates the abuse a portable product will experience in actual usage. The main disadvantage is that it is difficult to control the orientation of the product at impact and this affects the repeatability of the test results and ease of monitoring by instrumentation. There is little mention of experiments in the literature where free fall drop testing has been automated. Goyal et al. (1999) proposed that the object being tested be suspended onto the guided drop table in the precisely desired drop orientation and that, just before impact, the object be released from its suspension. The intended result, theoretically, is that although the required orientation at first impact is maintained, the object is free to move unconstrained subsequent to the first impact. Another variation of the quest for greater repeatability in product-level drop testing consists of using grippers to control the orientation of the product until just before impact (Lim and Low 2002).

Lim et al. (2003) surveyed the response of several commercial portable products (Nokia 8250 and 8310, Sony Ericsson T68i, Compaq 3850, HP Palm m105 and m505) using strain gauges and accelerometers to monitor the response of the PWB during drop tests. The maximum strain values ranged from 500 to 2,500 microstrain and varied considerably between the different drop orientations depending on the product. Although the horizontal drop orientations generally yielded the highest strains in the PWB and the highest accelerations, there was considerable variability, which indicates that the actual behavior is quite complex and eludes simple generalizations.

In a study of the role of the rigidity of the mobile housing in determining the impact tolerance (Goyal et al. 1999), it was found that thin-walled clamshell case constructions, currently favored for its size and weight advantages, may not provide sufficient rigidity to impact induced loads. Housing modifications to increase the stiffness improved the drop reliability. In addition, it is believed that the drop tolerance of the mobile phone would improve if the battery pack were to remain firmly attached to the phone, minimizing velocity amplifications and possible chattering.

The shock response spectrum (SRS) approach was applied in using compliant suspensions to reduce peak acceleration and increase drop impact performance (Goyal et al. 2000). Results from another study with a personal digital assistant (PDA) using accelerometers and strain gauges, located along both the longitudinal and transverse directions, suggests that although there is a reasonably good correlation between acceleration and strain, it is often very difficult to completely unravel the complex strain-time or acceleration-time data except in select orientations (Seah et al. 2002).

Board-level drop testing. Because of the complexities inherent in product-level drop testing, alternative ways of estimating the product reliability from simpler tests have received much attention. One such technique is the *board-level drop test,* where the PWB assembly is subjected to impact loads or high accelerations while measuring the acceleration, velocity, and strain on the assembly. Such board-level drop tests provide a common basis to evaluate the impact tolerance of electronic products if one assumes that the conditions during product-level drop impact can be reproduced adequately by dropping a test PWB assembly. The advantages of board level testing are the following:

- The shock pulse amplitude can be fairly well controlled
- The orientation of the PWB assembly is controlled closely
- The tests are relatively more repeatable

The primary disadvantage is that the test is not a true reflection of reality because it does not include the effect of secondary events such as clatter, chatter, or multiple impacts, which can have a significant bearing on reliability.

Ong et al. (2003) examined the relevance of a board-level drop tester by comparing it with the data collected from an instrumented drop of a Nokia 3210 model phone. It was found that, for the product-level drop test, depending on the orientation of drop, the impact force can vary by up to a factor of five. Their results indicate that an axial impact exerted the highest forces. Further, because of the possibility of multiple impacts, the damage induced in a single drop in a product-level test may actually be much higher than the damage induced due to a single drop in a board-level test. In addition, Ong et al. (2003) report that in board-level drop tests, flexure of the PWB can last much longer than in product-level drop tests.

Despite these differences, board-level drop tests are attractive for investigating package reliability and process quality issues. Mishiro et al. (2002) observed a correlation between solder joint stresses and PWB strains in a study where numerical analysis and strain measurements were employed to assess CSP reliability for three different package constructions. Even if the PWB strain is the same, the package structure played a significant role in controlling the solder joint stresses and hence drop impact reliability. In particular, the package structure with a 0.15-mm-thick elastomer between the die and polyimide substrate performed better than the package where the interposer consisted of a multilayer laminate, which in turn was better than the package with only a polyimide substrate. Further, non-solder mask defined (NSMD) pad structure was shown to be significantly better than solder mask defined (SMD) pad structure for drop reliability. With regard to PWB build-up layer, aramid–epoxy PWBs with low adhesive strength performed poorly because of premature delamination in the buildup layer. Underfilling the package-to-board interspace was found to improve the reliability when the Young's modulus was sufficiently high. When the underfill modulus was too low (5 MPa), drop test reliability was much worse. Similar results were also reported in another study, where underfilling the CSP improved the reliability significantly in drop loading and the degree of improvement depended on the underfill modulus (Canumalla et al. 2002). Thus, it appears there is an optimal range for the underfill modulus for a particular system. Also, the presence of even a small void encompassing the corner solder joint could magnify the stresses in the solder joint, effectively negating any anticipated benefit of underfilling.

Recognizing the relative complexity of a product-level drop test, the relatively simple board-level drop test has been used to quantify drop reliability in terms of the package structure, materials, and processing. For example, Hannan and Viswanadham (2001) evaluated the drop reliability of CSPs with reworkable underfills for reliability enhancement. Kujala et al. (2002) used a board-level drop test to compare the relative performance of land grid array (LGA) package and CSPs under both thermal cycling and drop impact. The board-level drop test was used as a means to study the reliability of a "corner-reinforced-only" CSPs for portable product applications (Toleno and Schneider 2003), where the CSP was held down only at the corners with epoxy, without actually having any underfill surrounding the solder joint in the package-to-board interspace. The drop reliability of such corner reinforced CSPs was lower than in the case of complete capillary underfill. However, the relatively modest three to four times improvement in the performance may be sufficient for some portable product applications (Tian et al. 2003). Board-level drop tests have also been used to investigate the effect of PWB and component pad surface finish, and concomitant interfacial strength, on drop test performance, and this is discussed in a later section.

Simulation of drop test behavior of PWB assemblies and products. Faced with the complexities of purely empirical product-level drop testing, there have been several attempts to complement experimental studies with finite-element simulation to better understand the drop phenomena. The key issues for a successful understanding of drop impact reliability are (a) sophisticated and consistent analysis tools, (b) test correlation for model validation and refinement, (c) specification to define reliability requirements, and (d) material property data, especially over a broad range of strain rates (Wu et al. 1998).

Simulation, when combined with board-level drop testing, can enable accurate prediction of not only the failure location but also durability to within 10% (Tee et al. 2004). It was found that drop orientation with the components oriented face down was a more stringent test condition than one with the components facing up. Results indicate that during the drop test, greater PWB bending induces larger stress to the solder joints. As aniticipated, it was found that the outermost solder joints have larger stresses and that smaller PWBs enhance drop performance. More importantly, it was reported that the lead-free solder studied had better board-level thermal cycling reliability but worse drop test reliability. It should be recognized that thermomechanical reliability alone does not assure product reliability under mechanical loads.

Relatively accurate correlation of model prediction and experimental data were reported using smeared property models (Lall et al. 2004). Significant error may be introduced due to aliasing of the experimental and computational data, and undersampled experimental data acquisition may mask the recognition of peaks in strain or displacement. In addition, smeared property models may not capture structural degradation during successive drops produced due to progressive delamination between materials. A validated modeling technique can be used to accurately predict failures observed in portable electronic products, such as disengagement of snap-fit housings and CSP solder joint cracking (Zhu 2003).

While state-of-the-art simulation was shown by various people to accurately predict different aspects of the drop test, a combination of simulation and experiments can be expected to be the most effective approach for improving and predicting reliability under drop or impact loading.

Analytical modeling of drop phenomena. In addition to numerical simulation, closed-form analytical modeling has been employed to understand the physics behind drop-related phenomena. Suhir (1991) obtained formulae to calculate the maximum displacements, velocities, and accelerations of surface-mounted devices when a shock load is applied to a flexible PWB at its support contour. Consideration of the nonlinearity of the PWB vibrations was found to be important in the case of large shock-induced deflections. The dynamic response of a rectangular plate element assembly subjected to drop impact was simulated as a box within a box, with one gasket between the outer and inner boxes and another between the PWB and the inner box. Results suggest that lower *g*-forces can be ensured by having the lower natural frequency considerably different from the higher frequency (Suhir and Burke 1994). For example, the inner cushioning gasket could be made substantially stiffer than the outer one. Probabilistic approaches could also be employed to ensure a low failure rate. The effect of the stiffness of a "spring" shock protector was also studied (Suhir 1995). Because the possibility of a "rigid impact" needs to be avoided at all costs, if the maximum drop height is not known, the advantages afforded by a soft spring cannot be fully utilized. The effect of viscous damping on the maximum displacement and the acceleration of a one-degree-of-freedom linear system subjected to a shock load during drop impact was also studied (Suhir 1996). Sometimes, the application of materials with high energy absorption can result in even higher acceleration levels, and this needs to be avoided by a careful consideration of the system's mass and spring constants.

Whether maximum acceleration is an adequate criterion of the dynamic strength of a structural element in an electronic product has been investigated using a simply supported beam and a cantilever with heavy end mass (Suhir 1997). Surprisingly, it was found that even if the accelerations experienced are not severe, one can expect significantly high dynamic stresses. These results are supported by observations during product-level drop tests, where the bending of the board plays a bigger role in controlling failure compared to purely inertial forces, especially for light components such as flip chips and CSP assemblies. Until recently, however, acceleration has been measured preferentially because it is easier to measure. Regarding alternatives to drop testing, it was found that the applicability of shock tests to replace product-level drop tests depends on whether the dominant frequency of the shock impulse (which is inversely proportional to duration) is sufficiently high in comparison to the fundamental frequency of the vulnerable structural element (Suhir 2002).

Bend or twist environment

Most portable electronic products experience more severe PWB-bending-related stresses than thermal stresses. PWB bending failure in the creep regime can be caused by localized bending near a screw location or in the high-cycle fatigue regime due to key press action. A third bending failure mode occurs when portable products are dropped (Darveaux and Syed, 2000). Recognizing the importance of understanding the reliability under bending loads, several studies in recent years have been aimed at characterizing the deformation and failure of solder joints.

Darveaux and Syed (2000) have used both three-point and four-point bending tests to examine the failure mechanisms under a range of conditions for different CSPs along with finite-element simulation of the damage processes. For displacement-controlled fatigue tests, life decreased with (a) reduction of span length, (b) increase in test board thickness, (c) increase in die size, and (d) increase in molding compound thickness. In load-controlled tests, which are more closely related to actual product reliability, opposite trends were observed. Simulation results indicate that the optimum component/PWB pad size ratio in bending is different than under thermal loading.

The failure modes observed can be summarized as (a) fracture in the solder or in the intermetallic layer at the component pad, (b) fracture in the solder or in the intermetallic layer at the PWB pad, (c) trace peeling and eventual laminate cracking of the PWB or the component, or (d) build-up layer fracture leading to trace cracks on the PWB. Since these are very similar to the failure modes frequently seen under drop or impact conditions, bend testing is generally perceived to be a relatively simple alternative to more complicated drop tests. Improving the strength of the weakest failure link can offer improvements in performance. For example, in 3-pt bend and drop impact tests on CSPs, anchoring the pads with via-holes improved the performance over having no-via-in-pads (Juso et al. 1998).

A few studies have been reported on the effect of strain rate during the bending test. For example, Geng et al. (2002) reported that the solder joint interconnection fails at approximately 50% lower board deflection when the test speed increases by two orders of magnitude (0.25– 2.54 cm/s). It is relatively well known that although solder strength increases with increasing strain rate, strain to failure decreases. In that context, as long as failure occurs in the solder, solder joints can be expected to fail at lower strain rates in high-displacement-rate bending tests. However, the data do not show a very distinct trend at higher strain rates (25.4 cm/s), possibly due to experimental artifacts. In a different study, with increasing ram displacement rate in a 4-pt bending test, strain gages mounted on the PWB showed increasing strain at solder joint failure sites (Harada et al. 2003). It was shown that Kirkendall voids at the intermetallic interphases between the Ni and the Ni–Sn–P layers degraded the interfacial strength enough to cause failure preferentially at these locations. It should be noted that Kirkendall-like voids were also reported at Cu–Sn interfaces in lead-free solder joints on organic solderability preservative (OSP) pads by Chiu et al. (2004), with severe drop performance degradation in strength upon thermal aging in the 100–150°C range. Some reliability studies also focused on testing methods for flexible or low stiffness PWBs. Rooney et al. (2003) reported an offset bend test configuration that is useful for testing assemblies with thin PWBs (0.5mm) having stiff components.

A planar 3-pt bending fatigue test method to assess the reliability of the CSP solder joints was recently proposed (Leicht and Skipor 1998). The applicability of this method was demonstrated for standard plastic ball grid array (PBGA) components mounted on a PWB. The same method to establish that via-in-pad structure by itself does not pose a reliability risk in bend fatigue (Jonnalagadda 2002). This is in accordance with the results reported previously by Juso et al. (1998). The applied load could induce dielectric (build-up layer) cracking, which in turn can lead to trace and via failures. Although lead-free solders have been found to be more durable than tin–lead solders in bend tests (Jonnalagadda et al. 2002), it should be remembered that different lead-free solders can be expected to behave differently, and some can perform worse than Sn–Pb solders depending on surface finish, test conditions, sample history, and several other variables. Moire interferometry coupled with 4-pt bend testing can reveal the localized influence of solder ball interconnections on chip carrier and PWB deformation (Stout et al. 2000). Large shear strains were found in solder balls across the entire array. It was found that maximum strains occur in the outermost row of the solder balls, which agrees with the observations from a study on underfilled CSPs with corner defects (Canumalla et al. 2002). In another study, the effect of cyclic bending on CSP assembly reliability was investigated in addition to monotonic bending (Shetty et al. 2000). The average overstress limit for a CSP studied was determined to be 2550 N-mm. It was concluded that the CSPs showed worse durability when the PWB assembly was subjected to negative curvature (CSP mounted surface of the PWB is convex). This is understandable since negative curvatures would subject the corner joints of the CSP to tension and lead to premature failure.

Portable electronic products were also evaluated for reliability under twist loads in addition to bend loads. For example, Perrera (1999) reported on the effect of twist loads of 9% and 12% and observed that solder joint failures occurred mostly by fatigue processes.

Shear tests

Interconnection failure is a common mode in portable electronic products, and it is widely accepted that interconnection strength and solder joint quality can play a central role in determining product reliability. Thus, measurements of the interconnection strength are useful in understanding reliability of the product. The term *interconnection strength* in this context denotes the effective strength of the package-to-board interconnection, and includes the strength of (a) the package–solder interface, (b) solder, (c) the solder–pad interface and (d) the build-up layer on the PWB.

Conventionally, the ball shear strength is used to denote the strength of attachment of a solder ball of a BGA or CSP to the component *prior to board assembly* (Erich et al. 1999). This measure of ball strength, although useful in measuring the quality of the ball attachment process, cannot be easily translated into a product level estimate of durability. This is because of the following reasons: (a) the bare component is no longer accessible for ball shear tests, and (b) the interconnection quality is determined not only by the solder/component bond but also by the solder strength, solder/PWB interfacial strength, and build-up layer quality.

Product-level tests such as mechanical drop, twist, and bend tests yield valuable information on the reliability in the field. However, the primary drawbacks of these tests in terms of targeted improvement actions are (a) the complexity and (b) the time required to analyze the results. Thus, there is a need for a product-level interconnection strength test that can yield relatively rapid results and simultaneously provide targeted quality improvement actions.

One candidate method is a recently developed product-level test, the package-to-board interconnection shear strength (PBISS) technique (Canumalla et al. 2004; Canumalla 2004). It was shown that the shear test is an effective tool to quantify the shear strength of CSPs and examine the effect of pad finish and build-up layer strength. Only low strain rate PBISS behavior was characterized because product-level twist and bend tests are performed at a low strain rate.

However, the strain rates experienced by the solder joint during drop tests are significantly higher. Therefore, the shear strength behavior measured at slow deformation rates is not directly applicable as a proxy for drop reliability of portable electronic products. Solder behavior changes significantly with the rate of deformation, and the damage to the CSP interconnection can be expected to be significantly different also during high rate of strain (Enke et al. 1989; Shohji et al. 2004; Darveaux 1992). In this context, it was demonstrated that high strain rate shear tests essentially mimic the failure mechanisms and relative performance observed in drop tests (Hanabe and Canumalla, 2004; Hanabe et al. 2006).

Corrosion and electrochemical environment

The failure mechanisms that are of importance in portable electronic products exposed to electrochemical environments can be described as follows:

- Corrosion
- Electrochemical migration (ECM)
- Conductive anodic filament (CAF)

The fundamental difference between corrosion and electrochemical migration is that corrosion involves the destructive attack of a metal by the environment as anodic oxidation without the necessity for electrical bias, whereas ECM involves the transport of metal ions from the anode to the cathode under the influence of an applied electric field. From a failure perspective, corrosion results in product failure primarily by causing electrical open or intermittent interconnections, while ECM results in failures primarily due to reduced surface insulation resistance, electrical shorts, or intermittent connections. Some factors affecting these failure mechanisms are the environment (temperature, humidity, presence of corrosive elements), operating conditions (bias voltage, current density, temperature, and conductor spacing), and materials (nature of metal or alloy, surface condition, ability to absorb humidity, coating composition, and thicknesses).

Corrosion

Corrosion, depending on the severity, results in the following failure pathways:

- Oxidative materials degradation resulting in loss of electrical continuity
- Partial degradation of materials accompanied by the formation of conductive oxidation product, such as easily ionizable salt(s), that could result in lower surface insulation resistance (SIR)
- Electrical shorts between adjacent conductive features
- Intermittent shorts or opens depending on the humidity levels and the ionic nature of the corrosion product.

Corrosion is often discussed in terms of half-cell reactions because all corrosion processes are essentially electrochemical reactions. The electrodes in question could be on the macro or micro scale. Macroscopic galvanic corrosion cells can occur when dissimilar metals are coupled electrically and exposed to a corrosive environment, while microscopic corrosion cells tend to occur on the scale of grains. In either case, oxidation occurs at the anode and reduction at the cathode. In other words, the metal dissolution occurs only at the anode. The medium or electrically conductive environment in which these chemical reactions proceed is usually referred to as the *electrolyte* even if the electrolyte may extend to a thickness of a few monolayers. Since all the cations produced by the anodic reaction are consumed by the cathodic reaction, both anode and cathode reactions proceed at the same rate for corrosion to occur in a continuous manner.

The propensity of a metal to undergo corrosion is described in terms of the standard electrode potentials, where the hydrogen electrode potential is arbitrarily assigned a value of zero. When two dissimilar metals are coupled, the less noble metal will corrode in relation to the more noble one. However, it is possible to promote corrosion of the more noble metal in a galvanic couple by electrical biasing, which makes the more noble metal the anode. Some forms of corrosion (Tullmin and Roberge 1995) that are relevant to portable electronic products are the following:

Uniform corrosion. This form of corrosion is evenly distributed over the surface, and the rate of corrosion is the same over the entire surface. A measure of the severity is the thickness or the average penetration.

Pitting and crevice corrosion. This localized form of corrosion appears as pits or crevices in the metal. The bulk of the material remains passive but suffers localized and rapid surface degradation. In particular, chloride ions are notorious for inducing pitting corrosion, and once a pit is formed, the environmental attack is locally autocatalytic.

Environmentally induced cracking. This form of corrosion occurs under the combined influence of a corrosive environment and static or cyclic stress. A static-loading-driven cracking is called stress corrosion cracking and a cyclic-loading-driven cracking is called corrosion fatigue. Residual stresses in electronic component leads from lead bending operations were observed to cause stress corrosion cracking failures in the presence of moisture (Guttenplan 1987). Stress corrosion cracking of package leads was also reported in the presence of solder flux residues (Raffalovich 1971).

Galvanic corrosion. This type of corrosion is driven by the electrode potential differences between two dissimilar metals coupled electrically. The result is an accelerated corrosive attack of the less noble material. Galvanic corrosion tends to be particularly severe if the anodic surface is small compared to that of the nobler cathode or in cases where a nobler metal is coated onto a less noble one. For instance, when a porous Au plating over a Ni substrate is exposed to a corrosive environment, the gold coating acts as a large cathode relative to the small area of exposed Ni. This sets up a galvanic cell at the exposed substrate which experiences intense anodic dissolution. It has been observed that pore corrosion can be enhanced by a galvanic corrosion process when the substrate metal is less noble than the coating, and vice versa (Yasuda et al. 1987).

Electrochemical migration

The distinguishing feature of ECM from corrosion is the formation of dendrites that cause a short between adjacent conductors. There are some similarities to corrosion as well, and the oxidation of the metal at the anode is common to both processes. ECM, which is also known as *migrated metal shorts* (Kohman et al. 1955; Shumka and Piety 1975), is probably best described as the transport of ions between two conductors in close proximity, under applied electrical bias and along an electrically conductive medium. In general, three conditions are necessary and sufficient for ECM failures to occur in PWBs and PWB assemblies: (1) presence of sufficient moisture (sometimes as little as a few monolayers), (2) presence of an ionic species to provide a conductive medium, and (3) presence of an electrical bias to drive the ions from the anode to the cathode. In the presence of sufficient moisture, the process is accelerated by temperature, and several mechanisms of ECM have been in vogue.

The first step in the classical model of ECM consists of metal ion formation by anodic oxidation (similar to corrosion), which may be either direct electrochemical dissolution or a multistep electrochemical process. At the anode, for example, where M represents a metal atom,

$$M \longrightarrow M^{n+} + ne^- \tag{8–2}$$

The second step is the transport of metal ions from the anode, through an electrolyte, towards a cathode. In the final step, at the cathode, the positively charged ions are reduced to a neutral metal atom. At the cathode

$$M^{n+} + ne^- \longrightarrow M. \tag{8–3}$$

Successive cationic reductions facilitate the growth of dendrites towards the anode along energetically favorable crystallographic orientations. Therefore, the surface insulation resistance of the material progressively decreases as the migration advances towards the anode.

Eventually, an electrical short results when the dendritic filament touches the anode. Ag (Kohman et al. 1955), Cu, Pb, Sn (Benson et al. 1988; DerMarderosian 1978; Ripka and Harsanyi 1985), Mo and Zn (Kawanabe and Otsuka 1982) have all been observed to form dendrites by this process. The presence of flux containing ionic species is a known contributor to ECM and has been studied widely using surface insulation resistance measurements (Turbini et al. 1992). Following the migration ability of pure metallization systems, the propensity for ECM may be ranked as follows: Ag>Pb>Cu>Sn (Harsanyi and Inzelt 2001; Takemoto et al. 1997).

A second mechanism of ECM was proposed to explain the migrated metal short formation involving noble metals such as Au, Pd, and Pt. Because of the relative chemical inertness of these metals, a halogen contaminant is needed to induce anodic dissolution (Grunhaner, Griswold, and Clendening 1975; Sbar 1976). In an acidic medium, a positively charged metal ion may form by the following route at the anode:

$$Au + 4\,Cl^- \longrightarrow AuCl^-_4 + 3e^- \tag{8–4}$$

$$AuCl^-_4 + H^+ \longrightarrow H[AuCl^-_4] \longrightarrow HCl + AuCl_3 \longrightarrow H^+ + 4Cl^- + Au^{3+} \tag{8–5}$$

These positively charged Au ions can migrate towards the cathode and form dendrites in a similar fashion as the classical model.

A third mechanism to explain the ECM of Ni starting at the anode involves the presence of a strongly alkaline electrolyte. The first step is the formation of a cation ($HNiO_2^-$) by anodic corrosion followed by a chemical process resulting in secondary ionic species (Harsanyi 1995).

$$Ni \longrightarrow Ni^{2+} + 2e^- \tag{8–6}$$

$$Ni^{2+} + 2OH^- \longleftrightarrow Ni(OH)_2 \tag{8–7}$$

It is suggested that instead of migrating to the cathode, the Ni^{2+} ions thus formed undergo the following reaction to form an anionic complex:

$$Ni^{2+} + 2H_2O \longrightarrow HNiO_2^- + 3H^+. \tag{8–8}$$

This anion complex migrates through the electrolyte under the applied electrical field. Finally, the metal atoms are deposited at the anode in the form of metallic dendrites due to the electrochemical reaction of the cationic species with the H^+, Ni^{2+} or OH^- ions. Similar process could be operative for Co and Cu ECM as well in cases where anodic deposits of the metal are observed.

Conductive anodic filament (CAF) formation

CAF is the type of electromigration failure mechanism where the loss of insulation resistance between neighboring conductors is caused by the growth of a subsurface anodic filament along delaminated fiber/epoxy interfaces (Lando et al. 1979). The first step in the formation of the CAF is the physical degradation of the fiber/epoxy bond. This is followed by an electrochemical reaction requiring both the presence of moisture and a potential gradient across the cathode and anode. The metal undergoes oxidation at the anode to yield a positively charged ion that migrates toward the cathode. As the metal species migrate toward the cathode, they precipitate at locations where the pH is thermodynamically conducive and, in time, the filament extends from the anode to the cathode causing a short. The CAF formation may occur along the surface of a PWB or between conductors in different layers separated by a dielectric or along the glass fibers in the weave (Viswanadham and Singh 1997).

Tin whiskers

Single-crystal whiskers of several metals including Sn, Cd, Zn, Sb, In, Pb, Fe, Ag, Au, Ni, and Pd have been reported (for example, McDowell 1993; Siplon et al. 2002; Zeng and Tu 2002). While the mechanism for the growth of whiskers of different metals may possess similarities, the mechanism of Sn whisker growth has been studied extensively. However, due to recent emphasis on the implementation of Pb-free solders and the consideration of Sn as a component in terminations and PWB finish, there has been an increased effort to study the reliability implications of Sn whiskers. Several reported field failures have been collected from medical, military, and space applications by Siplon et al. (2002). It is generally agreed that whisker growth occurs at the base of the whisker in response to imposed stresses or residual stresses below the surface. The formation of Cu_6Sn_5 or other intermetallic compounds at the interface between the tin and the substrate layer has been shown to result in a compressive stress in the Sn film (Zeng and Tu, 2002; Lee and Lee, 1998). Once the oxide layer covering the tin has ruptured, tin whiskers can be extruded as a means of releasing compressive stress. It has been demonstrated that the use of certain substrate-coating combinations, such as

Ni over Cu, significantly reduces whisker growth (Schetty 2000). It was also demonstrated that avoiding brighteners, annealing of any residual stresses, using thicker tin layers, and addition of Pb are beneficial in reducing the propensity for whisker growth. On the other hand, the use of brighteners, lack of annealing, tin layers thinner than 2 mm, copper-based substrates, and addition of Zn were shown to promote tin whisker formation (Wassink 1989). The study of Sn-whisker-related reliability issues in portable consumer electronic products is in its infancy insofar as published reports of whisker-related failures. Owing to considerable variations in Sn plating formulations and test methodologies, estimation of product failure risk has not been easy. However, decreasing pitch and increasing circuit density coupled with the drive towards Sn-rich solder compositions can be expected to elevate the risk of failure due to Sn-whisker-related issues in the near future.

Reliability comparisons in the literature

Reliability testing and accompanying failure analysis that are needed to fully understand the magnitude and nature of reliability concerns can be expensive in terms of time and resources. As discussed earlier in this chapter, there is a constant business-driven need to minimize or accelerate reliability tests. Therefore, it is only natural that every effort is made to utilize any available historical data to assess current reliability risks and minimize the reliability testing that needs to be performed. While the value of reliability comparisons is clear in terms of reducing the need for testing and saving time and money, comparison and utilization of reliability data from different sources is a difficult exercise at best, and one has to be cognizant of the multitude of factors that influence the final reliability projections. In this section, some of the relevant aspects in comparing reliability results from different sources are discussed.

Thermo-mechanical reliability

Effects of thermal fatigue are generally evaluated through accelerated thermal cycling tests. Test units, in statistically significant numbers, are subjected to a predetermined thermal profile over a number of cycles until all or 50% of the samples fail, and failure distributions are determined. In evaluating technologies, comparisons of failure data from a variety of sources are attempted to verify, substantiate, or discern significant variations in reliability and understand the mechanisms. There are several pitfalls in this approach. The first one is the definition of failure. Some regard a percent change in the resistance of total risk net consisting of a number of solder joints. Others may consider resistance spikes of a given magnitude and lasting over a specified duration, and still others may consider only an open joint as constituting a failure. The number of joints in a risk net may be different from study to study as well as in the same study depending on the I/Os of the packages being studied. The actual value of the resistance change can be significantly different in each case if only percent change in resistance is considered. In great many instances, the failure criterion is not even indicated or included. A comparison of the probability plots can lead to misleading conclusions if the failure criteria are not identical in all of them.

Test parameters are also crucial and need to be considered explicitly for meaningful reliability comparisons. For example, in a thermal cycling test, the important parameters are the ramp rates and dwell times at the temperature extremities. A ramp rate of 15°C/min and a dwell time of 10 min at each extremity are generally considered appropriate in many instances. However, the literature contains data with six cycles per hour all the way up to two cycles per hour. Differences in the dwell time at extremities can have significant influence on the thermal fatigue and creep behavior of interconnection alloys. The temperature that the package and board experience in a given profile can be different from the settings of the temperature chamber. Many studies indicate only the temperature values involved and do not provide the actual temperature the product under test actually sees. It is only prudent to compare temperature profile of the chamber versus the actual temperature experienced by the product under test as a function of time.

Other important factors that influence the discrepancies between the two are the number of layers, copper and epoxy content, thickness of the board and its heat capacity, the nature and size of the components, and presence and absence of heat sinks. For example, a high I/O large ceramic component may take a longer time to attain steady state temperature in comparison to a thin small package such as a CSP. If the cycle profile is not set correctly, it can alter the dwell time on some packages. Thus, a package of high heat capacity is more likely to experience a shorter dwell than a smaller package. The net result is that the solder joints in the bigger package may not experience the anticipated creep relaxation, and hence the failure may be altered by an unpredictable amount. In addition, during the ramp-up portion of the cycle, temperature can overshoot the preset values and it takes some time for the temperature to reach the set value. If a number of boards are being tested in the chamber, the location of the boards in the chamber and their disposition can influence the temperature each board or package experiences. Boards stacked together and aligned perpendicular to the direction of air flow in the chamber will result in the boards immediately facing the air flow experiencing a different profile than other boards in the stack. In addition, the likelihood of blind spots in the chamber cannot be ruled out. Thus, a complete characterization of the thermal chamber to ensure that packages and the board attain the equilibrium temperature is very important.

Comparison of failure distributions can be complicated if the statistical distributions are not properly chosen and failure mechanisms not well understood. The most popular solder joint failure distributions are the two-parameter Weibull distributions and occasionally three-parameter Weibull distributions. Even while using the two-parameter Weibull distributions, a single average line is often drawn through two apparently distinct distributions (discussed in chapter 7). This often leads to erroneous μ values. In addition, a tacit assumption is made that there is only a single failure mechanisms.

Sometimes, reliability results are reported without a failure analysis. Even when the failure mechanism is reported, the mechanism that is reported is based on the analysis done at the end of the test and not immediately following the detection of failure by electrical test. Thus, the understanding of the failure mechanism is corrupted or distorted by crack propagation, and microstructural changes occurred subsequent to the failure detection. When the distribution plots exhibit failures that indicate differing slopes, it is important to delineate them and conduct failure analysis to determine the exact failure mechanisms.

Thus, comparison of thermal cycling reliability tests has to be carried out with extreme caution taking into account all the factors that affect the inferences and conclusions. The current literature on reliability does not appear to readily lend itself to definitive correlations and accurate comparisons.

Mechanical reliability

Mechanical reliability comparisons for portable consumer electronics are more complicated and difficult than thermomechanical reliability comparisons because of dynamic and structural complexities. There are many more variables to be taken into account in the assessment of board-level mechanical reliability. These include package size, solder ball size, board structure and dimensions, drop height, orientation, impact duration, strike surface, etc. At the product level, reliability comparisons are even more complicated due to additional dependencies on the product form factor, weight distribution, impact orientation, occurrence of secondary impacts, and other test-related variables. Therefore, the ability for comparison of mechanical drop test reliability is at its infancy.

Consistent test procedures with consistent acceleration and impact and failure criteria are critical in ensuring that results from one reliability test can be compared with results from another. For example, peak acceleration and the impact energy attained by the product depend on the frictional forces induced by the guide mechanism in the test equipment. Therefore, the actual impact velocity can be different from the theoretically computed value.

The number of mounting screws and their location also has significant effect on the drop reliability. Boards mounted with only four screws can have a lower impact life compared to those mounted with six screws under the same loading conditions due to greater bending. The type of screws and the torque applied to them can have a pronounced effect on the drop performance. The likelihood of screws loosening after subsequent drops cannot be ruled out. The dislodged screws can dramatically alter the board response during the drop. In addition, it needs to be verified that the failure locations and mechanisms are identical before attempting to compare reliability values. For example, failures that occur during drop can be due to interfacial brittle fracture at the package pad/solder interface, PWB pad/solders interface, or the copper trace break at entry to the pad.

Location of the package on the PWB also plays an important role in determining reliability. Board bending and warpage can be very dependent on the board dimensions, and are usually greater along the longer dimension of the product. Typically, but not always, packages positioned at the center of the board are more susceptible to failure than the ones away from the center when the product is dropped on its face or back.

Package construction plays a significant role as well. Many portable electronics use low profile packages to accommodate the rather slim product form factors. These packages, such as ball grid array packages like very-thin-profile fine-pitch BGA (VFBGA), thin-profile fine-pitch BGA (TFBGA), and quad flat pack no-lead (QFN) packages, have low solder joint standoff, thinner die, and thinner molding compound. For example, VFBGAs have been shown to have slightly better performance than TFBGAs having the same I/Os (Tee et al. 2003). In a different study, the 208 QFP package solder joints were observed to fail in a relatively small number of drops due to their mass, and FLGA 300 (0.8 mm pitch) packages were relatively more durable (Xie et al. 2003).

Materials' aspects such as surface finish on the package and PWB pads can be expected to have a significant effect on the drop test reliability. Compatibility between PWB and component termination finishes, sometimes even inside the component module, can play a significant role in determining drop reliability. For example, incompatibility between Cu finish on resonators and ENIG finish on interposer PWB was found to severely degrade drop test performance (Saha et al. 2004). In this case, the copper from solder/component interface migrated to the solder/interposer interface during the reflow and impeded the growth of Ni–Sn intermetallics and, instead, promoted the formation of a ternary Ni–Cu–Sn intermetallic phase. In the absence of a strong metallurgical bond between the Ni on the interposer PWB and the solder, premature failures occurred in drop testing.

Although, in general, Sn–Cu interfacial bond has been found to be superior to the Sn–Ni interfacial bond, recent evidence seems to suggest that Cu–Sn intermetallic bond can have risks as well. For example, the Kirkendall type of voiding found at the Cu/Cu_3Sn interface, especially after thermal aging, has been shown to impair board level drop performance (Chiu et al. 2004). Modification of the intermetallic bond strength by addition of trace amounts of some elements also needs to be considered when comparing reliability results from different studies. For example, addition of 0.3% In and 0.04% Ni to the Sn–Ag–Cu solder was shown to improve drop test reliability by as much as 20% even after 150°C thermal aging in comparison to the Sn–Ag–Cu solder (Amagai et al. 2004).

Influence of Material Properties on Reliability

Printed wiring board

The proliferation of portable electronic appliances in the form of mobile phones, personal digital assistants, pagers, etc., has brought about a "density revolution" in the PWB technologies. Ever-smaller board features have necessitated new approaches to design, materials, fabrication, assembly, and testing. The consumer demand is for faster, cheaper, lighter, and more reliable electronic hardware. Conventional multilayer boards with 150 μm lines and 150 μm spaces with 325 μm drilled through-hole vias cannot always accommodate the wiring densities required for fine pitch high I/O area array devices such as ball grid array and CSPs. Therefore, weight reduction and high density requirements have resulted in the need for high-density interconnect (HDI) boards. For portable electronic hardware with high density, thinner boards with finer lines and spaces with very small vias are needed. Thus evolved the completely new PWB industry of HDI micro-via board technology featuring extremely thin laminates and multilayer microvias. Several techniques such as surface laminar circuitry (SLC), laser drilled micro-via techniques, any layer inner via hole (ALIVH) technology have evolved. Buried, blind, and through-hole vias were needed to accommodate the product functionalities. These features are significantly different from the conventional PWB technologies and are approaching those used in the semiconductor industry. A semiconductor technology attitude is being cultivated by the PWB industry to meet the new challenges. At the same time, the reliability requirements for portable electronic hardware are often more stringent than the conventional hardware. The only relaxation in the reliability requirement is one of product life; they are shorter than those required for desktop and business products. However, the mechanical and environmental requirements are more severe.

The complexity of the product varies considerably and may contain PWB assemblies that are either single-sided or double-sided. A double-sided assembly will be more rigid and display a different shock response. In some cases, depending on the product complexity, both buried and blind vias may be used simultaneously. The buried vias may be plated or filled with conductive paste and cured. The reliability of thin populated boards with blind and buried vias is inadequately understood under various mechanical loading conditions. Issues such as mis-registration of the buried vias in the individual layers can pose a reliability exposure.

Another important aspect of microvia technology is the shape of the vias, namely, square-well or bathtub, and the copper plating thickness and uniformity due to the variations in via shape. In addition, the registration of the microvia on the capture pad is very important and crucial for product reliability. In case of poor registration, the laser drilling may be partially off the pad and penetrate the adjacent laminate. This can result in voiding during reflow process due to the egress of the occluded moisture in the laminate, impacting the package-to-board interconnection integrity.

In a high-density PWB, different materials are used for the microvia layer including non-reinforced epoxies, woven-fiber reinforced resins, chopped fiber reinforced resins, such as aramid-reinforced materials, and resin-coated copper foils. The adhesion of the reinforcing material to the base resin can have a significant impact on reliability. Additionally, several Cu-to-laminate adhesion enhancement treatments, including mechanical abrasion, have been in vogue. Each of these aspects can impact the reliability, especially under mechanical loading.

Package

In portable electronic products, package size and style can influence product concepts, and vice versa. Packages have to fit the form factor of the product, which is usually very thin. Double-sided surface-mount assemblies with low-standoff low-profile packages are the order of the day. This limits the feasible options to CSPs, VSSOP, TSOP, lead-less packages, LGAs, and quad flat no-leaded package types, to name a few. With increasingly effective utilization of PWB real estate, an emerging trend is to explore the out-of-plane dimension to increase the packaging density within the constraints of the form factor and package height limitations. Device stacking and package stacking are becoming increasingly popular. An understanding of the failure modes and mechanisms of these packages on a variety of laminate materials and their construction under thermal and mechanical loading is still in its infancy. Package size, materials and construction, die size and thickness, the order of the stacking, and the bonding methods used can all have significant impact on the failure nature and mechanisms. Failures can range from package damage such as popcorning, to silicon die damage, interconnection failures, delamination, laminate cracking, etc. Industry trends indicate that with thinner die, such as 50–70 μm thin die, packages with as many as six to seven dies stacked together can be anticipated in the near future.

Surface finish

Surface finish of PWBs and the package termination plays a significant role in the integrity and reliability of an interconnection. Hot-air solder leveling, which has been the main PWB surface finish for well over half a century, has outlived its usefulness since the advent of high I/O fine-pitch surface-mount and area-array packaging technology. Several surface finishes have since come into use. OSPs and electroless nickel immersion gold (ENIG) have almost replaced solder leveling. ENIG has been used extensively owing to its long shelf life and excellent solderability wherever coplanarity requirements are stringent. However, as hardware integration and miniaturization continued, resulting in smaller feature sizes, problems related to defects in ENIG emerge. The hypercorrosivity of immersion gold plating composition and the attendant high phosphorous content can cause sporadic and unpredictable solderability problems (also referred to as black pad) (Biunno 1999). In addition, as portable electronic hardware is more subject to mechanical loading, intermetallic brittle fracture at the solder–pad interface is sometimes encountered. Also, as has been mentioned earlier, it is generally recognized that nickel–tin intermetalllics are more brittle than the copper–tin intermetallics. Often, dual surface finishes are employed, with OSP to preserve solderability and ENIG for electrical contact surfaces.

With ever-increasing emphasis on the implementation of lead-free solders as the interconnection material, surface finishes of PWB, package leads, and terminations are being reexamined to arrive at acceptable alternatives. Immersion silver, immersion tin, palladium, nickel–palladium–gold etc., are being looked at. There does not appear to be a consensus on surface finishes. While each surface finish has its merits, the industry has to weigh the alternatives in terms of cost, performance, and reliability for a given product group.

In the ensuing chapter, several failure mechanisms pertaining to PWBs, packages, and interconnections under a variety of loading conditions are described.

Reliability test practices

Accelerated thermal cycling test practices are influenced not only by the design life and the operating environment but also by the nature of the PWB assembly. In most cases, portable electronic hardware by its very nature has to be small, lightweight, and possess high I/O density. This implies the use of surface laminar circuitry or other HDI PWB technologies. Also, inherent is the use of small, low-profile packages.

It has been reported that the industry-standard temperature cycle profile, where the upper and lower temperature dwells are invariant, leads to an underestimation of fatigue life (Dishong et al. 2002). It is well known that inelastic strain accumulation is generally proportional to fatigue life. It has been suggested that temperature fluctuations during upper dwell times can reduce elastic strain accumulation, and, as such, using minicycles during dwell times will reduce the maximum inelastic strain. The magnitude of such inelastic strain reduction depends on the number of minicycles and their temperature ranges. Thus, selective superimposition of a judicious number of minicycles during the high-temperature dwell may enable a more realistic fatigue life prediction.

In addition, portable electronic hardware involving radio communication features can have components such as power amplifiers and RF devices that may run hotter during operation in addition to the thermal exposure imposed by the environment. The thermal effects in such cases can cause excessive growth of interfacial intermetallics, which may be deleterious to the interconnection integrity. Power cycling tests may be much more appropriate in such cases.

Thermal and mechanical stress exposures in portable electronic hardware are rather frequent and sometimes concurrent in contrast to desktop machines, and the effect on product performance can be significant. For example, the interfacial intermetallic growth, which by itself may not affect the solder joint integrity due to the compliance of the alloy, can progressively degrade the mechanical reliability. Thus, separate thermal and mechanical reliability assessments may not reflect the true product performance, as the synergistic effect is not taken into account. The effect of thermal aging on the mechanical reliability can be significant and should be considered in all reliability assessments.

Practical considerations in reliability evaluation

In the previous chapter, we discussed statistical models commonly used for portable electronic product reliability analysis, and the concepts of the physics of failure pertaining to portable electronic products were reviewed earlier in this chapter. The following discussion of accelerated testing development serves as a bridge to the failures discussed in the next chapter. The practical import of these disparate concepts is that they come together to help develop accelerated reliability tests for portable electronic products to mimic failures that are observed in the field, albeit at much shorter time frames so as to help in relatively quick assessment of the impact on field reliability. It is of paramount importance that the failure mechanisms in accelerated reliability tests are unchanged from field failures. In the following sections, a few relevant acceleration models are discussed with the goal of developing practical reliability tests for portable electronic products. The reliability tests discussed are generally applicable to portable products as a whole and not indicative of actual reliability tests performed at any particular OEM.

Acceleration models

The two primary goals of reliability testing are to (a) discover weaknesses in the design early in the design process and (b) ensure that the product meets the reliability requirements later before release to production. The flavors of reliability tests used for these two purposes are somewhat different because the audience and expectations are different. Reliability tests conducted to discover design weaknesses are geared to yield continuous data, while reliability assurance tests are often structured such that the end result is a clear pass/fail because a pass implies that the product can be released for production. The underlying basis for both these tests is the concept of acceleration, where failures are induced in the laboratory in relatively short times under harsher, but nonetheless prototypical, conditions than would be encountered in the field. The key to making this approach feasible is that the engineer knows how much faster the laboratory test would induce failure relative to the field. In other words, the cornerstone to this approach is a thorough understanding of the acceleration factor and the physical parameters that cause and accelerate degradation of the product. The acceleration model is the relationship or equation that makes this possible by relating the degradation of life to the degradation-inducing variable such as temperature, current, voltage, strain, etc.

In general, there are different ways in which failure or degradation may be accelerated (hastened) in the laboratory relative to field conditions.

1. Increasing the use-rate (Compress Time): This method of acceleration can be explained using the example of a mobile phone or other similar portable product. As explained earlier in this chapter, drop-impact-induced failures are a common occurrence in portable products. Consider, for example, that on average a mobile phone is dropped onto concrete twice a week from a height of 0.5 m (or 104 drops per year). If the product is designed for a median life of 5 years, assuming everyday usage, it would accumulate 520 drops over its lifetime. A simple acceleration of the drop test would then involve subjecting the phone to 0.5-m drops onto concrete repeatedly until 520 drops are completed, which can be done within a few hours. The damage induced over 5 years of normal product usage can be duplicated within a few hours by removing the "non-damaging usage time" and increasing the frequency of drops.
2. Increasing the aging-rate (Accelerate Failure): Increasing the level of variables such as temperature, humidity, etc. can enhance the progression of failure. For example, increasing the humidity can weaken anisotropic conductive adhesive film (ACF) and accelerate electrical failure by increasing the resistance. In solder creep, for example, increasing the temperature of the test will reduce the time to creep rupture.
3. Increasing the stress level (Reduce the Inherent Strength): Units operating at higher temperatures often fail earlier than those operating at lower temperatures because the strength at high temperature drops below the applied stress which may not vary with temperature. In this case, the failure mechanism itself is not accelerated but the strength of the sample is reduced instead.

Often, combinations of these methods are also employed in a single test to achieve higher acceleration levels. Variables like temperature and voltage, especially, can have dual impact: increasing the rate of failure or aging, and also reducing the strength relative to the stress, and there may not be enough knowledge of physical or chemical processes to adequately describe the acceleration. In such cases, empirical models can, if used judiciously, be useful for extrapolation to use conditions.

As mentioned earlier, the premise of accelerated testing is that the devices are subject to high levels of the accelerating variables to hasten the failure processes and then these results are extrapolated to lower usage levels using acceleration models. However, failure processes are often quite complex, and the relationship between the state variables and the actual failure mechanism may be very complicated. In such cases, it may suffice to have a simple model that adequately describes only the rate-limiting steps in the overall process as long as it is adequate for extrapolation. At other times, even when there is little understanding of the chemical or physical processes leading to failure, empirical models may be the last resort. With empirical models, extrapolations outside the available experimental data are fraught with inaccuracies and must be undertaken with great care. Some acceleration models commonly encountered in electronic packaging and reliability are described below.

1. Arrhenius Model (Temperature)

The Arrhenius equation is a widely used, simple but remarkably useful function relating the reaction rate to temperature. Typically, the relationship will not be applicable to all temperature–acceleration issues but may be an adequate descriptor over only a limited temperature range. The reaction rate (K) is given by

$$K = Ae^{-E\alpha/kT} = Ae^{\left(\frac{-E_\alpha \times 11{,}605}{T}\right)} \tag{8–9}$$

where A is the constant characteristic of the material and test,

E_α is the activation energy in eV,

k is the Boltzman constant ($8.6171 \times 10^{-5} = 1/11{,}605$), and

T is the temperature in Kelvin.

The Arrhenius acceleration factor between a use temperature T^{use} and reaction temperature T^{lab} is given by the relationship

$$\mathrm{AF} = \frac{k^{\mathrm{lab}}}{k^{\mathrm{use}}} = \exp\left[\mathrm{E}_\alpha\left(\frac{11{,}605}{T^{\mathrm{use}}} - \frac{11{,}605}{T^{\mathrm{lab}}}\right)\right] \tag{8–10}$$

2. Eyring Model (Temperature)

The Eyring relationship is based on physical concepts such as molecular collisions, etc. and can be considered an extension of the Arrhenius equation. The reaction rate (K) is given by the equation

$$K = A\left(\frac{kT}{h}\right)^{m} e^{-E\alpha/kT} \tag{8–11}$$

where A is the material and reaction constant,
k is the Boltzman constant,
h is Planck constant,
T is the temperature in K, and m is a constant (0 or 1).
The acceleration factor is given by the equation

$$AF^{\text{Eyring}} = \left(\frac{T^{\text{lab}}}{T^{\text{use}}}\right)^{m} \times AF^{\text{Arrhenius}} \tag{8–12}$$

As can be seen from Eq. (8–4), the acceleration factors predicted by the Eyring model are larger than predicted using the Arrhenius model, and, therefore, extrapolations to use temperatures are more conservative using the Arrhenius model.

3. Inverse Power Relationship (Voltage)

Increasing voltage or electric field is commonly used to accelerate failure of electrical insulation, light bulbs, capacitors, transformers, heaters, etc. The voltage stress across a dielectric is measured in units of V/mm or kV/mm. The inverse power relationship is commonly used to predict failure times at the use voltage based on experimental data collected at elevated laboratory voltages, and the AF relationship can be expressed as

$$T^{\text{lab}} = \left(\frac{V^{\text{lab}}}{V^{\text{use}}}\right)^{\beta} T^{\text{use}} \tag{8–13}$$

where T^{lab} and T^{use} are the failure times under lab voltages and use voltages, respectively,

V^{lab} and V^{use} are the voltages in the lab and at use conditions, β is generally <0, and the AF is given by

$$AF = \frac{T^{\text{use}}}{T^{\text{lab}}} = \left(\frac{V^{\text{lab}}}{V^{\text{use}}}\right)^{-\beta} \tag{8–14}$$

4. Lawson Model (Humidity)

Moisture has a significant effect on the degradation of electronic products and packages, as it gives rise to corrosion, electrochemical migration, leakage, etc. One of the models proposed to estimate the life of electronic devices is by Lawson (1984). The life of the device according to this model is given by

$$t^{fail} = Ae^{E_\alpha/kT} e^{bRH^2} \tag{8–15}$$

where t^{fail} is the time to failure,

E_a is the activation energy,

k is the Boltzman constant,

A is a material and test constant,

$b = 0.00044$,

RH is the relative humidity, and

T is the temperature in K.

The acceleration factor is given by

$$AF = \frac{t_{\text{use}}^{\text{fail}}}{t_{\text{lab}}^{\text{fail}}} = \frac{e^{E_a/kT_{\text{use}}} e^{bRH_{\text{use}}^2}}{e^{E_a/kT_{\text{lab}}} e^{bRH_{\text{lab}}^2}} = e^{E_a\left(\frac{11605}{T_{\text{use}}} - \frac{11605}{T_{\text{lab}}}\right)} e^{b(RH_{\text{use}}^2 - RH_{\text{lab}}^2)} \tag{8–16}$$

5. Eyring Extended Model (Humidity)

The Eyring temperature model has been extended to account for humidity in two different forms as given below:

$$t^{\text{fail}} = Ae^{E_\alpha/kT} e^{BRH} \tag{8–17}$$

$$t^{\text{fail}} = Ae^{E_\alpha/kT} e^{B/RH} \tag{8–18}$$

where t^{fail} is the time to failure,

E_a is the activation energy, 0.60 eV,

k is the Boltzman's constant,

A is a material and test constant,

$B = 0.304$,

RH is the relative humidity, and

T is the temperature in K.

6. Peck Model (Humidity)

The Peck model employs a logarithmic function of humidity and is given by

$$t^{\text{fail}} = Ae^{[E_\alpha/kT + B\,1\text{n}(RH)]} \tag{8–19}$$

where t^{fail} is the time to failure,

E_α is the activation energy, 0.54 eV,
k is the Boltzman constant,
A is a material and test constant,
$B = -4.55$,
RH is the relative humidity, and
T is the temperature in K.

The AF is given by

$$\text{AF} = \left(\frac{t_{\text{use}}^{\text{fail}}}{t_{\text{lab}}^{\text{fail}}}\right) = \left(\frac{RH_{\text{use}}}{RH_{\text{lab}}}\right)^{-4.55} \times \exp\left[\frac{E_\alpha}{\text{k}}\left(\frac{1}{T_{\text{use}}} - \frac{1}{T_{\text{lab}}}\right)\right] \tag{8–20}$$

7. Coffin–Manson Model (Fatigue)

The original model was proposed to describe the life under strain-controlled isothermal fatigue conditions based on the premise that the life is proportional to the plastic strain in each fatigue cycle, and is given by

$$Nf(\Delta\varepsilon_p)^2 = C \tag{8–21}$$

where Nf is fatigue life,

$\Delta\varepsilon_p$ is the plastic strain, and C is a constant.

The Coffin–Manson model, which was derived from the strain-controlled fatigue behavior of steels, is not directly applicable to thermomechanical fatigue of solder alloys because creep dominates the deformation of lead and tin alloys, even at room temperature.

8. Norris–Landsberg modified Coffin–Manson Model (Thermo-mechanical Fatigue)

To account for the relatively high homologous temperatures, Norris and Landsberg modified the Coffin–Manson equation as

$$N_f = f^{1/3}(\Delta T)^{1.9} e^{0.01/T} \tag{8–22}$$

where f is the frequency of thermal cycling (per day),
T is the maximum temperature in K,
and ΔT is the temperature range of thermal cycling.

The AF is given by the equation

$$AF = \frac{N_f^{\text{use}}}{N_f^{\text{lab}}} = \left(\frac{\Delta T^{\text{use}}}{\Delta T^{\text{lab}}}\right)^{1.9} \left(\frac{f^{\text{use}}}{f^{\text{lab}}}\right)^{1/3} e^{0.01(T_{\max}^{\text{lab}} - T_{\max}^{\text{use}})} \tag{8–23}$$

Constructing reliability tests for portable electronic products

In this section, we shall construct example reliability tests and calculate the acceleration factors using the discussion presented earlier. For example, consider that the portable product is subject to the user conditions given in table 8–3. Also, assume that it takes two days to drop the product repeatedly 150 times. Based on the conditions assumed in table 8–2 and the AF calculated using the equations above, a candidate reliability test plan to ensure that the product will survive for three years in the field might consist of the following tests:

- 150 drops from a height of 0.5 m onto concrete
- 440 thermal cycles from –20 to 60°C, and
- 82 hr of exposure to RH 95% at a temperature of 55°C.

The same tests could be used differently, depending on whether it is early in the design process or just before release to production. For finding weaknesses in the design, for example, the product drop test is conducted in an interrupted fashion, with read points every 50 drops where the functionality of the device is checked. In addition, the test may be carried out to 300 drops, which is well past the design life of the product. A Weibull plot is constructed from the failure data, and failure analysis is performed not only on the failed units but also on the surviving units in a process of discovery. Whatever weaknesses are found, the design, materials, or the manufacturing process is modified to eliminate the failures. On the other hand, if the product is close to production, a qualification test may be conducted in which the product is dropped only 150 times and a pass/fail result is obtained depending on the criterion. If a pass result is obtained, the product is released to production.

Table 8–3. Assumed conditions for a portable electronic product

Environment	Field Use Conditions	Number of hours or Cycles/year	Reliability Test Conditions	AF	Requirement equivalent to 3 years
Drop Impact	0.5 m	50	0.5 m	—	150 drops
Thermal Cycling	-10° to 40°C	365	-20° to 60°C	2.5	440 cycles
Temperature Humidity	RH = 55 T = 26°C	2080 (52x40)	RH = 95, T =55, Ea=0.54eV, B = -4.55	76.5	82 hours

Summary

As portable consumer electronic hardware becomes more complex with multitudes of functions and increased data handling capacity, further miniaturization and higher levels of integration at all levels of packaging will be a natural trend. The reliability demands will be higher to ensure customer satisfaction and product acceptance. The implications for reliability, failure, and root cause analysis will be significant. More functions will be integrated into the device. The silicon device thickness will be in the range of 40–50 μm. Stacked devices as well as folded and stacked packages will be more prevalent with a combination of multiple levels of wire bonding and/or flip chip interconnection. Another emerging trend in packaging is the three-dimensional integration at the wafer level. New materials that will have better mechanical properties and moisture resistance will be developed. More functions will be embedded into the PWB and these may include active, passive, and optical devices, together with new embedded interconnection schemes. The PWB technology itself will witness revolutionary changes with thinner and improved materials capable of 10–25 μm vias, 10–20 μm lines and spaces, and structures involving several layers of stacked vias. Consequently, hitherto unknown failure mechanisms are likely to be encountered. As the feature sizes diminish, the distinction between first- and second-level packaging becomes nebulous. Failure analysis, even at the PWB assemblies, will be a formidable challenge. With shorter product development cycles and faster-to-market business environment, the need for more automated analytical tools with minimal operator intervention for rapid and repeatable root cause analysis will increase. Innovative reliability test practices will be needed to shorten the test durations to accommodate faster development schedules.

References

Amagai, M., Y. Toyoda, T. Ohinishi, and S. Akita, 2004. High drop test reliability of lead free solders. Paper presented at the Proc. IEEE Electronic Components and Technologies Conf.: 1304–1309.

Benson, R. C., B. M. Romanesko, J. E. Weiner, B. N. Nall, and H. K. Charles. 1988. Metal electromigration induced by solder flux residue in hybrid microcircuits. *IEEE Trans. Components, Hybrids, and Manufacturing Technology* 11(4): 363–370.

Biunno, N. 1999. A root cause failure mechanism for solder joint integrity of nickel/immersion gold surface finishes. *IPC Printed Circuits Expo*, Long Beach, CA: 1–9.

Canumalla, S., S. Shetty, and N. Hannan. 2002. Effect of corner-underfill voids on the chip scale package (CSP) performance under mechanical loading. Paper presented at the Proc. 28th Int. Symp. Society for Testing and Failure Analysis, November 3–7, Phoenix, AZ: 361–370.

Canumalla, S., H. D. Yang, Viswanadham, and T. Reinikainen. 2004. Package to board interconnection shear strength (PBISS): Effect of surface finish, PWB build-up layer and chip scale package structure. *IEEE Trans.Components and Packaging Technologies*. 27 (1): 182–190.

Canumalla, S. 2004. Test fixture and method, U.S. Patent No. 6681640.

Chiu, T. C., K. Zeng, R. Stierman, D. Edwards, and K. Ano, 2004. Effect of thermal aging on board level drop reliability for Pb-free BGA packages. Paper presented at the Proc. IEEE Electronic Components and Technologies Conf.: 1256–1262.

Darveaux, R. 1992. Constitutive relations for tin-based solder joints. *IEEE Trans. Component, Hybrids, and Manufacturing Technology* 15 (6): 1013–1024.

Darveaux, R., and A. Syed. 2000. Reliability of area array solder joints in bending. Paper presented at the SMTA Int. Symp.

Dermarderosian, A.1978. The electrochemical migration of metals. Paper presented at the Proc. Int.Soc. Hybrid Microelectronics: 134.

Dishong, T., C. Basaran, N. Cartwright, Y. Zhao, and H. Liu. 2002. Impact of temperature cycle profile on fatigue life of solder joints. *IEEE Trans.Components, Packaging, and Manufacturing Technology, Part B: Advanced Packaging* 25 (3): 433–438.

Enke, N. F., T. J. Kilinski, S. A. Schroeder, and J. R. Lesniak. 1989. Mechanical behaviors of 60/40 tin-lead solder lap joints. *IEEE Trans. Components, Hybrids, and Manufacturing Technology* 12 (4): 459–468.

Erich, R., R. J. Coyle, G. M. Wenger, and A. Primavera. 1999. Shear testing and failure mode analysis for evaluation of BGA ball attachment. Paper presented at the IEEE/CPMT Int. Electronic Manufacturing Technology Symp.: 16–22.

European Telecommunication Standard (ETS). 1994. 300-019-1, Classification of Environmental Conditions.

Gere, J. M., and S. Timoshenko. 1990. *Mechanics of Materials*.

Goyal, S., S. Upasani, and D. M. Patel. 1999. The role of case-rigidity in drop-tolerance of portable products. *Int. J. Microcircuits and Electronic Packaging* 22 (2): 175–184.

Goyal, S., and E. B. Buratynski.. 2000. Methods for realistic drop testing. *Int. J. Microcircuits and Electronic Packaging* 23 (1): 45–52.

Goyal, S., E. B. Buratynski, and G. W. Elko. 2000. Role of shock response spectrum in electronic product suspension design. *Int. J. Microcircuits and Electronic Packaging* 23 (2): 182–190.

Greg, P., P. Chen, and X. Ling. 2002. Effect of strain rate on solder joint failure under mechanical bond; IEEE Electronic Components and Technol. Conf.: 971–978.

Grunhaner, F. G., T. W. Griswold, and P. J. Clendening. 1975. Migratory gold resistive shorts: Chemical aspects of failure mechanism. Paper presented at the Proc. Int. Reliability Physics Symp.: 99.

Guttenplan, J. D. 1987. Corrosion in the electronics industry. *ASM Metals Handbook*, 9th ed. 3 vol. Metals Park, OH: ASM International.

Hanabe, M., and S. Canumalla. 2004. Package to board interconnection shear strength (PBISS) behavior at high strain rates approaching mechanical drop. Paper presented at the IEEE Electronic Components and Technology Conf.: 1263–1270.

Hanabe, M., S. Canumalla, and A. Kujala, 2006. Enhancing drop test efficacy for product quality improvement through interconnection shear tests. Paper presented at the Electronic Components and Technologies Conference (ECTC), May 28–30, San Diego, CA: 72–77.

Hannan, N., and P. Viswanadham. 2001. Critical aspects of reworkable underfills for portable consumer products. Paper presented at the Proc. IEEE Electronic Components and Technologies Conf.: 181–187.

Harada, K., S. Baba, Q. Wu, H. Matsushima, T. Matsunaga, Y. Uegai, and M. Kimura. 2003. Analysis of solder joint fracture under mechanical bending test. Paper presented at the IEEE Electronic Components and Technologies Conf.: 1731–1737.

Harsanyi, G. 1995. Electrochemical processes resulting in migrated short failures in microcircuits. *IEEE Trans. Components, Packaging and Manufacturing Technology – A* 18 (3): 602–610.

Harsanyi, G., and G. Inzelt. 2001. Comparing migratory resistive short formation abilities of conductor systems applied in advanced interconnection systems. *Microelectronics Reliability* 41: 229–237.

International Electrotechnical Commission (IEC). 2002. 60721-3-7, Classification of groups of environmental parameters and their severities – Portable and Non-stationary use.

JMP ä v7, SAS Institute, NC, USA.

JESD94.01, Application specific qualification using knowledge based test methodology, May 2007.

Jonnalagadda, K. 2002. Reliability of via-in-pad structures in mechanical cycling fatigue. *Microelectronics Reliability* 42: 253–258.

Jonnalagadda, K., M. Patel, and A. Skipor. 2002. Mechanical bend fatigue reliability of lead-free PBGA assemblies. Paper presented at the Eighth Intersociety Conference on Thermal and Thermomechanical Phenomena in Electronic Systems, (ITHERM 2002): 915 –918.

Juso, H., Y. Yamaji, T. Kimura, K. Fujita, and M. Kada. 1998. Board level reliability of CSP. Paper presented at the IEEE Electronic Components and Technologies Conf.: 525–531.

Kanchanomai, C., Y. Miyashita, Y. Mutoh, and S. L.Mannan. 2003. Influence of frequency on low cycle fatigue behavior of Pb-free solder 96.5Sn/3.5Ag. *Materials Science and Engineering* A 345: 90–98.

Kanchanomai, C., Y. Miyashita, and Y. Mutoh. 2002. Low cycle fatigue behavior and mechanisms of a eutectic Sn–Pb solder 63Sn/37Pb., *Int. J. Fatigue* 24:671–683.

Kawanabe, T., and K. Otsuka. 1982. Metal migration in electronic components. Paper presented at the Proc. Electronic Components Conf.: 99.

Kohman, G. T., H. W. Hermance, and G. H. Downes. 1955. Silver migration in electrical insulation. *Bell Sys. Technol.* J. 34: 1115.

Krumbein, S. J. 1988. Metallic electromigration phenomena. *IEEE Trans. Components, Hybrids, and Manufacturing Technology* 11 (1): 5–15.

Kujala, A., T. Reinikainen, , and W. Ren. 2002. Transition to Pb-free manufacturing using land grid array packaging technology. Paper presented at the Proc. IEEE Electronic Components and Technologies Conf.: 359–364.

Lall:, D. Panchagade, Y. Liu, W. Johnson, and J. Suhling. 2004. Models for reliability prediction of fine-pitch BGAs and CSPs in shock and drop impact. Paper presented at the Proc. IEEE Electronic Components and Technologies Conf.: 1296–1303.

Lando, D. J., J. P. Mitchell, and T. L. Welsher. 1979. Conductive anodic filaments in reinforced polymeric dielectrics: Formation and prevention. Paper presented at the Int. Reliability Physics Symp.: 51–63.

Lawson, W. 1984. A review of the status of plastic encapsulated semiconductor component reliability. *British Telecom. Tech.* J. 2: 95–111.

Lee, B.-Z., and D. N. Lee. 1998. Spontaneous growth mechanism of tin whiskers. *Acta Materialia* 46 (10): 3701–3714.

Leicht, L., and A. Skipor. 1998. Mechanical cycling fatigue of PBGA package interconnects. Paper presented at the Proc. 30th Int. Symp. Microelectronics, October 14—16: 802–807.

Lim, C. T., and Y. J. Low. 2002. "Investigating the drop impact of portable electronic products. Paper presented at the Proceedings of the IEEE Electronic Components and Technologies Conf.: 1270–1274.

Lim, C. T., C. W. Ang, L. B. Tan, S. K. W. Seah, and E. H. Wong. 2003. Drop impact survey of portable electronic products, Paper presented at the Proc. IEEE Electronic Components and Technologies Conf.: 113–120.

McDowell, M. E. (1993). Tin whiskers: A case study. Paper presented at the Aerospace Applications Conf.: 207–215.

Meeker, W. Q., and L. A. Escobar. 1988. *Statistical Methods for Reliability Data*. New York: Wiley.

Meeker, W. Q., and M. Hamada. 1995. Statistical tools for the rapid development and evaluation of high-reliability products. *IEEE Trans. Reliability* 44 (2): 187–198.

MIL-HBK-310. 1997. Global Climactic data for Developing Military Products.

Mishiro, K., S. Ishigawa, M. Abe, T. Kumai, Y. Higashiguchi, and K. Tsubone, 2002. Effect of the drop impact on BGA/CSP package reliability. *Microelectronics Reliability* 42: 77–82.

Myers, R. H., and D. C. Montgomery. 2002. *Response Surface Methodology: Process and Product Optimization Using Designed Experiments*. New York: Wiley.

Ong, Y. C., V. P. W. Shim, T. C. Chai, and C. T. Lim, 2003. Comparison of mechanical response of PCBs subjected to product-level and board-level drop impact tests. Paper presented at the Proc. IEEE Electronic Components and Technologies Conf.: 223–227.

Perrera, U. D. 1999. Evaluation of reliability of mBGA solder joints through twisting and bending. *Microelectronics Reliability* 39: 391–399.

Raffalovich, A. J. 1971. Corrosive effects of solder flux on printed circuit boards. *IEEE Trans.Components, Hybrids, and Manufacturing Technology* 7 (4): pp. 155–162.

Ripka, G., and G. Harsanyi. 1985. Electrochemical migration in thick-film conductors and chip attachment resins. *Electrocomponents Science and Technology* 11: 281.

Rooney, D., N. T. Castello, M. Cibulsky, D. Abbott, D. J. Xie. 2003. Materials characterization of the effect of mechanical bending on area array package interconnects, *Microelectronics Reliability* 39: 463–477.

Saha, S. K., S. Mathew, and S. Canumalla. 2004. Effect of intermetallic phases in mechanical drop environment: 96.5Sn3.5Ag Solder on Cu and Ni/Au pad finishes. Paper presented at the IEEE Electronic Components and Technologies Conf.: 1288–1295.

Sbar, N. L. 1976. Bias humidity performance of encapsulated and unencapsulated Ti-Pd-Au thin film conductors in an environment contaminated with Cl_2. *IEEE Trans. Parts, Hybrids, and Packaging* 12: 176.

Schetty, R. 2000. Minimization of tin whisker formation for lead free electronics finishing. Paper presented at the IPC Works Conference, Miami, FL:

Seah, S. K. W., C. T. Lim, E. H. Wong, V. B. C. Tan, and V. P. W. Shim. 2002. Mechanical response of PCBs in portable electronic products during drop impact. Paper presented at the Proc. IEEE Electronic Components and Technologies Conf.: 120–125.

Shetty, S., A. Dasgupta, V. Halkola, V. Lehtinen, and T. Reinikainen. 2000. Bending fatigue of chip scale package interconnects. Paper presented at the ASME Int. Mechanical Engineering Congress and Exposition, Orlando, FL, November 5—10: .

Shohji, T. Yoshida, T. Takahashi, and S. Hioki, 2004. Tensile properties of Sn-Ag based lead-free solders and strain rate sensitivity. *Materials Science and Engineering A* 366: 50–55.

Shumka, A., and R. R. Piety. 1975. Migrated gold resistive shorts in microcircuits. Paper presented at the Proc.Int. Reliability Physics Symp.: 93–98.

Siplon J. P., G. J. Ewell, E. Frasco, J. A. Brusse, and T. Gibson. 2002. Tin whiskers on discrete components: The problem. Paper presented at the 28th Int. Symp. Society for Testing and Failure Analysis, November 3–7, Phoenix, AZ: 421–434.

Stout, E. A., N. R. Sottos, A. F. Skipor. 2000. Mechanical characterization of plastic ball grid array package flexure using moire interferometry. *IEEE Trans. Advanced Packaging* 23 (4): 637–645.

Suhir, E. 1991. Nonlinear dynamic response of a flexible printed circuit board to a shock load applied to its support contour. Paper presented at the IEEE Electronic Components and Technology Conf.: 388–399.

Suhir, E. and R. Burke. 1994. Dynamic response of a rectangular plate to a shock load, with application to portable electronic products. IEEE Trans. *Components, Packaging, and Manufacturing Technology, Part B: Advanced Packaging* 17 (3): 449–460.

Suhir, E. 1995. Shock protection with a nonlinear spring. *IEEE Trans. Components, Packaging, and Manufacturing Technology, Part A* 18 (2): 430–437.

Suhir, E. 1996. Dynamic response of a one-degree-of-freedom linear system to a shock load during drop tests: Effect of viscous damping. *IEEE Trans. Components, Packaging, and Manufacturing Technology, Part A* 19 (3): 435–440.

Suhir, E. 1997. Is the maximum acceleration an adequate criterion of the dynamic strength of a structural element in an electronic product? *IEEE Trans. Components, Packaging, and Manufacturing Technology, Part A* 20 (4): 513–517.

Suhir, E. 2002. Could shock tests adequately mimic drop test conditions? Paper presented at the IEEE Electronic Components and Technology Conf.: 563 – 573.

Takemoto, T., R. M. Latanison, T. W. Eagar, and A. Matsunawa. 1997. Electrochemical migration tests of solder alloys in pure water. *Corrosion Science* 39 (9): 1415–1430.

Tee, T. Y., H. S. Ng, D. Yap, X. Baraton, and Z. Zhong. 2003. Board level solder joint reliability modeling and testing of TFBGA packages for telecommunication applications. *Microelectronics Reliability* 43: 1117–1123.

Tee, T. Y., H. S. Ng, C. T. Lim, E. Pek, and Z. Zhong. 2004. Impact life prediction modeling of TFBGA packages under board level drop test. *Microelectronics Reliability* 43: 1131–1142.

Tian, G., Y. Liu, R. Lall W. Johnson, S. Abderrahman, M. Palmer, N. Islam, and J. Suhling, 2003. Drop reliability of corner bonded CSP in portable products. Paper presented at the Int. Electronic Packaging Technical Conf., July 6–11, Hawaii:

Taguchi, G. 1986. *System of Experimental Design: Engineering Methods to Optimize Quality and Minimize Cost.* White Plains, NY: UNIPUB.

Toleno, B. J., and J. Schneider. 2003. "Processing and reliability of corner bonded CSPs. Paper presented at the Int. Electronics Manufacturing Technology Symp.: 299–304.

Tullmin, M., and Roberge: R. (1995). Corrosion of metallic materials, *IEEE Trans. Reliability*, Vol. 44: 271–278.

Turbini, L.J., J. A. Jachim, G. A. Freeman, and J. F. Lang. 1992. Characterizing water soluble fluxes: surface insulation resistance vs. electrochemical migration. IEEE International Mfg. Technol. Symp. 80–84

Vakevainen, K., J. Hannula, and K. Ojala. 2001. Determination of Usage Environment of Mobile Phones. Paper presented at the Proc.ESTECH 2001 Symposium, Institute of Environmental Sciences and Technology, Phoenix, AZ:

Viswanadham:, and P. Singh. 1997. *Failure Modes and Mechanisms in Electronics Packages.* New York: Chapman and Hall.

Wassink, R. J. K. 1989. *Soldering in Electronics.* Great Britain: Electrochemical Publications, Ltd.

Wu, J., G. Song, C. P. Yeh, and K. Wyatt. 1998. Drop impact simulation and test validation of telecommunication products. Paper presented at the Intersociety Conference on Thermal Phenomena: 330–336.

Xie, D., M. Arra, S. Yi, and D. Rooney. 2003. Solder joint behavior of area array packages in board level drop for hand held devices. Paper presented at the IEEE Electronic Components and Technology Conf.: 130–135.

Yasuda, K., S. Umemura, and T. Aoki. 1987. Degradation mechanisms in tin- and gold-plated connector contacts. *IEEE Trans. Components, Hybrids, and Manufacturing Technology* 10 (3): 456–462.

Zeng, K., and K. N. Tu. 2002. Six cases of reliability study of lead-free solder joints in electronic packaging technology. *Materials Science and Engineering Reviews* 38: 55–105.

Zhu, L. 2003. Modeling technique for reliability assessment of portable electronic product subjected to drop impact loads. Paper presented at the Proc. IEEE Electronic Components and Technologies Conf.: 100–104.

9

Failures and Prevention

Introduction

Failure analysis (FA), in general, refers to the process of analyzing failures with a view to identifying the root cause for eliminating failures. Physical failure analysis (PFA) of portable electronic products is the focus of this chapter. A good understanding of the taxonomy of failure analysis investigations is essential for effective failure analysis. The surest way to remove the effect (the failure) is to eliminate the root cause of that failure. It is not sufficient to understand what has failed (failure mode) or even how a failure has occurred, namely, the failure mechanism. What is really needed is a very clear understanding a root cause of why a particular failure has occurred (root cause).

Often, the handoff in failure analysis activity occurs at or slightly prior to the failure mode being determined. PFA is sometimes conducted on samples that have not yet failed, to understand the state of damage and construct the failure mechanism progression even though "electrical failure" has not yet occurred.

Electronic packaging is a multidisciplinary field and involves coordinating team members from the branches of electrical engineering, mechanical engineering, materials science, physics, chemistry, reliability, and mechanics. Likewise, successful failure analysis also requires people with diverse competencies working together.

Definitions in Failure Analysis

The taxonomy of failure analysis is as follows:

1. Failure Symptom: This is the first indication to the user that the product is not performing its intended function. For example, a blank display on a mobile phone is a failure symptom.

2. Failure Mode: This is the manner in which the failure has occurred. Often, the failure mode is arrived at by asking the question "What?" In any particular product, there may be only a few different failure modes operational at the same time. For example, the symptom of a blank display can be a result of a number of different failure modes such as broken display, broken display driver (Si), broken chip-scale package (CSP) solder joint on the printed wiring board (PWB), corroded indium tin oxide (ITO) traces on the display module, fracture of the display flex connection, etc. However, only ITO corrosion and broken display can dominate the failure process, whereas other possible modes can be recessed. As can be seen from the failure modes listed (partial list), more than one failure mode can be operative simultaneously in a particular product and still cause the same failure symptom. However, it should be remembered that knowledge of a failure mode is not sufficient to remedy the issue.
3. Failure Mechanism: The physical, chemical, or other phenomenon that controls the failure is known as the *failure mechanism*, and there are often many failure mechanisms for each failure mode. The failure mechanism is often arrived at by asking the question "How?" As in the earlier example of a blank display in mobile phone, the broken display can be a result of several failure mechanisms such as impact fracture of the glass display, fatigue fracture of the display, monotonic bending fracture, etc. While the failure mechanism gets us one step closer to the cause of the failure, even knowledge of the failure mechanism alone is not enough to fix the issue.
4. Root cause: The root cause of failure is arrived at by repeatedly asking the question "Why?" This is the reason or circumstances by which failure has occurred. Therefore, fixing the root cause necessarily has to eliminate the failure mechanism. In this case, lack of design margin, lack of protection against environmental stresses, or use of inappropriate materials or processes can be the root cause. For example, using a display glass with poor edge-cutting quality that has microcracks larger than a critical size will reduce the fracture resistance of the display module to usage loads. If the failure mechanism of the broken glass corresponds to cracks emanating from the edges, the root cause is often the poor glass-cutting quality.

Failure Analysis of Portable Electronics Products

Computing and communication have become mobile and, indeed, pervasive. Increasingly, these portable consumer electronics devices are characterized by a higher density of packaging and by higher levels of integration. Although these devices sometimes possess the computing power and capabilities of desktop and office devices, they are vastly different from a technological and reliability perspectives. Concomitantly, some of the failure mechanisms are also different than those encountered in conventional communication or office computing realms.

The technology and business drivers influencing the failure analysis needs of the mobile and handheld products are discussed briefly first. From the PWB technology point of view, the trend is towards (a) thinner layers, (b) extremely high-density circuit lines and spaces (100-μm or finer line and spaces), and (c) finer vias (75–100 μm) that are drilled with nonmechanical means. This generates a need for embedded devices and stacked multilayer vias. In terms of component technologies, lower standoff and lower profile packages with thinner silicon are being used. As the PWB real estate demands grow, three-dimensional device stacking with 50–75-μm-thick silicon will be used more and more. In addition, circuit card assemblies will be double-sided to maximize the circuit card density. The industry will be challenged to eventually migrate to concepts such as system-on-package (SOP) and system-on-chip (SOC) through elimination of some levels of packaging.

Owing to their portability, the environment in which hand-held electronic devices operate is often more severe in terms of exposure to thermal excursions, humidity, and corrosive environments. Additionally, portable products are much more prone to mechanical loads that include drop, bend, twist, etc. Further, owing to rapid developments in information processing technology, the product life of hand-held products is shorter than that of conventional office machines. The number of power ON cycles, the operational voltages, and other conditions will be different. Among the various loads that portable electronic hardware experience, the mechanical and humidity loads are perhaps more severe than others. Mechanical drops from heights of up to a meter and half on hard surfaces are not uncommon. Also, the products are more prone to exposure to high-humidity noncondensing and condensing atmospheres.

In a high-volume, low-cost, competitive business environment, time taken to achieve sufficient product quality and reliability take on added significance. Cost-efficient manufacturing and time-to-market need to be considered in addition to product quality owing to high customer expectations. In such an environment, the speed and effectiveness of failure analysis has a direct business impact, and the analysis processes need to be optimized to meet business needs effectively.

Failure Analysis Process

The first step in the process of failure analysis often begins with defining the goal of the investigation and gathering background information. This process, sometimes known as *pre-failure analysis,* defines the scope of the failure analysis investigation. A typical sequence of steps in this pre-failure analysis process is shown in figure 9–1. An adequate field failure data collection process and statistical analysis of the field return data are essential for estimating the product reliability. However, they are not sufficient to improve product reliability unless coupled with a thorough understanding of the underlying failure mechanisms.

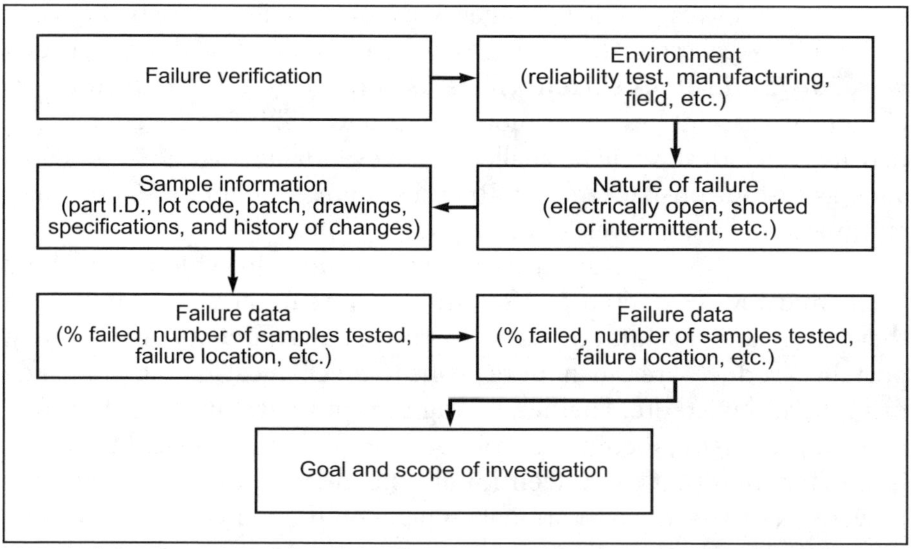

Fig. 9–1. Pre-failure analysis process to determine the goals and scope

First, the statistical validity of the failure is examined to ensure that the effort spent in identifying a root cause will result in financial payback. Root-cause analysis is sometimes severely convoluted in enterprises that are not vertically integrated because the root cause for failure may reside a few steps up or down the supply chain. In such cases, root-cause analysis may have to be redefined in terms of boundaries of direct influence and the burden of further analysis transferred to the next level in the supply chain. Often, this means that first-tier vendors will have to carry the responsibility of failure analysis even if their suppliers perform the actual analysis. For example, an original equipment manufacturer (OEM) (level 1) may find that the cause of failure is the poor quality of a component surface finish on a ball grid array (BGA) module and refer the investigation to the contract manufacturer (CM) (level 2). The CM may determine that the cause of the poor surface finish is lack of adequate cleaning at their supplier (level 3). It is not uncommon to find several levels of the supply chain involved in a failure analysis investigation. The success of such a venture is dictated by good communication up and down the supply chain and the degree to which each level player understands the technologies and processes of other players in the value chain.

A basic failure analysis process is described in figure 9–2. The tools involved in failure analysis include, but are not limited to, microscopes (optical, X-ray, acoustic, infrared, and electron); decapsulation and material removal tools such as ion milling; chemical analysis tools for volumetric and gravimetric analysis; spectroscopy tools (UV/visible, infrared) electron beam analysis tools (Auger, X-ray wavelength, and energy); and laser-induced mass spectroscopy. In addition to these experimental tools, simulation tools are also important in analyzing complex geometries and materials. Some representative failure mechanisms are discussed in this chapter. The recognition that failure analysis is multidisciplinary in nature is important.

Expertise drawn from different disciplines of science and engineering needs to be utilized for successful determination of root cause. A brief introductory review of the techniques commonly employed in failure analysis is presented next to help answer the following questions that typically come up in planning and executing a failure analysis investigation:

1. Do I need bulk analysis or surface analysis?
2. Do I need more than one analysis and will one analysis interfere with a subsequent analysis?
3. What kinds of samples are needed? Do I need “golden” samples? What sample sizes are needed?
4. What analysis methods do I need to answer the questions facing us?
5. Are the techniques selected sensitive enough?

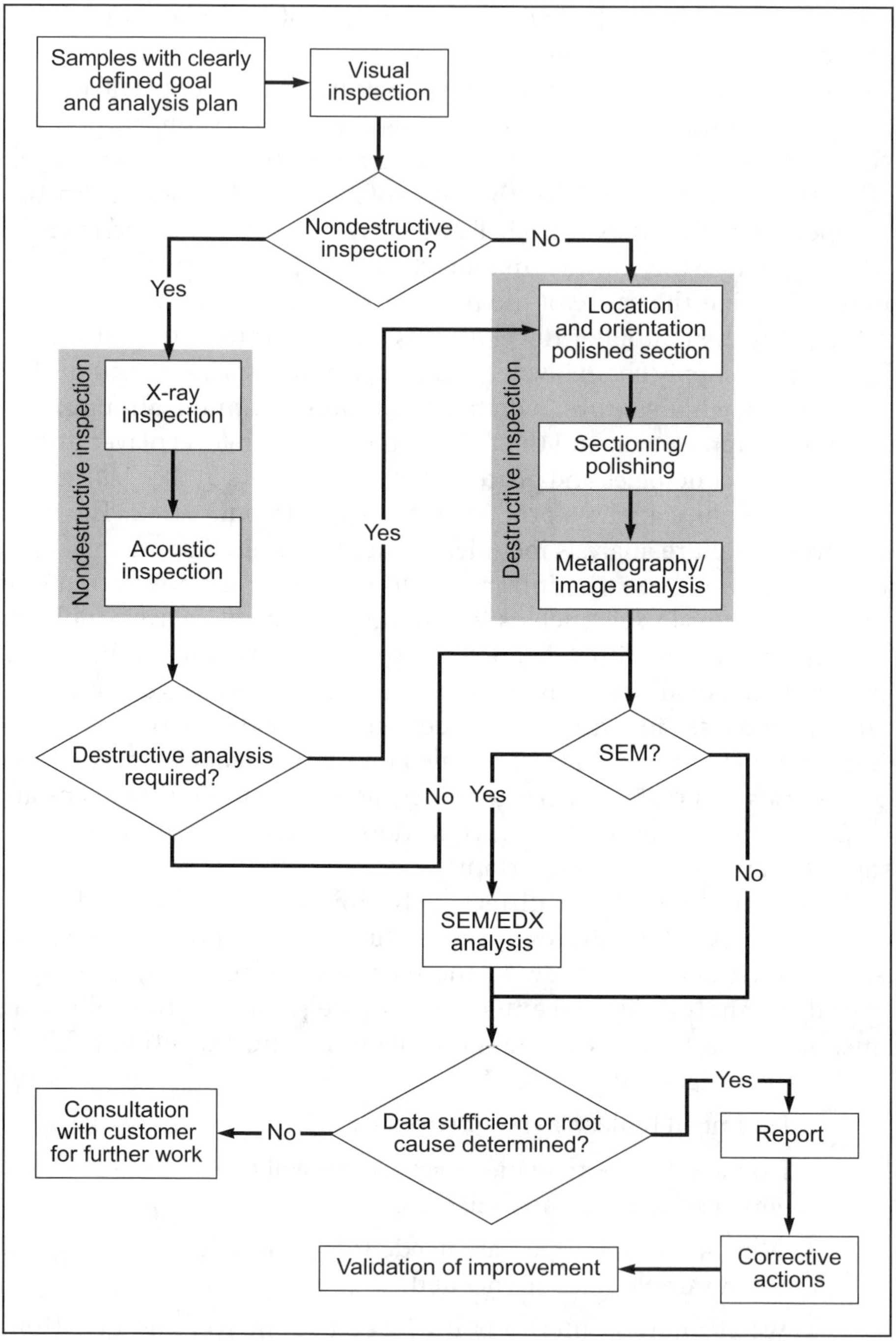

Fig. 9–2. Process flow showing destructive and/or nondestructive analysis steps preceded by visual inspection

An appreciation of the capabilities of the analytical technique also enables a meaningful dialog with the analyst. For example, when a volatile species is involved, using a vacuum-based technique will be inappropriate because the species of interest is scavenged before it can be analyzed and subsequent analysis will only indicate its absence. If the chemical species of interest is not in an ionized state and is covalently bonded, using ion chromatography will indicate the absence of that species, contrary to reality. Therefore, a detailed discussion should take place between the parties seeking the analysis and the analyst. The role of good communication in effective failure analysis cannot be emphasized enough. It is also important that the analyst be provided as much pertinent information on the sample as necessary.

Several sources for in-depth treatment of the individual techniques described below are listed in the references section.

Imaging

Imaging, which is almost always a first step in the analysis, is extremely important in recognizing the most obvious defects, such as cracks, delaminations, blisters, measling, burnouts, deformation, discoloration, contamination, etc. Imaging can be classified into several categories based on the energy used: light (optical), X-ray, ultrasonic (acoustic), and infrared (thermal).

Optical microscopy. The unaided human eye is capable of discerning features as small as 10–15 μm in size, and light microscopes are employed to see smaller features at magnifications ranging from 10× to 5,000× using either transmitted or reflected light. The magnification of most modern compound microscopes is the product of the individual magnifications of the objective and eyepiece lenses. An optimal illumination system is important to obtain the best results with optical microscopes. Stereomicroscopes, which offer magnifications ranging from 10× to 150× have reasonably good depth of field and are suitable as a first inspection tool. They are often used to study the presence of flux residues, corrosion products, metal migration, solder ball shape, etc. Higher magnification optical microscopes, such as metallurgical microscopes, offer magnification up to 5000× and other ancillary techniques such as polarized light, bright field/dark field, and differential interference contrast but at the expense of depth of field. Metallurgical microscopy is used to study polished cross-sections and, when coupled to sophisticated image-analysis systems, they can be used for intermetallic thickness, coating thickness, and area-fraction measurements.

X-ray microscopy. This is a nondestructive method for carrying out inspections of package interiors. The property of different materials to absorb X-rays differently is used for generating contrast in the image. In general, a highly focused beam of electrons impinges on a target block made of tungsten, which gives out a beam of X-rays that irradiate the object of interest. The contrast in the image arises from the fact that regions of lower absorptivity, such as voids, cracks, and organic materials such as molding compounds, resins, and coatings, transmit much of the impinging X-rays in comparison to metals such as bond wires, traces on the PWB, solder balls, bumps, and ceramics. The intensity of transmitted X-rays is captured by a detector array and displayed on a monitor after suitable image enhancement to yield a real-time image. This technique is often called two-dimensional (2D) or real-time X-ray imaging. While extremely useful for detecting voids and suitably oriented cracks, the planar representation makes it difficult to understand the true shape and size of a 3D defect or feature.

In recent years, computed tomography (CT) scans have become popular in medical diagnostics. One drawback of that technique is the requirement that the object be approximately equiaxed in the three dimensions. In conventional CT, the object is rotated on an axis perpendicular to the X-ray beam, and data consists of slices perpendicular to the axis of rotation. However, since the thickness dimension of most electronic products is an order of magnitude smaller than the other two dimensions, obtaining a 3D representation of the object requires modifications to the scanning technique.

In laminography, data are collected by using an X-ray source and detector that rotate in synchronized motion, 180° out of phase with one another, while the subject remains stationary. Images are collected continuously throughout the rotation. The images obtained for an arbitrary number of revolutions are stitched together into a final image. Object features that are in the focal plane are prominent in the final image; object features that are out of the focal plane are blurred. A series of such "laminograms" of neighboring planes can be assembled to yield a 3D representation of the subject. Features as small as 25 μm can be imaged (Moore, Vanderstraeten, and Forssell 2002).

Nondestructive reconstruction of solder joint geometry and defects is a difficult but worthy goal in X-ray inspection. Electronic assemblies are characterized by planar dimensions that are 2–3 orders of magnitude larger than the thickness, and this makes it difficult to reconstruct the solid by conventional CT. A further complication is that many of the defects that affect reliability, such as controlled collapse chip connection

(C4) bump cracks, have a much larger planar dimension than the through-thickness dimension (perpendicular to the PWB). CT reconstruction requires more than 100 images projected from uniformly spaced radial directions around the cross-section with fan-beam X-ray, which is hard to implement in PWB inspection. X-ray digital tomosynthesis, a variation on laminography (Siewert and Mark 1994), has been proposed to address some of these difficulties, and the method consists of using Korhonen neural networks to implement shape correction and intensity correction in two different steps (Roh, Park, and Cho 2003). There is increasing interest in the capabilities offered by 3D X-ray imaging, and this will ensure that 3D X-ray inspection will become more commonplace in the near future.

Infrared/thermal imaging. Thermal or infrared imaging is an imaging technique that operates at wavelengths longer than visible light, and both transmission and reflection modes are used for failure analysis. Further, the inspection can be classified as either active, where the sample is illuminated, or passive, where the sample under inspection is powered on. A reflective infrared microscope (or camera) is useful in bond pad damage detection and facilitates identification of bond pad cracks, corrosion, cratering, etc. (Shell and Golwalkar 1991; Yasuda et al. 1991).

The infrared emission microscopy (IREM) is widely used for the localization of subsurface defects within ICs (Bailon et al. 2003). At the silicon level, backside photoemission imaging of failure sites offers several advantages:

- Failure sites are often hidden beneath metallization, and this is increasingly a problem as more levels of metal are used. Even where emission is visible, the metallization can obscure its origin and complicate interpretation. Hence, backside imaging is better.
- Imaging from the backside may not require potentially destructive decapsulation techniques to be used on the surface of the die.
- For flip-chip mounted devices, backside imaging is the only way of imaging the device without destroying the connections to the outside.

Thermal imaging is also finding increasing use as a failure analysis technique at the board level. Dynamic image subtraction algorithms have been employed in conjunction with passive, large-area infrared imaging of entire portable electronic product PWBs, where the defective sample will show anomalous thermal profiles in comparison to known good units. For example, barrel via cracks in the interposer of power amplifiers would manifest as hot spots due to the higher resistance of the crack faces. Most IR thermography systems use one of two types of detectors: indium antimonide or mercury cadmium telluride (Barton and Tangyungong 2005). Indium antimonide (InSb) detectors are sensitive in the wavelength range 1.5–5.5 μm. Mercury cadmium telluride (HgCdTe) detectors are sensitive over the range of 8–12 μm.

IR thermal imaging systems have acceptable temperature resolution but suffer from a limitation on spatial resolution governed by the wavelength of the radiation. IR systems can easily detect hot areas or spots on integrated circuits at relatively low power densities but may have difficulty resolving features less than about 15 μm. Because of their speed of inspection and fault localization potential, IR systems are attractive for failure analysis of multichip modules, circuit boards, and IC packaging issues, sometimes even entire portable products.

Acoustic microscopy. This ultrasonic-energy-based imaging technique has become an integral part of failure analysis at the semiconductor and package level, and less so at the system level. It is also known as *acoustic microimaging* or *ultrasonic imaging* or *C-scan imaging.* The fundamental basis for contrast in this imaging technique is that ultrasonic energy is reflected, transmitted, or scattered differently by defects than by the surrounding "good" material.

Ultrasonic waves, i.e., sound waves at frequencies higher than 20 kHz, are essentially elastic waves that are affected by the properties of the materials they travel through. Also, unlike sound at audible frequencies, ultrasound does not propagate well in air. So, whenever a sample with a material discontinuity is subjected to insonification at ultrasonic frequencies, every interface reflects some energy back and transmits the rest. Thus, electronic packages, which contain numerous material interfaces such as Si/mold compound, mold compound/Cu, etc., are ideally suited for inspection by acoustic microscopy. Moreover, packaging defects such as delaminations, cracks, void, etc., reflect all of the incident ultrasound and show up as high-contrast features in the images, making acoustic microscopy ideal for defect and failure analysis. A number of articles have been written on the fundamentals of acoustic microscopy for electronic package inspection because of the popularity of the technique in failure analysis (Hartfield and Moore 2005; Kessler and Yuhas 1979, Gordon et al. 1993). Quantitative acoustic microscopy is a relatively recent development where the elastic properties of materials have been measured nondestructively using the acoustic microscope (Canumalla et al. 1997).

The use of quantitative acoustic microscopy in electronic package failure analysis is a relatively young and growing field. For example, it can help measure the elastic properties of underfills (Canumalla, Oravecz, and Kessler 1998) and "fingerprint" molding compounds (Semmens and Canumalla 1999, Canumalla and Kessler 1997). The use of quantitative acoustic microscopy for metrology has also been demonstrated for measuring bond-line thickness of interior thermal interface materials nondestructively (Canumalla and Schackmuth 2000). The success of acoustic microscopy depends on the ability of the acoustic microscopist and the selection of the best possible transducer and imaging parameters. For example, electronic materials such as molding compounds affect the frequency-dependent attenuation of ultrasound, and the selection of an optimum transducer requires both experience and expertise (Canumalla 1999).

The estimation of the delaminated area in acoustic images is relatively straightforward in most cases. Whenever ultrasound is incident at an interface, the reflection coefficient, i.e., the proportion of energy reflected back, is given as

$$R = \frac{Z_2 - Z_1}{Z_2 + Z_1} \tag{9–1}$$

where Z_1 is the acoustic impedance of the material from which the acoustic energy is incident and Z_2 is the acoustic impedance of the second material at the interface. Table 9–1 lists the acoustic impedances and velocities of some materials common in electronic packaging. In acoustic microscopy, by convention the incident pulse is considered positive. From basic principles, when a pulse is incident from a lower acoustic impedance material to a higher acoustic impedance material, the reflection pulse has the same polarity as the incident pulse. Conversely, when the pulse is incident from a higher impedance material to a lower impedance material (solid to air), the pulse reflected from that interface will be inverted in polarity. The amplitude of reflection is governed by the difference in acoustic impedances between materials across the interface. Thus, a delamination will be indicated by a strong reflected pulse with an inverse polarity as compared to the incident one. In common practice, amplitude images obtained by the scanning of the acoustic beam are pseudo-colored by this phase inversion information such that delaminations are colored in red. This simple interpretation, however, breaks down when the delamination is filled with water or in the presence of thin layers at the interface. In such cases, the potential confusion can be cleared by using appropriate broadband pulse propagation models (Canumalla 2005).

Table 9–1. Approximate ultrasonic properties of common electronic materials (Selfridge 1978; Canumalla and Kessler 1997; Canumalla 1999)

Material	Compressional Velocity (Longitudinal) mm/usec	Shear Velocity (Transverse) mm/usec	Density (g/cm^3)	Acoustic Impedance (MRayls)	Attenuation (dB/mm)
Silicon	8.43	5.84	2.34	19.7	
Tin	3.3	1.7	7.3	24.3	
Lead	2.2	0.7	11.2	24.6	
Nickel	5.6	3.0	8.84	49.5	
Copper	5.01	2.27	8.93	44.6	
Gold	3.24	1.20	19.7	63.8	
Aluminum	6.42	3.04	2.7	17.3	
Alumina	10.52		3.86	40.6	
Silicon Carbide	13.06	7.27	3.2	42	
Silicon Nitride	11	6.25	3.27	36	
Silver	3.6	1.6	10.6	38	
6300 HJ (OCN EMC/Sumitomo)	3.29		1.7	5.7	$0.0111f^2 + 0.3859f$
7320CR (BP EMC/Sumitomo)	3.88		1.7	6.8	$0.0113f^2 - 0.0153f$
MP8000CH (OCN EMC/Nitto)	4.07		1.85	7.5	$0.0003f^2 + 0.2277f$
MP190ML (OCN EMC/Nitto)	3.59		1.6	5.5	$0.0276f^2 + 0.0991f$
Silica	5.7	3.75	2.2	12.55	

Decapsulation

Failure analysis of packages sometimes requires the removal of the plastic encapsulation materials to probe the area of interest. The removal of the plastic encapsulant is termed *decapsulation*. In wet decapsulation, the package is subjected to suitable chemical solvents that dissolve the plastic molding compound. Because many of the molding compounds are either glass- or alumina-filled epoxies, the solvents chosen should dissolve only the organic matrix and not affect the Si chip or any other metallurgies so as not to detract from the analysis. The package is immersed in the solvent and is refluxed at a suitable temperature until all the plastic is dissolved and the area to be analyzed is exposed. Solvents used for decapsulation are often a mixture of sulfuric and nitric acid or N-methyl pyrrolidone. There also exist dry decapsulation methods without the use of wet chemicals. The package is etched in a vacuum chamber where an etch gas such as oxygen or carbon tetrafluoride is introduced into the chamber. When RF energy is applied to the gas at reduced pressure, reactive atoms, ions, or free radicals called *plasma* are formed. The plasma reacts with the package material producing volatile species due to particle bombardment resulting in etching of the package. The product gases and vapors are scavenged by the vacuum system attached to the chamber (Beall 1992). The plasma etching technique is also used at the chip level to remove passivation layers or specific metallurgies to arrive at the defect site.

Recently, a laser-assisted technique was developed to reduce the time for decapsulation (Schwindenhammer 2006), in which a Nd:YAG laser at 1,064 nm was used for ablation prior to chemical deprocessing. The result was a much shorter decapsulation time because the "harder" mold compound near the surface of the package was removed using laser decapsulation leaving the "softer" molding compound for the conventional sulfuric acid/nitric acid wet etching. Using the chemical process closer to the die also minimized any ablative damage to the active circuitry from the laser decapsulation process.

Moiré interferometry

This technique is for the analysis of in-plane displacements caused by thermomechanical and mechanical loads (Liu et al. 2004). Moiré interferometry provides whole field maps of in-plane deformation contours with submicron resolution. This technology also provides valuable information on both normal and induced shear strain deformation values. Such a capability is extremely useful for studying the driving forces for delaminations, cracks, and solder fatigue. Moiré interferometry is thus attractive to failure analysis of portable electronic products and electronic packages in elucidating the root cause as well as in failure mechanism description.

Dye penetrant

The dye penetrant method is used extensively to quickly detect cracks in packages. The package or product is immersed or treated with a solution containing a fluorescent or bright-colored dye. The liquid vehicle for this dye is usually selected to have a low viscosity and low boiling point, and the surface tension/capillary forces are taken advantage of to get the dye into the crack or crevice. The sample is then cleaned to remove the excess dye and heated at about 100°C to dry the dye. Subsequently, the sample is examined under UV light to detect the presence of cracks into which the dye has penetrated. Alternatively, the CSP or other soldered component is pulled or pried off the PWB and the fracture surface of the solder joint is examined for dye coverage. This technique is useful in cases where the cracks are open to the surface and the dye can penetrate into the crack. The limitations of this technique are that cracks that are not accessible to the dye are difficult to detect and little metallurgical information can be obtained from the fracture surface other than the area fraction of the cracked joint. However, dye penetrant can be a useful technique when electrical fault isolation is unsuccessful or difficult.

Cross-sections and metallurgical analysis

Metallurgical analysis of cross-sections of solder joints is one of the mainstays of failure analysis, especially at the first- and second-level interconnections. Because the electronics industry employs several metals and alloys, metallurgical analysis has become a very useful technique to understand the failure mechanism and arrive at the root cause of failure. Metallurgical analysis of polished cross-sections is much more than simply identifying the presence of a crack or fracture, and requires a sample with minimal surface damage due to the polishing technique.

The specimen under examination is first cut several millimeters away from the actual plane of interest using a water-cooled diamond abrasive wheel such that the area of interest is not adversely affected by the cutting process. In conventional cross-sectioning, the sample under investigation is potted in an epoxy material that provides support and facilitates handling without causing damage. This slug containing the sample is ground down to the exact location of interest using progressively finer grit papers and often using diamond particle suspension or submicron alumina particle suspensions. Each polishing or grinding step introduces damage in the sample, which is approximately proportional to the size of the particles used for grinding/polishing. The goal of each grinding/polishing step is to remove the damage from the previous steps, such that at the end of the final fine-polishing step the damage in the sample is minimal. The reader is referred to several excellent treatises on sample preparation and metallurgical analysis (Samuels 2003; Vandervoort 2004).

Sometimes, electropolishing or ion milling is needed to remove the layer of damaged material from the finely polished surface, especially for soft metals. In this context, the focused ion beam (FIB) is also employed to clean the surface of damage.

Chemical analysis

In general, chemical analysis is a bulk analysis technique rather than a surface analysis method. It comprises either volumetric or gravimetric analysis, which depends on measurement of volumes or measurement of weights, respectively. Both kinds of chemical analyses depend on balanced chemical reactions between the reactants in the sample and products arising from the chemical reaction. Some examples include the determination of copper content in a plating bath and sulfate determination in a bath of persulfate etch. As such, while these techniques are very important, they are used only in a small fraction of the failure analysis jobs in a typical product level failure analysis lab.

Volumetric Analysis

In this method, a known quantity of sample is brought into solution and a known volume is titrated against a reagent solution of known concentration to a discernible end point indicated by distinct color change. Acid–base, oxidation–reduction, or complexation reactions occur, and from the concentrations of the reagent, the volumes of the solutions, and the stoichiometries of the chemical reactions, the concentration of the unknown ingredient is computed using chemical equilibrium equations.

Gravimetric analysis

In contrast, in gravimetric analysis, a known quantity of the sample is brought into solution and the analyte is precipitated using suitable reagents. The precipitate is digested, filtered, heated, and weighed. From the weight of the precipitate, the stoichiometry of reactions and the amount of sample, the composition is determined using standard equations (Viswanadham and Singh 1998). Several modern analytical techniques exist to supplement the basic volumetric and gravimetric procedures, and these have become relatively quick and efficient methods.

Atomic absorption/emission spectroscopy

The basic procedure in this technique consists of introducing the sample solution into a flame as a fine spray where it is desolvated, vaporized, or atomized. Radiation from a light source corresponding to the electronic transition state of the analyte atom under study is passed through the flame. The element in question absorbs radiation in characteristic wavelengths, which depends on the concentration of the analyte atoms. A relative measure of the transmitted signal with and without the sample in the flame is obtained, and a calibration curve is obtained with known concentration samples in the range of interest. The concentration data in the sample is estimated from the trend of absorbance versus the known concentrations. A significant limitation of this technique is that only one element can be analyzed at a time, although simultaneous analysis of several elements has been reported in the literature (Lawson et al. 1982).

Conversely, in emission spectroscopy, in principle, the analyte atoms are raised to an excited electronic state by thermal collisions in the emission zone. As the excited atoms return to the ground state, the excited state atoms emit radiation characteristic for each element. This radiation passes through a suitable monochromator to isolate the desired spectral wavelength and is detected by a photomultiplier detector. The intensity of the signal gives a measure of the concentration that can be estimated using calibration curves. One caveat is that both absorption and emission spectroscopy data are influenced by matrix interference effects (Hageman et al. 1982).

UV/visible spectroscopy

Visible and ultraviolet comprise radiation in the 200–800 nm range, and electronic transitions in materials produce emission/absorption spectra in this range. When a particular wavelength of incident radiation coincides with an allowed transition to a higher energy in a given molecule, absorption occurs. The absorbance A of radiation is governed by Beer's law, which is given by the equation

$$A = abc \tag{9–2}$$

where a is the absorptivity of the material, b is the path length traveled by the radiation in the sample, and c is the concentration of the absorbing species in the sample. Therefore, knowing the wavelength at which the absorption takes place, and the values of a and b, the concentration of the species under consideration can be obtained. Information about the peak absorption wavelength is obtained by measuring the absorbance as a function of the wavelength, and calibration curves are generated with solutions of known concentrations at the chosen wavelength. The concentration of the species in the unknown sample is obtained by using these calibration curves. If the calibration plot of the absorbance versus concentration is not a straight line, the analyst must consider complexation reactions, polychromatic radiation, and interferences. Water is used as the solvent for most inorganic species, while cyclohexane or 95% ethanol is used for organic substances. Among the instrument types, double-beam spectrometers are preferred over single-beam spectrometers, and grating dispersive elements are superior to single-beam prism monochromators. Because the wavelength can be determined relatively accurately, the technique is capable of good precision, but the accuracy depends on the quality of the standard solutions used in generating the calibration curves.

Infrared spectroscopy

This is a very important technique in the identification of organic compounds, and this technique operates in the infrared region of 0.7–500 μm of the spectrum. When a sample is irradiated with infrared radiation, specific functional groups such as OH, NH, etc., in the sample absorb radiation corresponding to their vibrational frequencies and result in a spectrum with characteristic absorption peaks. Since each compound has a unique infrared spectrum, this technique can help identify the fingerprint of an unknown substance by comparing against a library of spectra. In Fourier-transform infrared spectrometers, a frequency transformation is performed accompanied by signal averaging and signal conditioning. This technique is analogous to fingerprint analysis of crime scenes, where a fingerprint by itself is of little value unless a match is obtained with a known fingerprint. Even when a complete identification is not possible, specific functional groups can be compared to draw inferences about the family of compounds. Identification of samples can be performed in all three states (solid, liquid, and gas) and either from the surface or from the bulk of the sample.

Thermoanalytical methods

There exist several techniques in this category where heat is applied to the sample and a change in physical property is measured to characterize the material.

Differential Scanning Calorimetry (DSC): In this technique, the test sample is heated along with a known reference material at a constant rate. When a phase change occurs in the test sample, there is a change in the heat required to keep the sample at the same temperature as the reference sample. Thus, when a solid melts in an endothermic reaction, more heat flow to the test sample is required to keep up with the change in temperature of the reference. Another property that is of great importance to the reliability of electronic packages is the glass transition temperature (where the amorphous polymer starts to soften as the temperature is raised). As a polymeric underfill is heated, the glass transition temperature would appear as a step in the baseline of the recorded DSC signal. This is due to the sample undergoing a change in heat capacity; no formal phase change occurs in the polymer. DSC is widely used in electronics industry as a quality control tool due to its applicability in evaluating sample purity and for studying polymer curing. Since a fully cured polymer generally has a well-defined glass transition temperature (T_g), the determination of this property also sheds light on the degree of cure and cross-linking.

Thermomechanical analyzer (TMA). The thermomechanical analyzer allows determination of changes of mechanical properties that result from application of heat. Properties such as strength, coefficient of thermal expansion (CTE), etc., can be estimated as a function of temperature. In electronic packaging, the CTE mismatch between the various packaging materials is a major cause of package and assembly failures, and the TMA is widely used to measure this property. In addition, by analyzing the rate of change of CTE with temperature, properties such as the glass transition temperature can be estimated because the CTE below the T_g is significantly lower than the CTE above the T_g.

The sample is enclosed in a chamber that can be heated, and a quartz probe is located on the surface of the sample and the changes in the material as a result of the heating such as dimensional change, softening, etc. are reflected in the probe movement. When no force is exerted on the sample, the technique is also called *zero-force TMA* or *dilatometry*. Variations of this technique also exist involving a static force, dynamic force, and modulated temperature for studying stress relaxation (creep), linear viscoelastic region of the stress–strain curve, and reversing CTE data, respectively.

Thermogravimetric analysis (TGA). In thermogravimetric analysis, the sample is suspended from the arm of a microbalance capable of weighing to a microgram inside a programmable oven. Weight changes as a function of temperature are recorded. If a material stability at the operating temperature or test temperature is suspected as cause of a given failure, it can be verified by this technique. Weight loss due to solvent evaporation, material decomposition at a given temperature, loss of water of crystallization, etc., can be determined by this method.

Chromatography

Chromatography depends on the separation of mixtures by color, and is a very powerful technique for the separation and quantification of mixtures. Organic liquid mixtures are injected into long resin columns, vaporized, and transported by an inert gas such as helium. In the column, different constituents of the mixture travel at different rates depending on their respective affinities to the column. As they emerge at different times, they are collected separately. Alternatively, the constituents are detected by various types of detectors and are recorded. The principle is extended to develop high-performance liquid chromatographs where the sample cannot be converted into vapor without the risk of decomposition. High molecular weight polymers are separated by this technique and are identified by reference to standard materials. Another variation of this technique is ion chromatography wherein both anions and cations in a mixture can be detected, separated, and quantified. Examples of applications of this technique include the identification and quantification of ions such as bromide, chloride, etc., in flux residues (Viswanadham et al. 1982). Chromatographic techniques are powerful tools in the analysis of ionic contamination, determination of molecular weights of organic polymers, etc.

Electron beam analysis

Electron beam analysis is an umbrella term for several techniques that depend on probing the sample with an incident electron beam. Techniques such as scanning electron microscopy (SEM), electron spectroscopy for chemical analysis (ESCA), X-ray photoelectron spectroscopy (XPS), and Auger electron spectroscopy (AES) all fall under this umbrella.

Scanning electron microscopy. This is by far the most popular failure analysis technique used in the industry. It extends the limits of optical microscopy by utilizing a beam of electrons instead of visible radiation to generate images. The beam is scanned across the sample in a raster pattern, and the resultant output is used to generate an image. Electrons emitted from the cathode of an electron gun are accelerated by the application of up to 30 kV potential between the anode and the cathode. The accelerated electron beam travels through a series of magnetic lenses that alternatively converge and diverge the electron beam along the axis. As the beam passes through these filters, electrons of greater or less than a specific energy are filtered, essentially forming a monoenergetic beam. The beam is focused using magnetic coils and is scanned across the sample. When the electron beam impinges on the sample, it undergoes both elastic and inelastic scattering. In elastic scattering, the beam is deflected with essentially no energy loss, whereas in inelastic scattering the beam produces backscattered radiation. The backscattered radiation can be collected by a suitably oriented detector, and the resultant image shows strong contrast with the atomic number of the material under the beam. Thus, in board-level and solder-joint analysis or samples, especially of polished cross-sections, the backscattered electron image is very widely used to obtain a high-contrast image of a sample that is relatively free from topographical contrast.

Figure 9–3 shows the electron beam and sample interaction. The beam-sample interaction produces Bremsstrahlung X-ray radiation (electron beam/nuclei of sample atoms) or secondary electrons and elemental characteristic X-rays. The secondary electrons can be collected by a suitably positioned detector and imaged to create high-resolution, high-depth-of-field secondary electron images. The secondary electrons escape from a depth of 10 A for metals and 100 A for nonmetals. A variety of imaging modes and display techniques are used to enhance the analytical capability, such as cathode luminescence, voltage contrast, channeling contrast, wavelength dispersion and stereo pair, superposition, mixed modes, dual magnification, etc. The reader is referred to several excellent references at the end of the chapter (Lifshin et al. 2002).

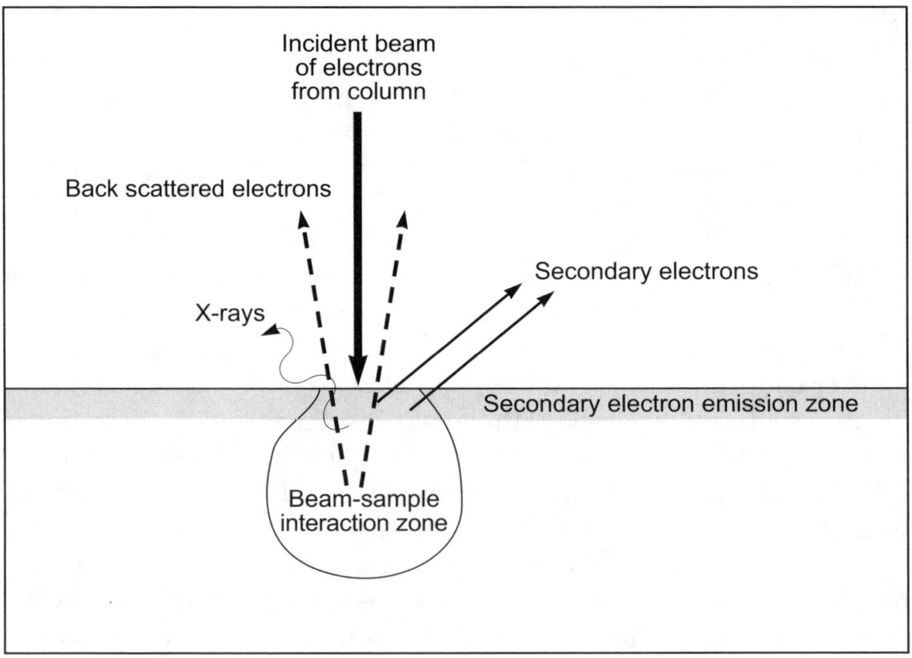

Fig. 9–3. Electron beam and sample interaction producing Bremsstrahlung X-rays

The X-rays produced by the electron beam as it travels through the bulk of the sample are specific to the element producing them, and an often-used attachment called energy dispersive X-ray analyzer (EDX) enables the elemental analysis of the sample. So, contamination, composition, etc., can be determined and also mapped to form an image of different elements in the picture to study distribution, migration, segregation, etc. The analyzer typically consists of a lithium-doped silicon wafer. The X-rays penetrate the silicon wafer to different depths creating hole–electron pairs and, thus, current pulses. The magnitude of the current pulses depends on the energy of the X-rays, which in turn is characteristic of the element from which they are generated. A typical SEM-EDX spectrum is shown in figure 9–4.

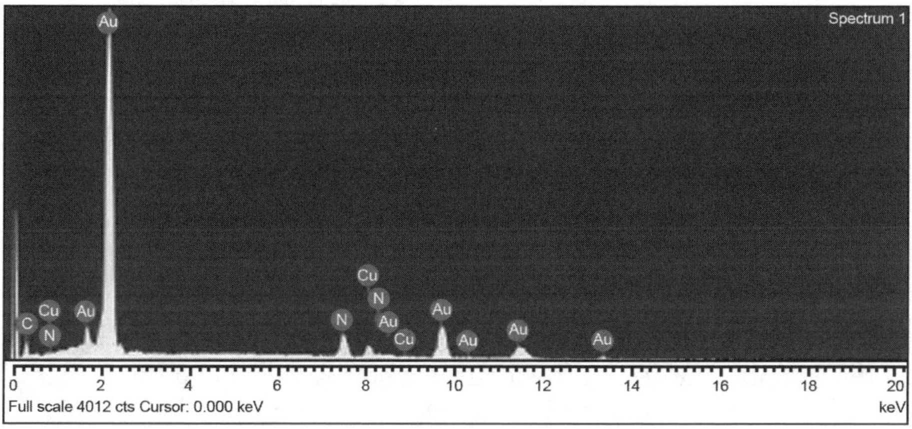

Fig. 9–4. An example of energy dispersive X-ray spectrum (EDX) from an SEM.

The SEM has a magnification range of 20–100,000 with a depth resolution of 20 A or better. Its depth of focus far exceeds that of any optical microscope, almost by a factor of 200 or 300 at comparable magnifications. Higher resolution is achieved by high accelerating voltages but at the cost of increasing the interaction volume. Lower voltages are often employed for enhanced contribution from the surface for studies involving contamination, thin films, etc. In addition to the beam voltage, sample tilt can have a significant impact on the nature of images obtained since tilt alters the depth penetration.

The quality and the amount of information is influenced by a number of factors, and it is important that the analyst be aware of several artifacts due to electron beam damage, specimen preparation, chemical process, contamination from the pumps used to evacuate the chamber, etc.

Auger electron spectroscopy (AES). Pierre Auger discovered the Auger effect in 1925 when he was working with X-rays, which was initially used for studying compositional variation with depth close to the surface of the sample. As in most electron beam analysis techniques, the sample is enclosed in a high-vacuum chamber and bombarded with a focused beam of high-energy electrons in the 2–50 keV range to remove the core K level electrons in the target atom. The resultant instability due to the hole in the inner shell is quickly remedied by an electron from one of the outer shells (L_1) filling the hole in the inner shell, and in the process releases energy equivalent to the difference in the orbital energies. The energy equivalent is the difference between the incident energy (EK_1) and the energy of the electron in the L_1 level (EL_1). This energy is released either as characteristic X-rays (the basis for X-ray fluorescence) or transferred to another electron at one of the higher levels (L_2), which is ejected as a characteristic Auger electron. The measured energy in this case will be $E_K - E_{L1} - E_{L2} - w$, where w is the work function of the target atom. Several transitions such as KL_1L_1, KL_1L_2, etc., are possible with different probabilities, and these are characteristic of a sample material. Figure 9–5 shows a schematic describing Auger transitions.

The attractiveness of this technique is related to the fact that, because of the very short mean free path of electrons in solids, only those electrons originating in the first few atomic layers of the surface escape the sample. Thus, the analysis results pertain to a very thin surface layer of the sample. The size of the focused electron beam and the ability of the detector help achieve a lateral resolution of 0.1 μm. The elemental sensitivity is between 0.1 and 1%.

A significant portion of the failures in semiconductors and microelectronics arise out of surface contamination because of the multitude of interfaces and the critical role played by adhesion between interfaces.

X-ray photoelectron spectroscopy (XPS/ESCA). In the basic electron spectroscopy for chemical analysis, a beam of monochromatic X-rays impinges on the sample surface enclosed in a vacuum chamber. The photons are absorbed by the sample surface and an electron is emitted from the outer valence shells. Generally, electrons from the different orbitals that have different bonding energies less than that of the incident X-ray beam and kinetic energies consistent with the energy conservation principle are, however, emitted with different probabilities:

$$hv = E_b + KE \tag{9–3}$$

where $h\nu$ is the energy of the incident X-ray beam, E_b is the binding energy of the ejected electron, and KE is the corresponding kinetic energy of the electron.

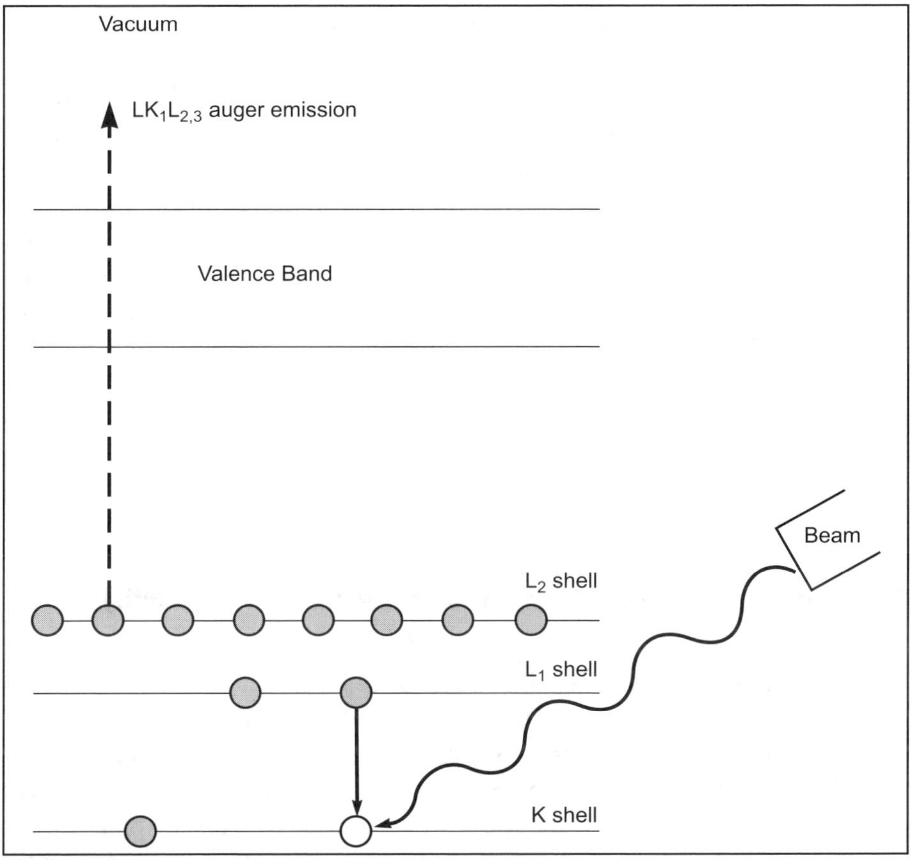

Fig. 9–5. Schematic describing Auger transitions

Since valence electrons are involved in the process, the technique provides valuable information on the oxidation states of the species involved in the process. The photon sources for ESCA are usually Mg or Al targets that are bombarded with electrons. The method provides organic and inorganic structural information. Another advantage of this technique is that it can be performed with moderately low levels of vacuum compared to other vacuum techniques. ESCA has a spatial resolution of 10 μm and the detection limits for most elements are in the order of 0.1–1.0 at.%. The special ability of this technique is that the chemical bonding can be understood (Vander Wood et al. 1988)

Secondary ion mass spectrometry (SIMS)

SIMS is an excellent tool for the analysis of solid samples. Similar to other electron beam techniques, the sample is bombarded with Cs, Ar, or O ions of approximately 10–20 keV. The ions penetrate the surface and lose their energy in a series of inelastic collisions with the sample surface atoms. The transferred energy causes particle ejection from the sample surface. These are generally mixtures of neutrally charged atoms, positive and negatively charged ions, and sometimes clusters of atoms in addition. These are extracted into a mass spectrometer where they are separated and detected, giving a mass spectrum (a plot of the number of ions detected as a function of mass). These are thus secondary particles and the analysis of these ions enables one to draw inferences about the composition of the material being investigated. The escape depth of these particles is dependent on the energy of the bombarding radiation, and SIMS has high sensitivity for most elements including low atomic number and isotopes. It also has fairly good resolution for depth profiling and lateral characterization. The detection limit for many elements is in the parts per billion range, and a depth resolution of 5–10 nm is achieved (Vander Wood 1988).

Laser-induced ionization mass spectrometry (LIMS)

This technique is an excellent complement to XPS/ESCA for a more complete characterization of organic surface layers and is based on determining the mass and intensity of elemental and molecular ionic species that are characteristic of the material's chemical composition. A finely focused Nd:YAG pulsed laser ($\lambda = 266$ nm) with a pulse time of 10 ns with an energy beam of 10^7 to 10^{12} W/cm^2 irradiates the sample. The laser-ablated ionic species are analyzed by a time-of-flight mass spectrometer. The species are separated in time as they travel to the analyzer. At lower energy levels one would observe laser desorbed species, while at higher energies the laser ionized species are observed. The technique can provide identification of all elemental species including isotopes. Molecular structure information is also obtained. The technique provides both positive and negative ion spectra. Examples of the use of this technique include identification of solder mask residue on copper pads, potential flux residues on the circuit card assemblies, epoxy residues, contamination, etc. The technique should be used with ESCA/XPS to obtain the maximum benefit (Vander Wood 1988).

Focused ion beam (FIB)

High-precision fail-site isolation in submicron dimensions can be achieved by utilizing a FIB as a milling machine. A 30–50-kV gallium beam is generated from a liquid gallium source and focused electrostatically into a fine beam. By scanning the specimen surface with an appropriate ion dose, the material can be ablated precisely. If a cross-section right through the center of a narrow via is required, a combination of mechanical cross-sectioning and FIB milling can be used. First, a relatively rapid mechanical polishing is performed until close to the perimeter of the interconnect. In a second step, trenches right through the center of the interconnect are milled by using an FIB tool. Unlike the conventional top-down FIB cross-section, the combined mechanical and FIB milling technique enables the analyst to inspect the cross-section exactly perpendicular to the interconnect, enabling precise measurements on cross-sections with a minimum of surface damage because the damage from the mechanical polishing is removed by the ion beam. This is analogous to the ion-milling operation used earlier in the industry, except that the FIB milling is much more precise. Apart from the sample preparation aspects for board and package level, FIB is an attractive technique for fault isolation at the semiconductor level using techniques such as voltage contrast and channeling contrast.

A summary of the different analytical techniques used in failure analysis is presented in table 9–2.

Table 9–2. Synopsis of the analytical and material characterization capabilities of different techniques

Technique	Probing radiation	Detecting radiation
Optical Microscopy	Visible	Visible
Fourier Transform Infrared Spectroscopy (FTIR)	Infrared radiation 7000–14,000 Å	Infrared radiation 1–1000 microns
Photoluminescence	Visible and infrared	Visible and infrared
Raman Spectroscopy (RS)	Optical visible	Optical visible
Scanning acoustic microscopy (SAM)	Acoustic waves 10-500 Mhz	Reflected acoustic waves
GC-MS & GC-FTIR	Injected fluid	GC-MS molecular fragments GC-FTIR-optical infrared
Differential scanning calorimetry (DSC)	Heat	Temperature
Thermogravimetric analysis (TGA)	Heat	Weight, expansion
Ion Chromotography (IC)	Ion-exchange column	Current/light absorption
Electron microphobe (EM)	1–40 keV	Secondary electrons & x-rays
High resolution electron microscopy (HREM)	100–300 keV	X-rays & electrons
Scanning tunneling Microscopy (STEM)	Tunneling electrons	Tip to sample current
Photoelectron Spectroscopy (UPS, XPS)	X-rays (Al, Mg, Ag, K-alpha) or UV light	Photoelectrons
X-ray fluorescence	X-rays	X-rays
X-ray diffraction	Cu-Kα or Mo Kα x-rays	X-rays
X-ray topography (XRT)	Cu-Kα or Mo Kα x-rays	X-rays
Secondary ion mass Spectrometry (SIMS)	Oxygen or cesium ions	Sputtered atomic molecular ions
Ion channeling & particle induced x-ray emission	1–3 MeV H_2 or He ions	Hydrogen or helium ions or X-rays
Nuclear reaction analysis (NRA)	High energy ^{19}F ions	Gamma rays (5–10 MeV)
Rutherford back scattering (RBS)	1–3 MeV H_2 or He ions	Hydrogen or helium ions
Neutron activation analysis (NAA)	Thermal neutrons 1.5×10^{13} atoms/cm^3	Gamma rays, beta particles

Detectability	Surface sensitivity	Sample size	Comments
Images	1 micron	10 × 10 cm	
0.5 ppm 2.5 × 10^{14} atoms/cm^3	Bulk measurement	1 × 1 × 0.5 mm or larger	Micro IR and gas can be done
1 ppb or 5 × 10^{10} to 2 × 10^{15} atoms/cm^3	1 micron	1 × 1 × 0.1 mm	
5 × 10^{19} atoms/cm^3, 0.1 %	0.1 micron	1 mm ×1 mm	Molecular structure of insulating films
Images	Probes beneath surface	15 cm ×15 cm	Nondestructive subsurface imaging
GCMS-5 × 10^{11} atoms/cm^3, 1 ppt. GC-FTIR-2x10^{17} atoms/cm^3 5 ppm	1 micron with chemical etching	1 mL of fluid	Acids degrade the columns
Molecular Properties	Bulk measurement	5-100 mg	Solids & liquids analysis
Molecular Properties	Bulk measurement	5-100 mg	Solids & liquids analysis
5 × 10^{13} atoms/cm^3, 1ppb	None	10 mL	Liquid samples only
5 × 10^{19} atoms/cm^3, 0.1%	1 micron	Up to 5 in. wafers	Quantification possible
Images(0.01%)	Thickness of the sample	<100 nm; 3 mm dia.	Atomic resolution possible-sample prep. difficult
Single monolayer	Single atoms	1cm × 1cm	Atomic force microscope for insulator
5 × 10^9 atoms/cm3, 0.1%	1 to 3 nm with 1 nm depth resolution	5 × 5 × 1 mm or larger	150 micron spot analysis possible; depth profiling with sputter etching
5 × 10^{18} atoms/cm^3, 0.01%	1 micron	2 × 2 × 1.5 mm or larger	Fast quantitative screening
Phases 1% lattice parameters	0.1–5 microns	Powder particles, crystals	Must be single crystals or polycrystalline material
Strains; in part in 10^8	0.1–50 microns with chemical etching	1 × 1 cm up to 8 in. wafers	Sample must be crystalline
1 × 10^{15} atoms/cm^3; 20 ppb	10 nm with 3 nm depth resolution	2 × 2 × 1cm	Best with conducting samples
5 × 10^8 atoms/cm^3, 0.01%	1% of a monolayer	5 nm across or larger	Crystalline sample
1 × 10^{20} atoms/cm^3, 0.2 %	20 nm	7 mm × 7 mm	
5 × 10^8 atoms/cm^3, 0.01%	1 micron with 20 nm depth resolution	5 nm across or larger	Nondestructive depth profiling
1 × 10^{10} atoms/cm^3, 2 parts per trillion to 1 parts per million	0.5 micron depth resolution with etching	1 cm × 1 cm or larger	Radiotracers available

Typical failure mechanisms

The failure mechanisms in hand-held electronic products are different from those commonly encountered in desktop or mainframe business machine environments. Broadly, they may be categorized by use environment as those caused by (a) thermal loading, (b) mechanical loading (including mechanical drop, vibration, bending, and twisting loads), and (c) electrochemical environments that induce corrosion and metal migration.

Thermal environment

Failures induced due to thermal stresses in portable electronic hardware are, in general, similar to those in other electronic products. In portable electronic hardware, where use of high density interconnect (HDI) with multiple microvia layers is prevalent, the shape of the microvia, copper thickness, and the voids in the microvia influence the nature of the interconnect failure. In general, interconnect failures tend to occur on the package side of the solder joint, and are influenced by the CTE of the package and sometimes aggravated by the solder-mask-defined pad geometry on the package side.

In conventional Sn–Pb solders, the fracture generally occurs in the solder adjacent to the intermetallic layer, where the region is Pb-rich in composition. For Pb-free solder alloys, the interconnect failure mechanisms may show different kinds of deviations from the previously known mechanisms for Pb–Sn alloys. Depending on the surface finish and the pad metallurgy, the interconnection can have multiple types of intermetallic phases dispersed in the bulk joint. In the case of tin–silver–copper system with OSP and ENIG surface finish, Cu–Sn, Ag–Sn, and Au–Sn intermetallic compounds (IMC) were found to be dispersed in the bulk of the joint or near the pads (Dunford et al. 2004). Solder joint failures due to thermal cycling are influenced by shear forces induced by CTE mismatch between the component and PWB, with both fatigue and creep damage mechanisms operative at the same time. A damage accumulation map for Pb-free solders is discussed next.

In Pb-free solders, there are several significant differences in the microstructure compared to the Sn–Pb eutectic or near-eutectic solders, and these microstructural differences result in a very different damage evolution process. The primary microstructural differences are as follows:

- The intermetallic morphology is more complex and a multitude of small spheroidal Ag_3Sn IMCs are observed at the Sn dendrite boundaries. These could serve as initiation sites for voids and microcracks.
- In addition to the increased presence of small particles in the interdendritic spacing, several large Cu_6Sn_5 intermetallics and Ag_3Sn plates are distributed throughout the solder ball, which can effectively constrain the solder joint during shear deformation. A partially etched Sn3.5Ag0.7Cu solder ball on an unassembled CSP is shown in figure 9–6 to illustrate how the intermetallics in this solder system are distributed throughout the bulk of the solder joint to a much larger extent than previously observed in Sn–Pb solders. A completely etched solder ball microcrostructure in figure 9–7 reveals the presence of Cu_6Sn_5 scallop-shaped intermetallic phases adjacent to the Cu pad in addition to the IMCs distributed in the bulk of the solder.
- These microstructural features can bring out damage mechanisms in Pb-free solders that were not a significant contributor to final failure in Sn–Pb solders under thermal fatigue/creep environments.

It should be noted that failure in thermal cycling in solders involves both fatigue and creep failure mechanisms. The relevant mechanisms of creep deformation are as follows:

1. Dislocation creep involves the movement of dislocations which overcome barriers by thermally assisted mechanisms involving the diffusion of vacancies or interstitials ($10^{-4} < \sigma/G < 10^{-2}$).
2. Diffusion creep involves the flow of vacancies and interstitials under the influence of applied stress ($\sigma/G < 10^{-4}$).
3. Grain boundary sliding involves the sliding of grains past each other.
4. Dislocation glide, which normally requires very high stresses, is probably not a major contributor to creep during thermal fatigue. Diffusion creep causes vacancies from grain boundaries experiencing tensile stresses to flow towards those that are experiencing compressive stresses.

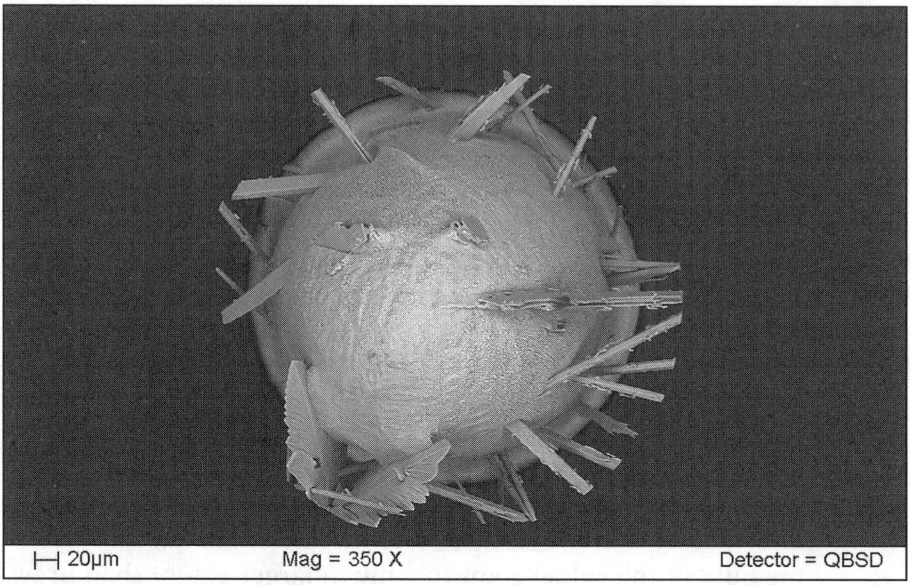

Fig. 9–6. A partially etched solder ball shows the distribution and shapes of different IMCs in the solder.

Fig. 9–7. Completely etched solder ball of an unmounted chipscale package revealing the scallop shaped Cu_6Sn_5 IMC phases on the Cu pad along with rod shaped Cu_6Sn_5 IMCs. In addition, planar dendrites of Ag_3Sn can also be seen interspersed throughout the surface.

In a solder joint with the microstructure and IMC morphology described in the earlier section, the driving force for failure is the imposed cyclic shear stress due to CTE mismatch and creep under this stress. Because of the low homologous temperature of solder, fatigue damage mechanisms are accompanied by creep damage mechanisms.

Consistent with previously reported damage mechanisms for Sn–Pb solder, inhomogeneous shear stress fields can result in recrystallization at pads, corners, and voids. Additional damage mechanisms not widely reported for Sn–Pb solder but observed for Pb-free solders are described next (Dunford et al. 2004).

Zones of recrystallized material were observed at locations with high strain gradients and strain incompatibilities, such as grain boundaries. These recrystallized zones grow with imposed cycling, and a multitude of smaller recrystallized grains form to relieve the strain. In parallel, creep-driven damage mechanisms were observed to a degree not reported in previous studies. Another creep-driven damage mechanism is the initiation of voids and cracks at locations of high strain incompatibility. For example, triple-point grain junctions and IMC–grain boundary junctions in the interior of solder balls and grain boundary (GB) junctions at the surface of the solder ball appeared to be the favored sites for crack initiation.

Further damage evolution is governed by the interaction of the localized damage (in the form of recrystallized zone) with the distributed damage (in the form of microcracks and voids). The severity of damage of all three types, namely, (a) recrystallization zones, (b) microcracks and voids at recrystallized grain boundaries (RGB), and (c) cracks and voids at grain boundaries (GB), grows with increased cycling.

Final failure, however, is dominated by the weakening of the material due to recrystallization and distributed microcracking in the damage zone. A macrocrack forms by the coalescence of the microcracks, primarily in the recrystallized zones. The propagation path of these macrocracks is very different from that observed for Sn–Pb solders. The IMC plates and rods sometimes serve to deflect the propagating macrocrack so that several macrocracks may exist in a solder joint without significantly impacting electrical continuity. These macrocracks coalesce with each other through the distributed damage, changing direction depending on local damage geometries and microcracks at the RGB or the cracks at the GB. Final failure occurs by propagation of the most dominant macrocrack traversing the solder ball, primarily near the pads on the board or the component. For example, in the solder joint of the CSP shown in figure 9–8, one can see the tortuous path taken by the propagating macrocrack and the distribution of the microcracks near the fracture plane. Near the bottom of the solder joint, away from the component pad, an elongated void formed due to creep-related damage enlarging an initially small crack or void is also seen. In the right half of the picture, the grain morphology with Ag_3Sn and Cu_6Sn_5 IMC particles interspersed in the interdendritic spaces is seen. A higher magnification picture of an elongated void caused by creep damage at grain boundaries in a different solder joint is shown in figure 9–9.

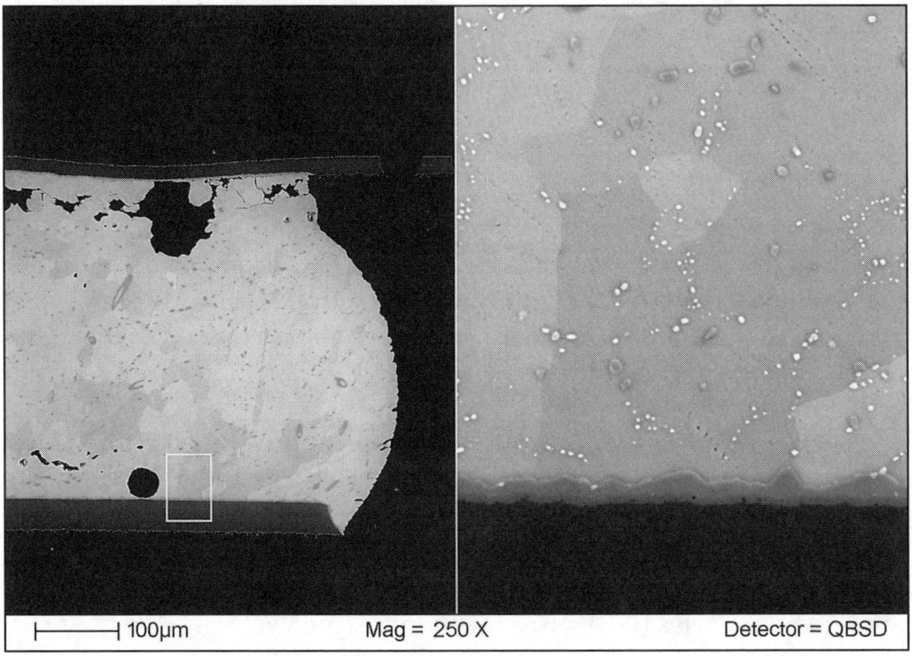

Fig. 9–8. Interconnection fracture due to thermomechanical fatigue loading in Sn3.5Ag0.7Cu solder joint of CSP. The backscattered electron micrograph reveals the fatal crack near the component pad in addition to voiding and other damage near the PWB pad.

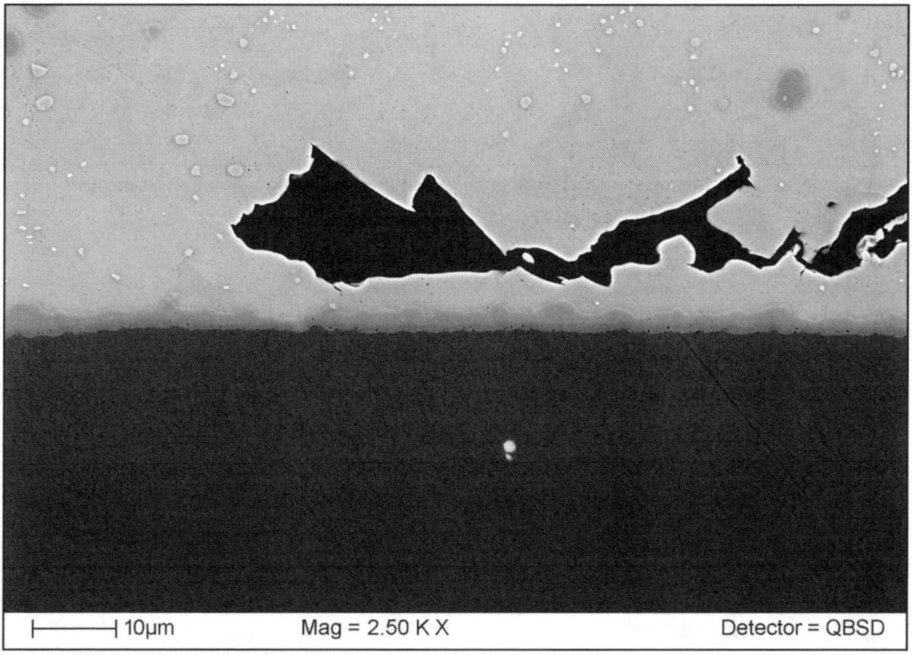

Fig. 9–9. Creep driven damage resulting in elongated voids at grain boundaries

The damage evolution map for thermomechanical loading that brings together the different operative mechanisms just described is shown in figure 9–10 (Dunford et al. 2004). The left-hand side of the damage map describes the fatigue damage process, while the right-hand side describes the creep-driven damage mechanism. Crack growth encompasses contributions from both the creep and fatigue processes. In addition, creep can also cause solder ball spalling, where a portion of the solder ball separates from the rest, and hasten final fracture.

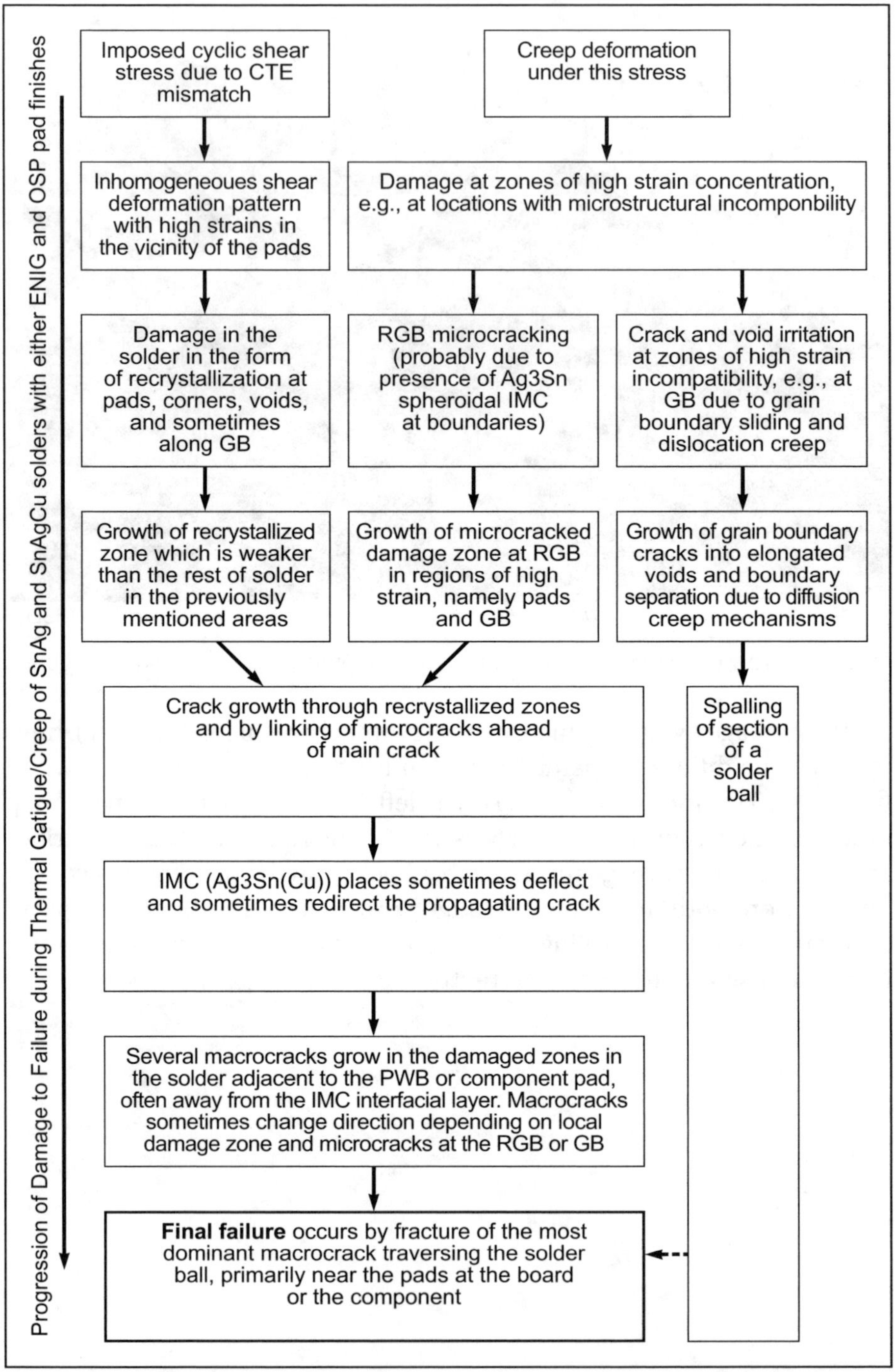

Fig. 9–10. Damage mechanism evolution map for Sn3.5Ag0.7Cu solder under thermomechanical loading

Mechanical environment

It is instructive to review the construction of a generic package mounted on a PWB before discussing the failure mechanisms. The PWB in portable electronic products serves not only as a carrier for the different electrical subsystems but also provides mechanical rigidity to the assembly. A typical PWB can have 4–12 electrical planes laminated between woven glass-fiber-reinforced epoxy layers that serve both as a dielectric and mechanical support. Electrical connection between these layers is often achieved through plated-through-hole vias, blind vias or buried vias. The outermost layer of the PWB, sometimes called the *build-up layer* or *redistribution layer,* is the first interconnection layer between the solder joint and the PWB. Interconnection failures can be found at different levels as shown schematically in figure 9–11, and can be classified as follows based on the location of the crack:

- Die fracture within the package (1)
- Interposer level failure within the package (2)
- Solder joint fracture (3)
- Crack initiation inside the component and subsequent damage to the solder joint
- Interfacial failure—at the solder/PWB pad interface (4)
- PWB-related failure—trace fracture (5)
- PWB-related failure—microvia fracture

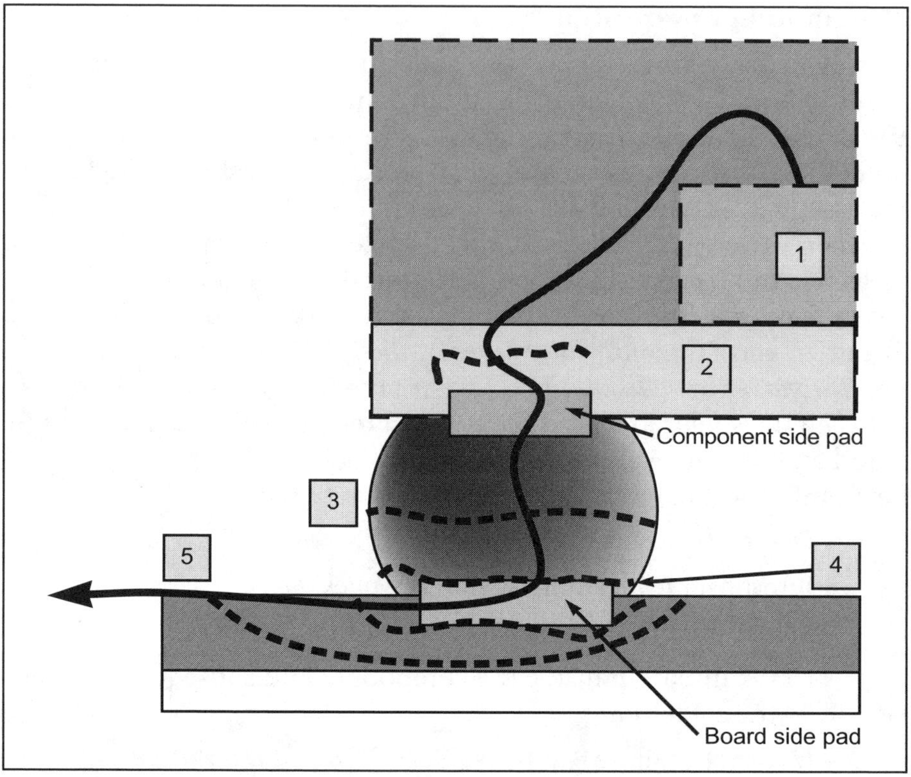

Fig. 9–11. Simplified schematic of electrical interconnection from the Si die to multilayer PWB through different levels of packaging. The dashed line represents possible crack or open.

Sometimes, when the die is not supported optimally inside the package, the flexure of the PWB can be transmitted to the relatively brittle semiconductor die inside the package. Cleavage fracture of the die can occur, causing electrical failure. An example of this kind of failure is shown in the optical micrograph in figure 9–12. The wire bonds on the die can also be seen along with the vertical crack in the die. The cracks at the inactive side of the die (bottom) are attributed to polishing damage during the grinding stage. Such artifacts have previously been observed in samples where excessive normal force was exerted on the sample, and should not be confused with cleavage type of cracking on the active side of the die. In this particular case, the die inside the land grid array (LGA) package was subjected to excessive bending during reflow, thereby cracking it. One factor driving this failure mechanism was the collinearity of the solder mask strip separating the LGA pads with hollow vias in the interposer. A second factor was the asymmetric laminate stackup of the interposer, which resulted in bending strains on the die during the reflow.

A third factor was that the resin system used for encapsulation had a CTE that was an order of magnitude higher than the FR4 in the interposer. The confluence of these three factors resulted in excessive bending of the die during board assembly, thereby causing die fracture. The problem was addressed by design and material changes to completely eliminate the failure mechanism during normal reflow operation.

Fig. 9–12. Die cracking due to mechanical loading shown in an optical micrograph

Interposer level package failure: The Cu circuitry inside the interposer can sometimes fail if the process conditions in the fabrication of the interposer are not optimal. The example shown in figure 9–13 illustrates the particular case where suboptimal adhesion between the via barrel and the via cap failed upon exposure to mechanical loading at the PWB level. In this particular case, the interposer PWB fabrication lacked some key cleaning steps after drilling of the vias. Therefore, due to the presence of a thin layer of organic contaminant, the subsequent Cu plating did not have adequate adhesion to the Cu via in the laminate. Since the joint was not a complete open, the electrical tests after the package assembly did not reveal the presence of the defect. However, when the phone was assembled, the thermal stresses due to the reflow resulted in a yield loss at the manufacturing line. Electrical failure analysis revealed that the intermittent open was located in the interposer. However, the device would function normally when pressure was applied to the top of the package. The defective units could be discerned clearly using an infrared camera, where the normal device heating pattern was significantly different from the thermal profile of a package with cracked interposer.

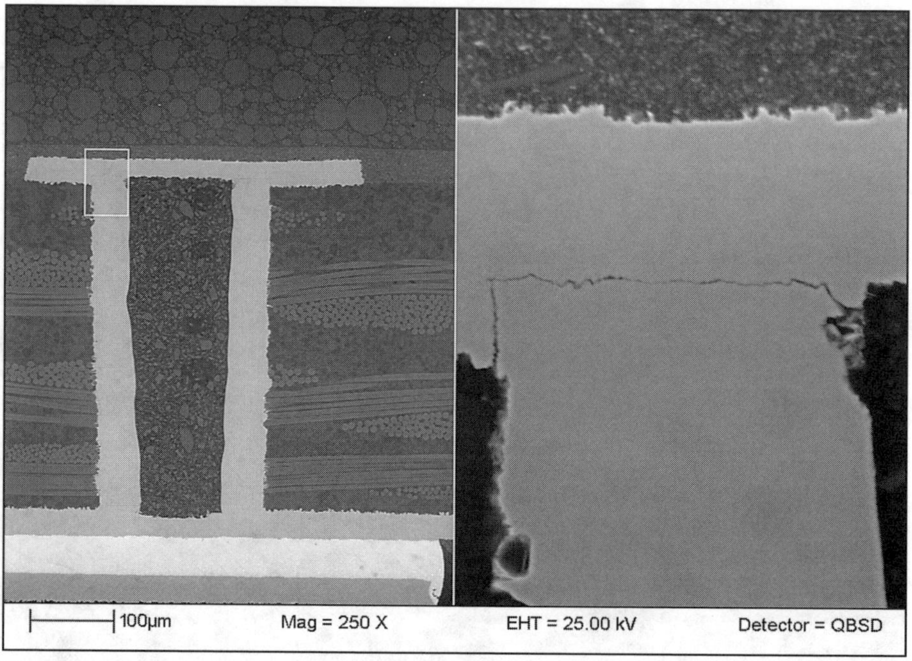

Fig. 9–13. Sub-optimal adhesion of the via-barrel/via-cap manifesting as via barrel cracking due to PWB level assembly causing electrical failure

Crack initiation inside component leading to solder joint damage: Ceramic components, due to their weight and lower fracture toughness, are particularly susceptible to failure when the product is dropped. Local stress concentrations on the ceramic component, such as those created by machining, can serve as crack initiation sites and cause premature failure as shown in figure 9–14. The crack that originated at the machining groove caused an electrical open upon propagation. Apart from the machining on the ceramic component, a second factor contributing to the crack originating in the component is the relatively high strain rate of deformation during drop loading. Since solder deformation characteristics are highly strain rate dependent at room temperature, the solder joint is stiffer and stronger under higher deformation rates, thereby subjecting the ceramic component to proportionately higher stresses.

For components operating at RFs, a relatively minor partial crack, as shown in figure 9–14, can sometimes cause parametric shift-induced failures rather than a hard open. In cases where parametric shift is the failure mode, conventional pass/fail testing may need to be augmented to account for performance degradation.

Fig. 9–14. Crack in solder joint and ceramic component after mechanical shock drop reliability testing

Solder joint fracture due to PWB level twisting: Bending and twisting are commonly encountered end-use environmental hazards for hand-held products. The deformation rates are much lower than those observed in mechanical drop. In such cases, the solder joint strength and stiffness are proportionately lower and promote fracture at the solder joint in contrast to locations within the ceramic component. An illustrative example is shown in figure 9–15, where the solder joint is completely fractured without the damage extending into the ceramic. The lack of machining damage near the solder joint was probably a secondary factor in limiting damage to the solder joint without cracking the component.

Fig. 9–15. Crack in solder joint after twist testing

Failures generated under bending and twisting loads often present intermittent electrical opens, and board-level fault isolation sometimes requires the exertion of a small bending or twisting load to "open up" the suspect solder joint. Special fixtures to power portable products and monitor the functionality of the product during the application of the external load are also good ideas to recreate failures.

Solder joint failure related to underfill process. It is a relatively common practice to provide additional reinforcement to a solder joint to improve its reliability under thermal and mechanical loads. For BGA and CSP packages soldered onto PWBs, this reinforcement can be achieved by the use of a suitable underfill material in the package-to-board interspaces. This constrains the assembly against bending and thermal strains. One of the more commonly used procedures for underfilling a CSP soldered onto a board consists of dispensing liquid underfill along one or more edges of the CSP perimeter such that capillary action forces the underfill to fill the entire space between the CSP interposer and the PWB. Upon curing, the liquid underfill hardens and encapsulates the solder joints completely, thereby providing additional reliability by mitigating the deleterious effect of either thermal or mechanical strains.

The quality of the underfilling process is dependent on several variables such as temperature of the PWB or liquid underfill, cleanliness of the surfaces, speed of dispensing, etc. When the quality of the underfill is non-optimal and voids are present at the CSP corners, the benefit of the underfill is not realized even if the size of the void exposes only the corner solder joint (Canumalla et al. 2002). An example of a partially underfilled CSP is shown in figure 9–16(a) and an optical micrograph of a more severe underfill defect is shown in figure 9–16(b).

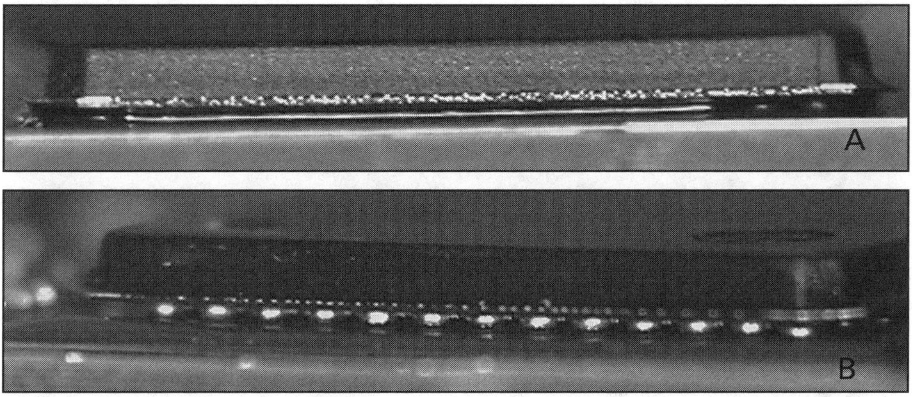

Fig. 9–16. (a) A partially underfilled CSP with a corner underfill void, and (b) a more severe underfill defect exposing a whole row of solder joints

The true extent of an underfill defect cannot be ascertained by either visual or X-ray inspection. For example, figure 9–17 shows a representative X-ray microscope picture of a CSP that does not reveal any underfill defect although visual inspection showed a substantial underfill defect at the perimeter. The scanning acoustic microscope, on the other hand, is very sensitive to voids and underfill defects. Difficulties encountered in acoustic inspection of CSP or BGA underfill include the signal-to-noise ratio due to material attenuation and uncertainty about the specific depth that the data includes. Both these problems are particularly severe for CSP and BGA underfill, unlike in flip-chip underfill inspection. A judicious selection of transducer frequency (ratio of focal length to diameter of the transducer), depth of focus, and gating is essential for successful inspection. The acoustic image of the CSP in figure 9–16(b) is shown in figure 9–18, and the areas of incomplete underfill can be clearly identified in the top half of the acoustic image. A virtual cross-section along the dashed line is shown in the bottom half of the acoustic image, and the relative depths of the die, the interposer, and the void can be seen. In addition, the bond wires extending from the die to the interposer are also visible. It is also useful to present the acoustic waveform along with the image to clarify the nature and location of defects.

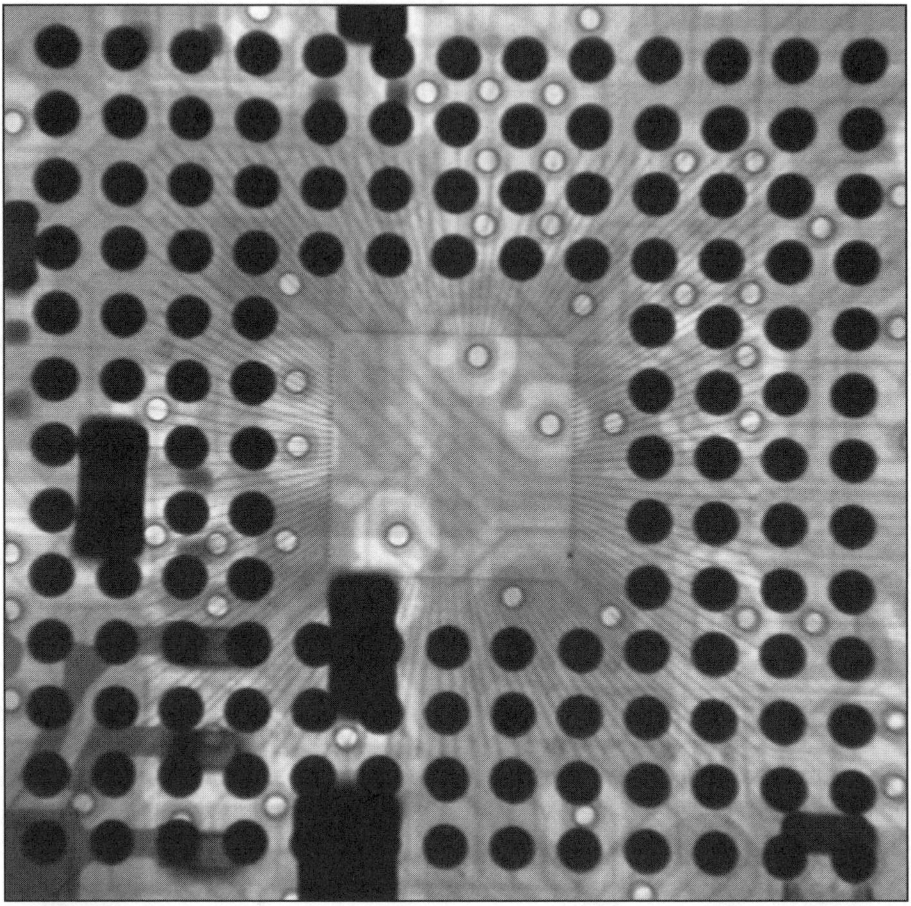

Fig. 9–17. X-ray microscope image of a poorly underfilled CSP incorrectly indicating the lack of underfill defects

An acoustic image of a different improperly underfilled CSP is shown in figure 9–19. The waveforms from three locations are presented alongside the acoustic image for ease of interpretation. The waveform from locations 1 and 2 shows how the die and the interposer lie above the depth of the defect shown in location 3. The positive (upward) reflection from the top of the die and the Cu pads on the interposer is in contrast to the negative (downward) pulse from the underfill void. Thus, there is no ambiguity in concluding that the void lies below the interposer, where underfill would normally be expected in an underfilled sample. The lack of support for the solder ball can lead to failure of the interconnects that are now exposed to higher levels of loading. When exposed to adverse environment such as mechanical loading, the solder joints or the build-up dielectric layer below the Cu pad on the PWB can develop cracks as shown in the scanning electron micrograph of the polished cross-section in figure 9–20.

Fig. 9–18. Acoustic image of the same CSP as in fig. 9-16(b) showing voiding in the underfill below the interposer of the CSP. A virtual cross-section (QBAM along the dashed line in the image) in the lower half of the image reveals that the underfill defect is below the interposer.

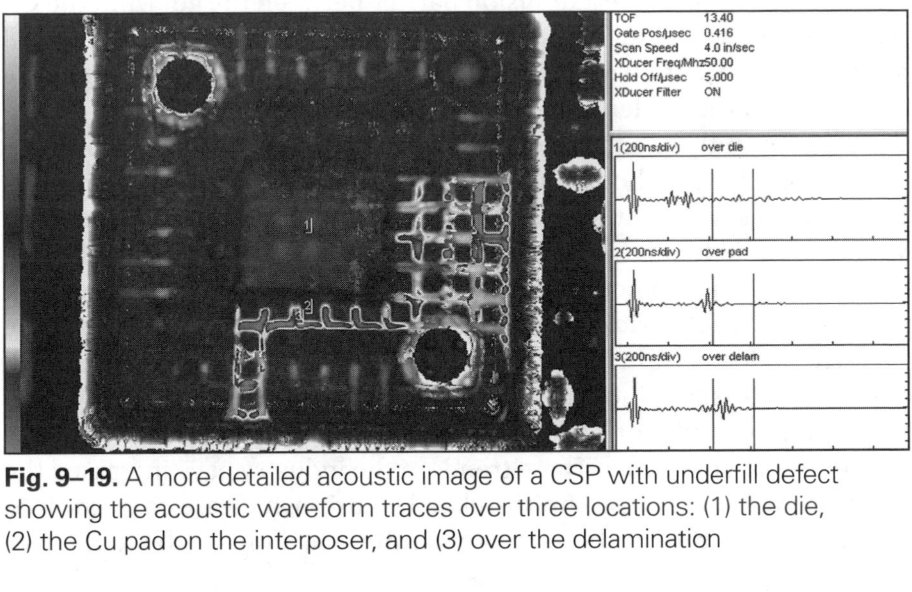

Fig. 9–19. A more detailed acoustic image of a CSP with underfill defect showing the acoustic waveform traces over three locations: (1) the die, (2) the Cu pad on the interposer, and (3) over the delamination

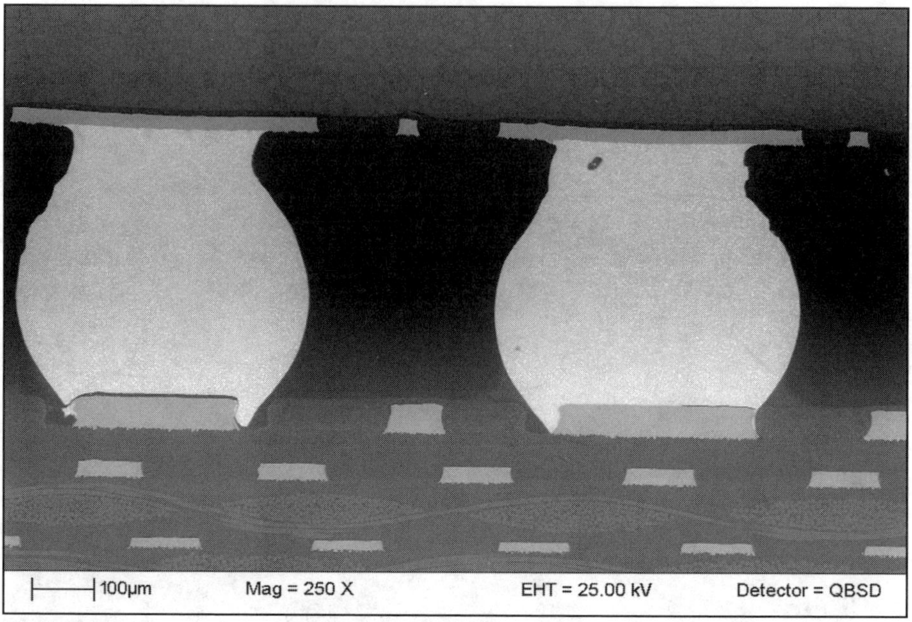

Fig. 9–20. Scanning electron micrograph showing the fractured solder joint and concurrent damage at a neighboring solder joint

PWB quality related fracture at solder/PWB pad interface. ENIG plating of the Cu pads on the PWB gained considerable popularity as a pad surface finish in recent years. This is because it provides a cost-effective means of ensuring coplanarity, which is a crucial requirement in high-density, fine-pitch, and area array package assembly. The Ni layer was intended to provide a diffusion barrier between the gold and the Cu pad. The very thin gold layer (<0.5 mm) was intended to protect the Ni surface from oxidation and preserve its solderability until reflow. During reflow, when the solder melts and wets the gold surface, the gold layer dissolves instantly into thc solder leaving a clean, solderable Ni surface for Ni–Sn metallurgical bond formation. Not so long ago, the interconnection pad sizes were relatively large because fine-pitch packages were not widely used. When pad sizes were relatively large, quality variations in the ENIG plating did not immediately or always result in interconnection failures because the pad size to defect size ratio was substantial. Now, the pad size to defect ratio is smaller due to higher density of packaging. In addition to this, the increased use of less aggressive organic solvent or water-soluble fluxes or no-clean fluxes can result in a pad surface that may not be as solderable. These trends in the industry have increased the risks due to ENIG surface-finish-related failures which are characterized by a brittle fracture at the solder/pad interface along with a dull, dark

Ni surface exhibiting "mud crack" type of surface morphology (Biunno 1999; Bradley and Banerji 1996).

One example of the brittle interfacial crack at the solder–PWB pad interface is shown in figure 9–21. The fracture occurs below the intermetallic layer at the solder/ENIG pad, which indicates that the interfacial bond between the Ni–Sn intermetallic layer and the underlying Ni layer is weak. Indeed, the fracture surface on the PWB side pad is often devoid of any adhering solder as seen in the backscattered micrograph in figure 9–22. The only evidence of solder adhering to the surface is seen in locations where a void in the solder offered easier crack propagation than fracture at the solder/Ni pad interface. The characteristic "mud cracking" types of features are also visible on the Ni fracture surface. A high-magnification micrograph of the polished cross-section figure 9–23 reveals the high-P Ni-layer and a transverse view of the hypercorrosion (black-pad) trenches in the Ni layer (just below the crack) into which the solder has ingressed.

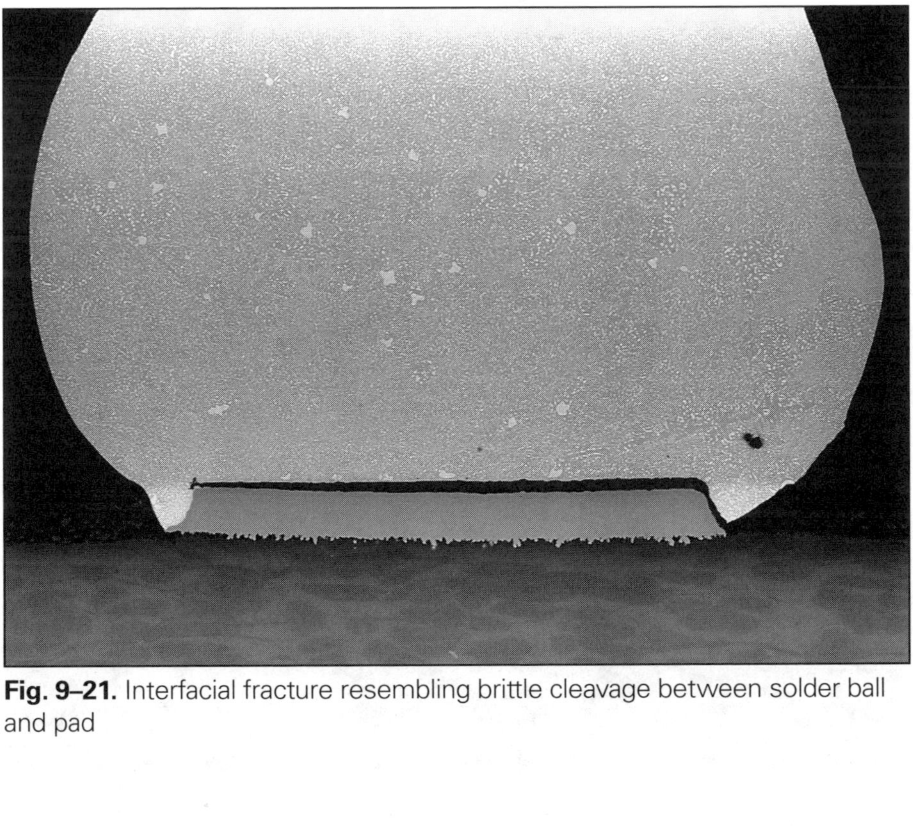

Fig. 9–21. Interfacial fracture resembling brittle cleavage between solder ball and pad

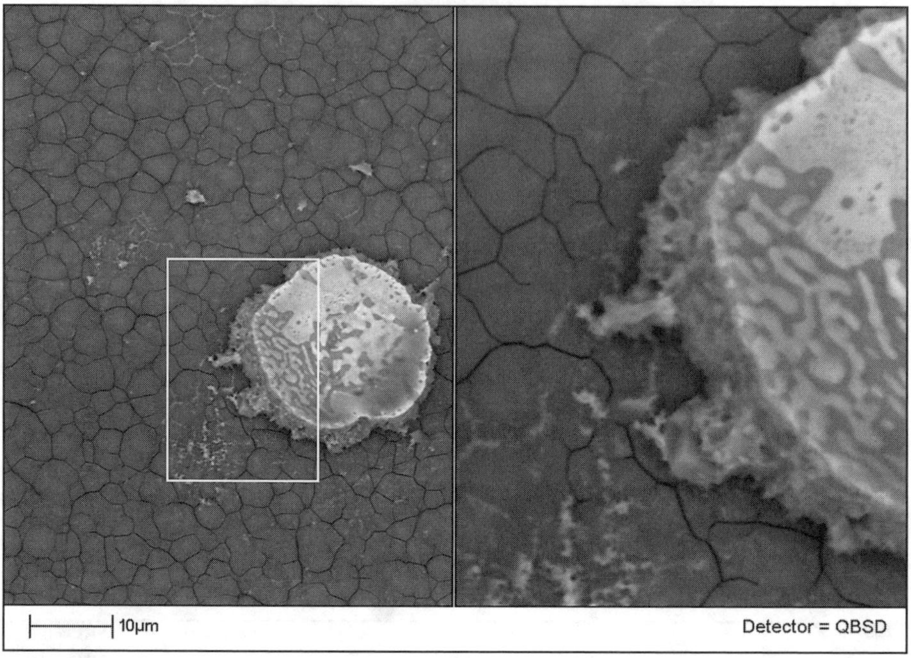

Fig. 9–22. "Mud crack" appearance of Ni fracture surface showing the poor bond quality of solder to the Ni pad

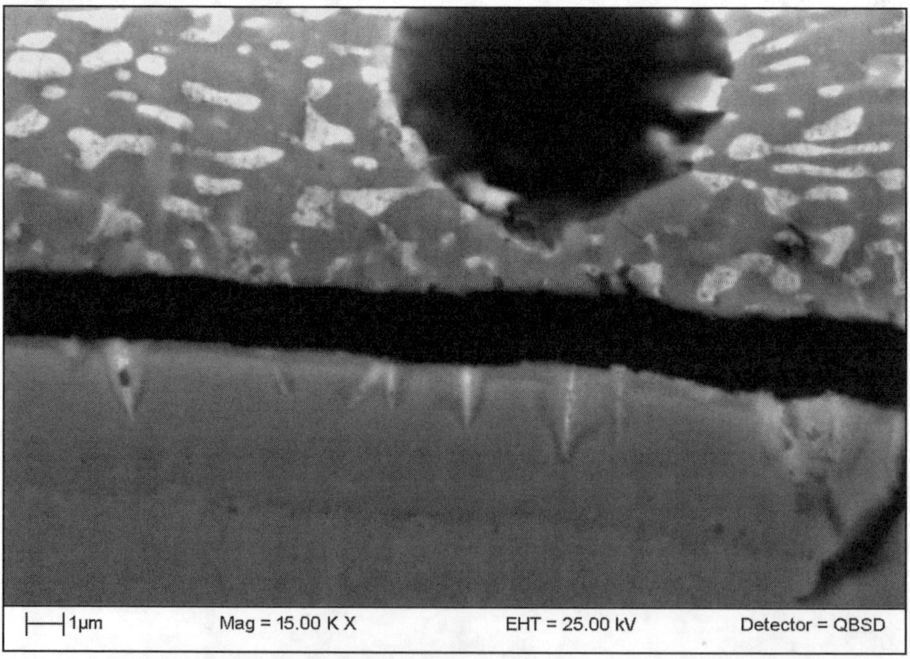

Fig. 9–23. Hypercorrosion of Ni layer observed on a microsectioned sample with Ni hyper-corrosion defect

PWB build-up layer cracking leading to trace fracture. The PWB is usually a multilayer laminate made up of layers of continuous, woven glass-reinforced epoxy, and Cu circuitry. The outermost layers sometimes referred to as the *build-up* or *redistribution layers,* serve both as the dielectric material and mechanical support for the Cu traces during PWB flexure or extension. Upon subjecting the assembled board to mechanical loading, such as encountered in mechanical drop, damage accumulates in the build-up layer in the form of cracking. Subsequent damage accumulation and electrical failure will depend on the redistribution method employed.

The progression of damage for the case when redistribution is achieved through traces is described below. Initially, the damage in the build-up layer accumulates until the trace is no longer supported because of extensive cracking of the laminate under the Cu pad. The damage in the build-up layer is exemplified by the backscattered electron micrograph in figure 9–24(a). This weakening of the build-up layer forces the trace to sustain an ever-increasing share of the mechanical loads imposed on the PWB, which eventually causes trace fracture by fatigue processes. An example of the fractured trace is shown in the top-view optical micrograph in figure 9– 24(b). The fracture process can be seen more clearly after a second microsectioning operation along the dotted line in figure 9–24(b). The backscattered electron micrograph of the double-polished sample is shown in figure 9–25(a) and a schematic explaining the crack under the pad causing the trace fracture is depicted in figure 9–25(b).

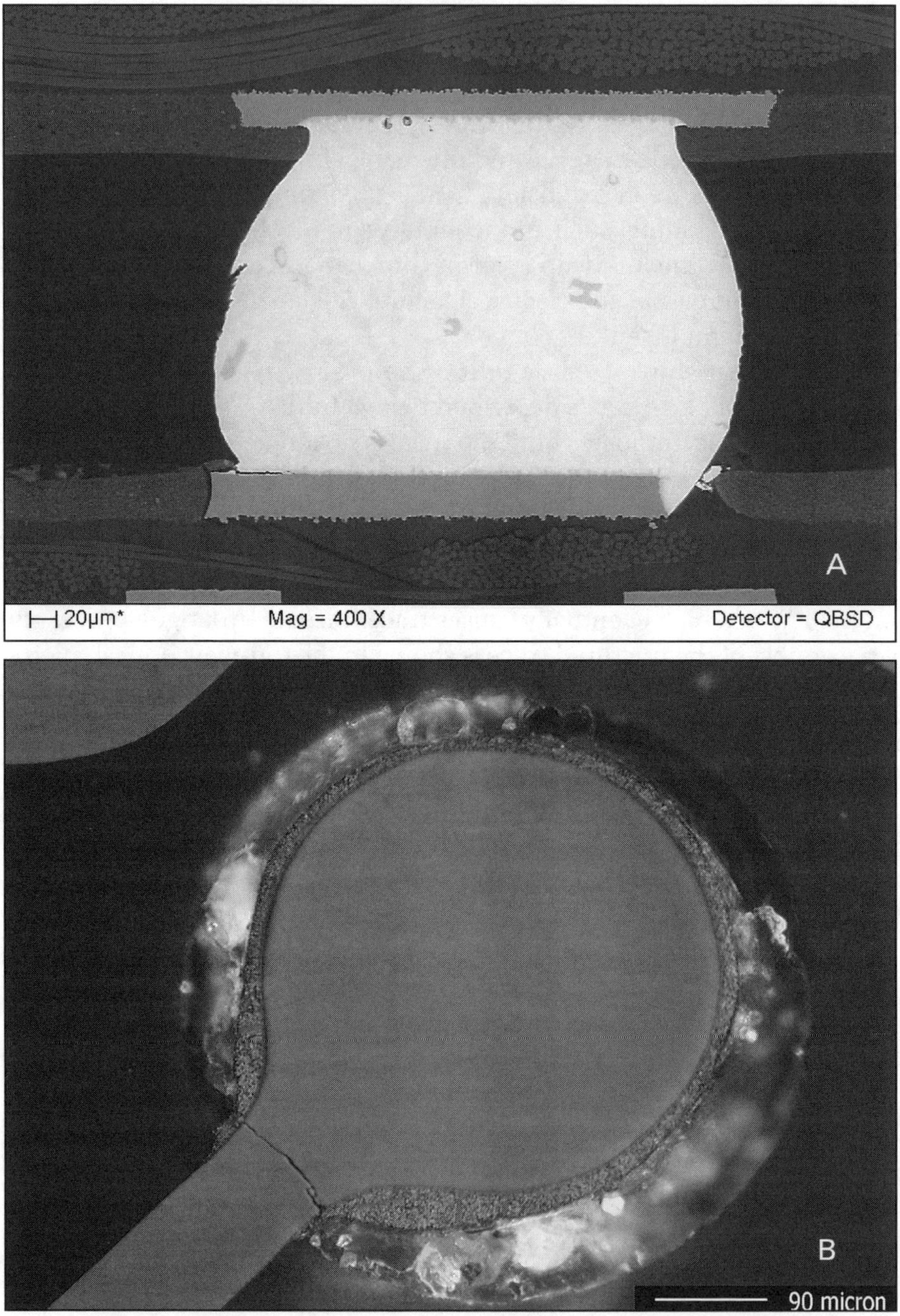

Fig. 9–24. (a) BSD electron micrograph of build-up layer cracking in a solder joint with trace and (b) optical micrograph (top view) of a sample suspected to have a broken trace after the solder ball was removed by mechanical polishing.

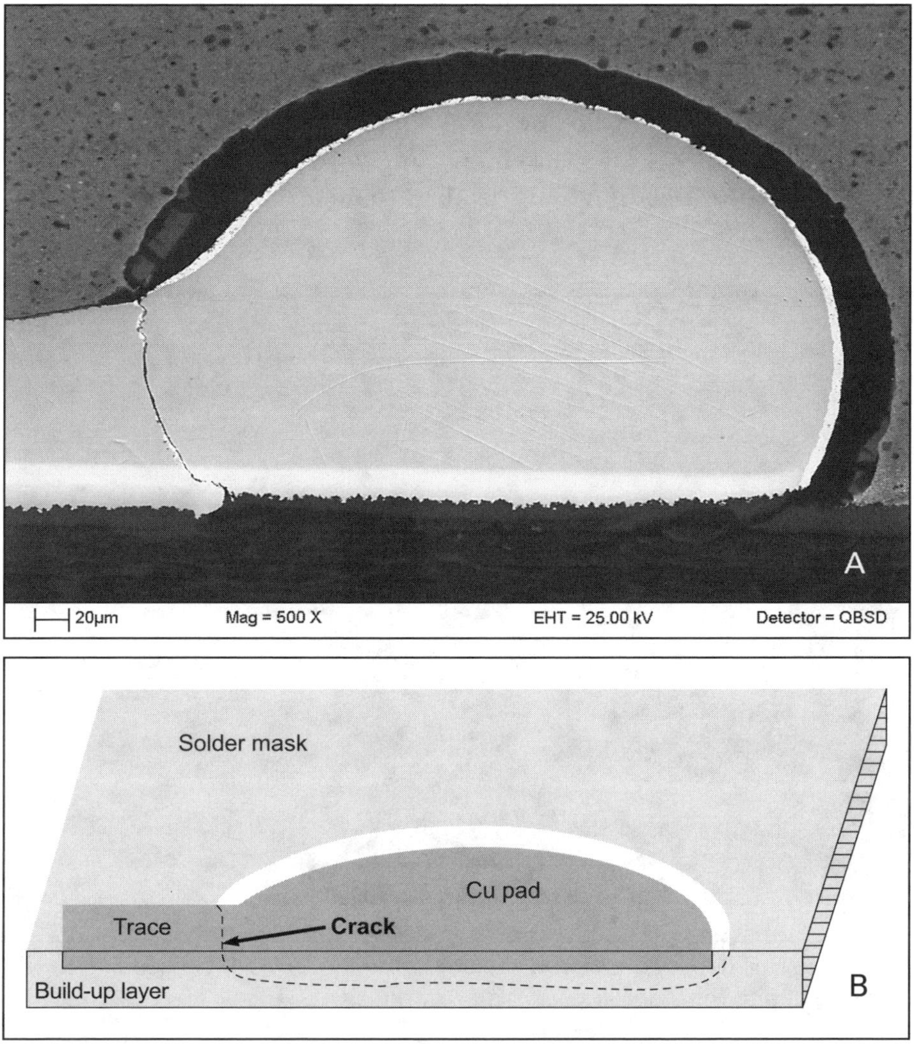

Fig. 9–25. (a) Trace fracture accompanied by build-up layer cracking revealed in a double-cross sectioned sample (sample shown different from that depicted in fig. 9–24, (b) schematic showing the location of the crack in the build-up layer

The damage progression is similar in cases where a via-in-pad redistribution method is employed. The damage in the build-up layer, again, accumulates in the form of cracks. Once the support afforded by the build-up layer is diminished by the cracking, further mechanical flexure of the PWB subjects the via to increasingly higher stresses. Eventually, the via fractures due to fatigue, leading to an electrical open as shown in figure 9–26.

Fig. 9–26. Build-up layer cracking in a solder joint with via in pad leading to via cracking due to mechanical drop related stresses

Electrochemical environment

For portable and hand-held electronic devices, two failure mechanisms related to electrochemical environments are of particular relevance: corrosion and electrochemical migration.

Corrosion. Gold plating of connectors is a common practice designed to protect the underlying Cu and Ni layers from corrosive attack and promote good electrical contact. However, under the action of friction, the relatively thin and inert Au coating can be removed locally, thereby exposing the Cu and Ni layers underneath. In such cases, fretting corrosion, pitting corrosion, and localized galvanic corrosion can occur simultaneously, especially in the presence of ionic species such as chlorides. This corrosion product, which is usually nonconductive, can cause electrical failure due to opens or intermittent. An example of gold-plated connector corrosion is shown in figure 9–27. The EDX elemental map for Au indicates that the coating is intact over the major portion of the area of interest. However, in the central portion of the image, the Au coating appears to have been removed completely, and the underlying Cu is exposed. This Cu surface, identified as a bright area in the Cu elemental map, also shows significant presence of O and Cl. The absence of any areas with high concentrations of Ni indicates that the mating surface of the connector has probably worn through the Ni layer in the area of contact.

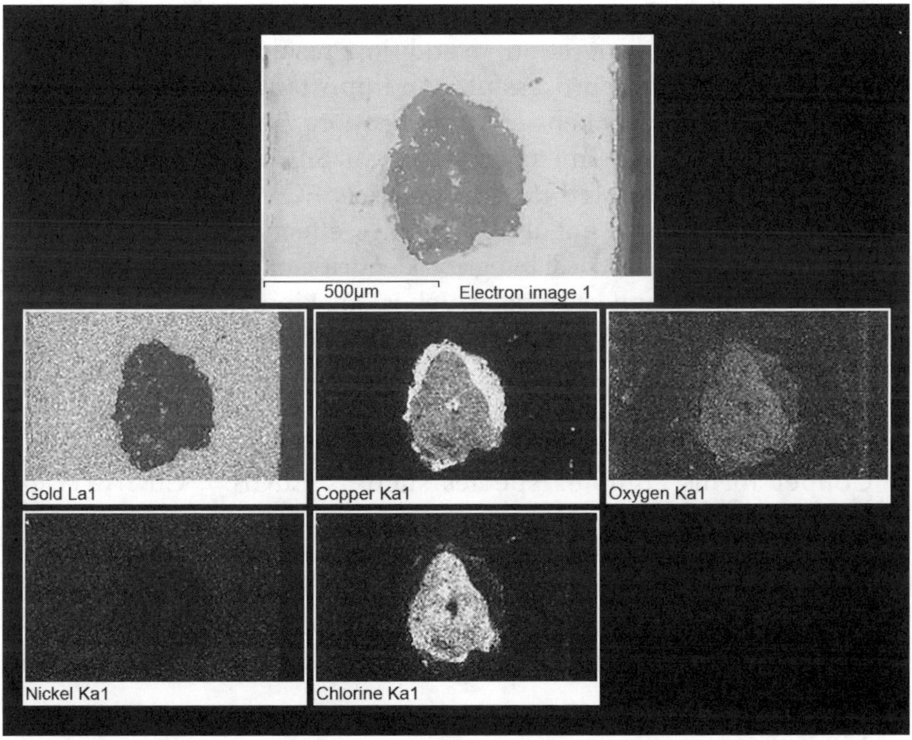

Fig. 9–27. Corrosion of Au plated connector along with EDX elemental maps of Au, Cu, O, Ni, and Cl

Electrochemical migration (ECM). In several studies comparing the propensity for ECM, different metallization systems can be ranked as follows: Ag>Pb>Cu>Sn (Harsanyi and Inzelt 2001). Although ECM phenomena have been observed with many metals, only Ag (Kohmann et al. 1955; DerMaderosian 1978), and Cu to a limited extent (Lahti et al. 1979), and perhaps Sn (Dumoulin et al. 1982), have been found to exhibit this behavior in the presence of humid but noncondensing conditions. Indeed, Dumoulin et al. (1982) concluded that silver migration presents the greatest risk because dendritic growth can occur whether Ag is outside the package or only partly exposed to humid air, on ceramic as well as on plastic substrates. Although Dumoulin et al. (1982) suggested that Cu migration and Sn migration did not pose as big a risk, based on their experimental data, in mobile electronic products which see a wide range of corrosive species during their lifetime, ECM of Cu and Sn can be as prevalent as Ag migration. In addition, residues on the substrates that originate from the process play an important role through water adsorption/conductivity behavior modification. One example of each is provided for ECM phenomena involving Cu, Sn, and Ag.

Krumbein (1988) noted that, in practice, ECM can manifest itself as two separate, though not always distinct, effects that lead to impairment of the circuit's electrical integrity. Dendritic or filamentary bridging between the anode and cathode, which is one kind, has been discussed at length before. Colloidal staining is the second manifestation of ECM, which can also cause a short. Deposits of colloidal Ag, Cu, or Sn have been observed to originate at the anode without necessarily remaining in contact with it. An example of this effect is also provided below.

Copper forms complex species such as $CuCl_4^{2-}$, $CuCl_2 (H_2O)$, $Cu(H_2O)^{2+}$, etc., in the presence of halide-containing species and moisture. An example of Cu electrochemical migration resulting in Cu dendrite formation is shown in figure 9–28. If plated-through hole vias or conductor pads are too close, Cu ECM can occur when the product is exposed to humid environments in the presence of an ionic contaminant.

Fig. 9–28. Cu electrochemical migration on a PWB subjected to damp heat exposure

The Sn electrochemical migration mechanism is similar to that of Cu, but is much more prevalent because Sn constitutes a major portion of several commercial solder compositions such as 62SnPb2Ag, 10SnPb, Sn3.5Ag0.7Cu, etc. In addition, exposed Sn is more widespread on an assembled PWB as compared to Cu. The particular example shown in figure 9–29 is from a test vehicle that failed upon exposure to damp heat testing. In this case, the potential difference between the terminals of a capacitor with Sn termination resulted in the migration of Sn from the anode towards the cathode. The right half of the picture shows a higher magnification view of the Sn dendrites at the cathode end of the termination. Elemental analysis mapping data of the surface of the capacitor is shown in figure 9–30, where the Ba, Ti, and O from the capacitor dielectric material can be seen clearly. In addition, the Sn map shows the presence of Sn between the terminations, where there should be none. In several passives, Ni is used as a barrier layer between the silver adjacent to the dielectric and the tin termination. In this particular case, the Ni barrier layer at the anode is visible in areas where the Sn from the surface has been consumed by the ECM process. Another example of Sn ECM is shown in figure 9–31, where colloidal form of ECM can be observed in addition to dendrite formation.

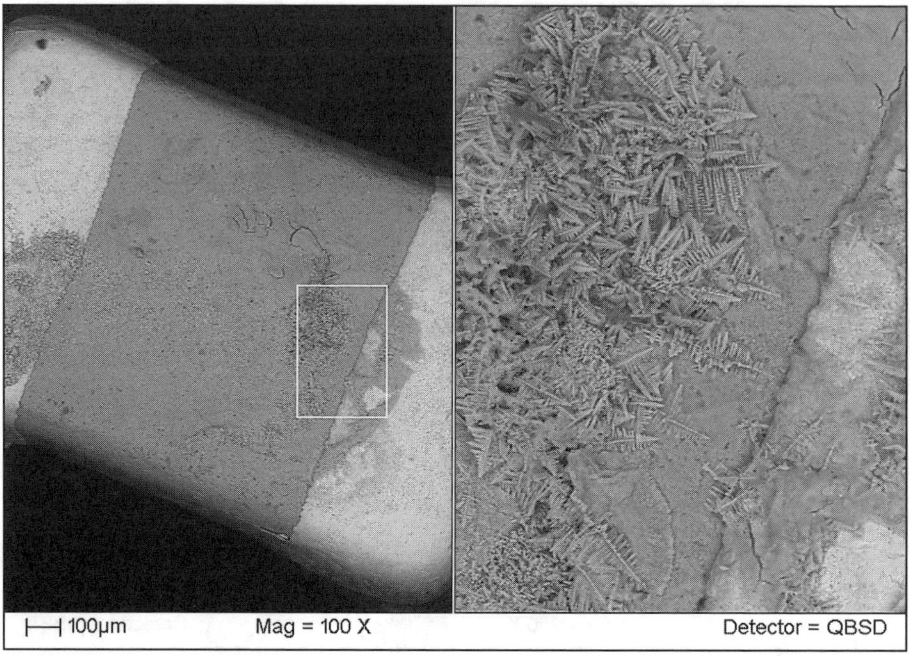

Fig. 9–29. Tin electrochemical migration on a capacitor with tin termination. The right half of the picture shows a higher magnification view of the Sn dendrites at the cathode end of the termination.

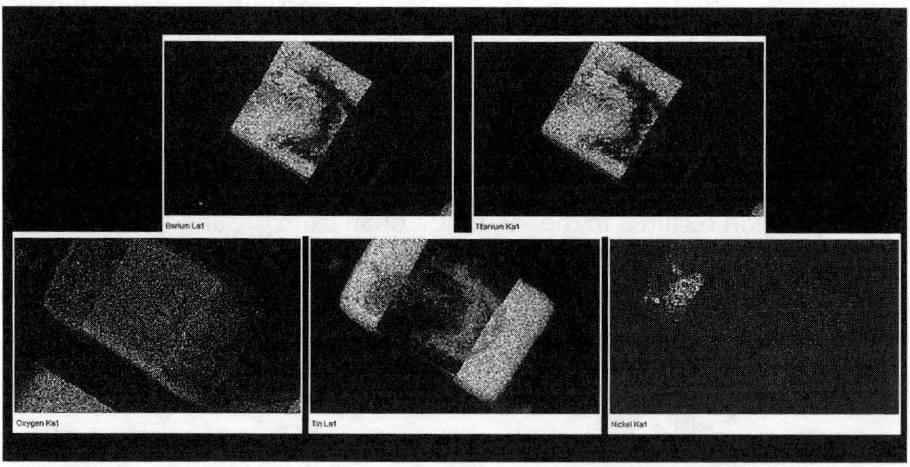

Fig. 9–30. Elemental maps for the capacitor shown fig. 9–25 for (a) Ba, (b) Ti, (c) O, (d) Sn, and (e) Ni showing the presence of Sn ECM between the terminations and exposure of the Ni barrier layer under the consumed Sn surface at the anode

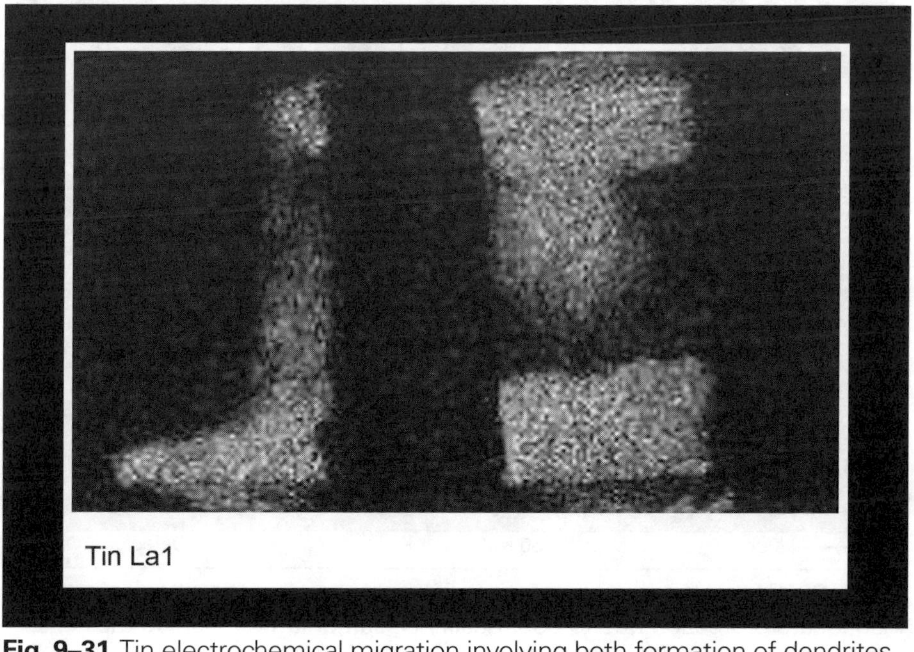

Fig. 9–31. Tin electrochemical migration involving both formation of dendrites and colloidal form of ECM on a resistor with pure tin termination

Silver ECM can occur on the PWB if there is exposed metal in the termination or pad finish, or it can occur on the surface of passive devices separate from the surface of the PWB. The occurrence of ECM on the surface of passive devices can potentially be a more serious reliability risk because of the current trend towards smaller size passives, which provides a ready site for ECM. A coating of Ag is commonly employed at the ends of the passive device to ensure that there is a good contact between the electrodes in a capacitor. However, since Ag is prone to ECM, it is advisable to isolate this Ag from the environment. Therefore, Ni is used as a barrier layer between the Ag base and the Sn outer layers. To be effective, this Ni layer should be continuous and free of cracks or gaps. In the event that the Ni layer is discontinuous, Ag can be exposed to the environment leading to dendrite formation as illustrated in figure 9–32. Here, dendrites of Ag can be seen growing on the surface of the passive component after damp-heat reliability tests.

Fig. 9–32. Silver electrochemical migration on a resistor with tin termination. Inadequate protection due to poor quality Ni barrier layers enabled the Ag to exhibit electrochemical migration.

Display failures

The display is a crucial subsystem of most portable electronic products. Since the display is one of the primary interfaces with the user and display failures are visually striking, portable electronic product field failure due to display-related issues constitute a significant fraction of the overall failure rate. It is reported that 2%–3% of all mobile phones suffer display damage as a result of glass breakage (Inoue and Fukuchi 1999).

The dominant failure symptoms in displays in mobile phones are shown in figure 9–33, and the dominant failure modes are shown in figure 9–34. It is seen that blank displays, missing lines, and missing lines/pixels constitute the dominant symptoms, while the dominant failure mechanisms are glass fracture, broken flex, and moisture uptake-related anisotropic conductive adhesive film (ACF) failure.

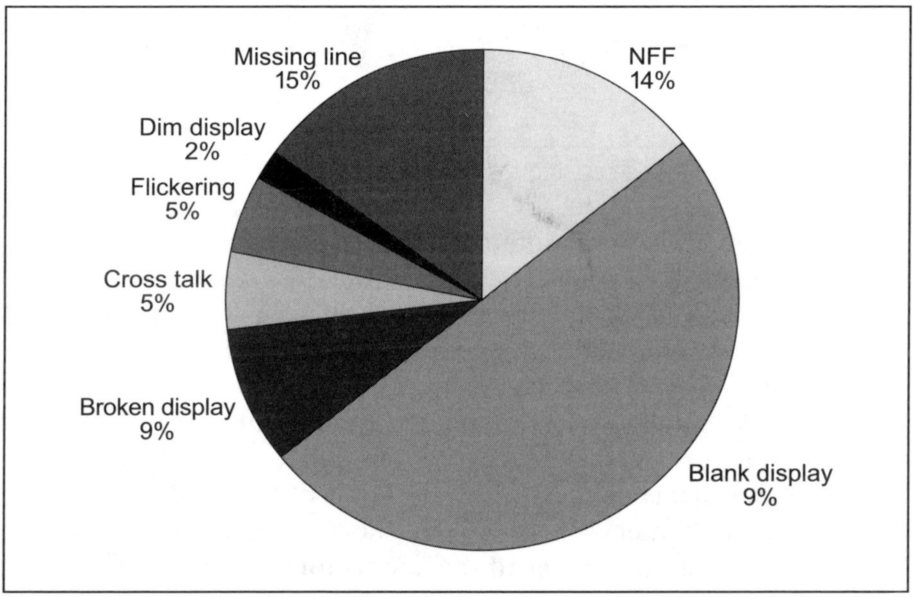

Fig. 9–33. Failure symptoms related to mobile phone displays

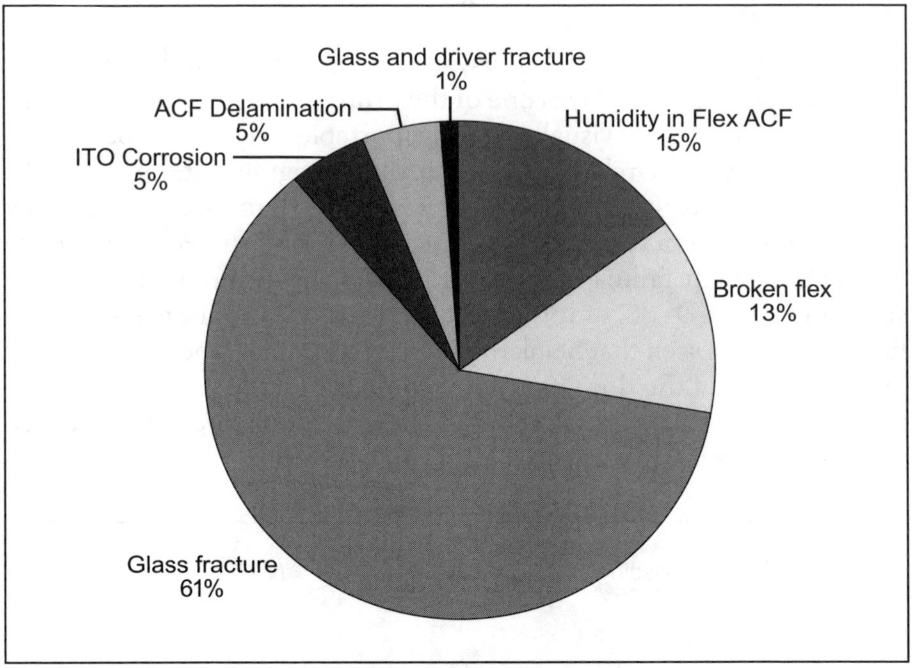

Fig. 9–34. Failure modes causing the failure symptoms in fig. 9–33

While there are several different kinds of construction in vogue in display modules, it is instructive to examine the generic construction of a display module, as shown in figure 9–35. The glass display itself typically consists of a sandwich of polarizing films, reflecting films, and two glass plates between which the liquid crystal is present. In addition, one of the glass plates also has indium tin oxide (ITO) traces to and from the driver (Si) flex tape on which numerous passives are soldered, and a flex connector that connects the display to the circuit card. Failures are, therefore, a function of electrical opens, shorts, or intermittent connections starting from the driver to the connector on the flex cable such as broken flex or broken copper traces on flex, corroded copper traces, broken solder joint, broken component termination, etc.

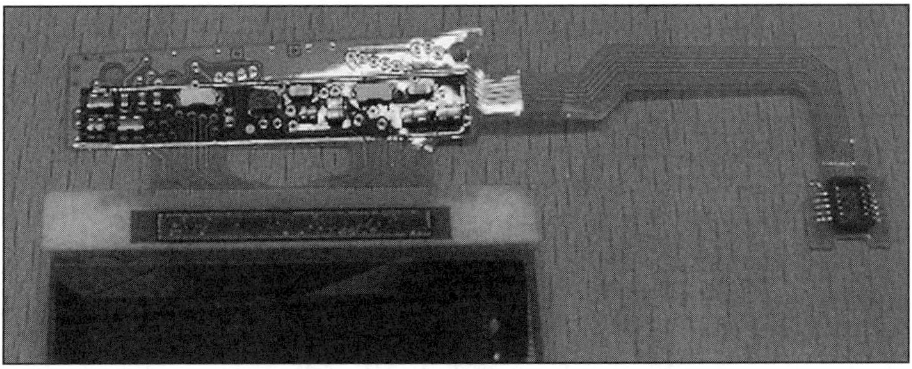

Fig. 9–35. Generic construction of a display module

An example of broken traces is shown in figure 9–36, and this can occur if the flex cable is bent with an excessively small radius of curvature or fatigued repeatedly beyond its elastic limit. Sometimes, these failures can occur during product assembly because the flex cable is manually handled during manufacture. Another issue with conductors on flex cables is the surface finish. If ENIG finish is used on the flex portion, care should be taken that it does not crack during product assembly because ENIG finish can crack during mechanical deformation causing fallout at line. Also, the thickness of the Cu trace, size and structure of Cu grains, and the residual stress in the trace after flex manufacture are key variables to minimize fracture of traces during use or assembly.

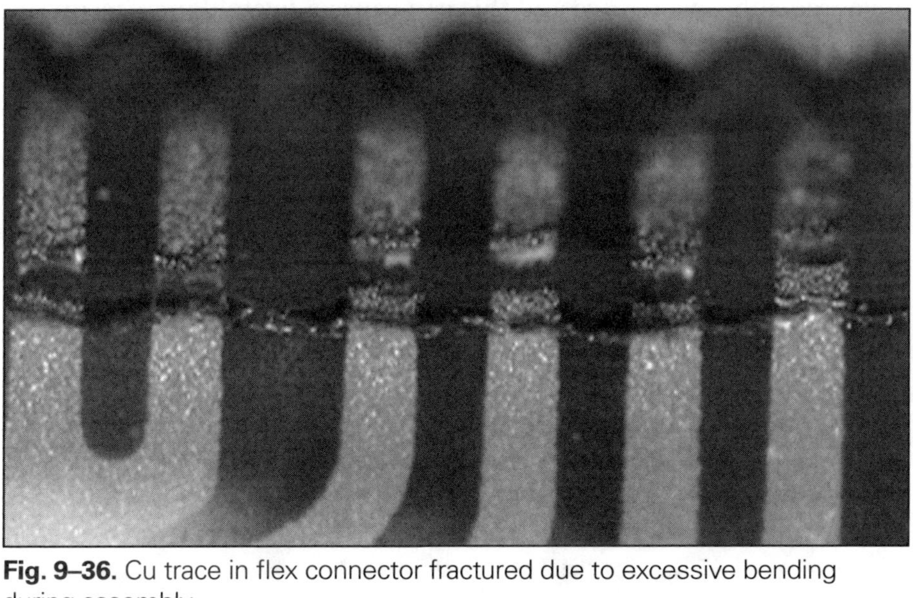

Fig. 9–36. Cu trace in flex connector fractured due to excessive bending during assembly

The ACF is composed of a nonconductive polymer matrix with Au-coated polymer spheres making the electrical connection between the ITO traces and the gold-plated pads of the driver IC when the chip is attached to the glass display under force and heat. Upon curing the polymer adhesive, the connection offers a stable resistance except when subjected to humidity. The adhesive film absorbs humidity and swells up, thereby increasing contact resistance or an open in extreme cases. This failure mechanism is well documented in the literature for chip-on-glass and other applications (deVries 2004). To extend the useful life of this joint, the ACF bond is often sealed with a silicone or other humidity barrier. At a portable product level, display failure due to humidity is a common failure mechanism and is often controlled by designing the enclosures (mechanics and plastics) in such a way that water does not collect near the display. Another factor controlling the performance of the display when subjected to humidity is the extent of delaminations or voids in the ACF which in turn is dependent on variables like surface contamination and process parameters. It is difficult to determine whether a particular display has failed by this failure mechanism, because optical and acoustic microscopy are not sensitive enough to reveal the difference between a failed and functional display, but electrical resistance measurements are a more reliable indicator. An example is shown in figure 9–37, illustrating the voids commonly seen in ACF and the presence of Au–Ni coated polymer spheres forming the conductive path. Upon exposure to high temperatures, the compressive forces from the cured adhesive reduce and this can cause a fivefold increase in the contact resistance (Rizvi et al. 2008). Further, the thermal stresses can crack the Au–Ni coating on the polymer spheres, causing an increase in resistance.

The phone or portable product is often subject to mechanical loads in the use environment, and these mechanical loads are sometimes transmitted to the display module. In the case of an improperly designed structure, or if the forces are excessive, the display driver (fig. 9–38) or the glass display itself can fracture (fig. 9–39), leading to display failure. The display driver fractures are relatively easy to identify once the module is disassembled. Often, the remedy to this failure mechanism is to redesign the display to ensure that the externally imposed forces are not transmitted to the display via cushioning or compliance in the module seal.

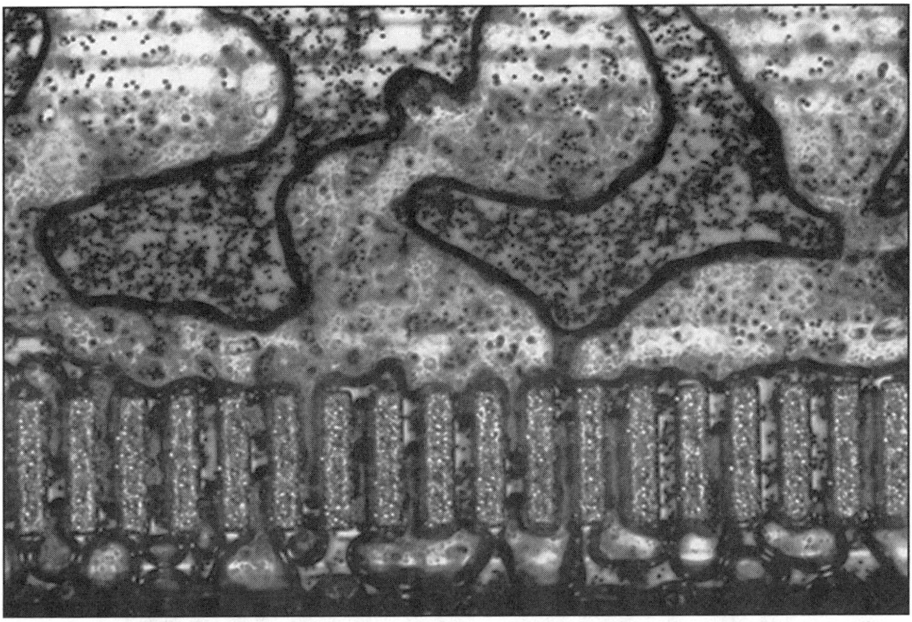

Fig. 9–37. Optical micrograph of a functional display revealing the conductive polymer spheres in the ACF (dark and bright spots) and the voiding away from the pads of the driver

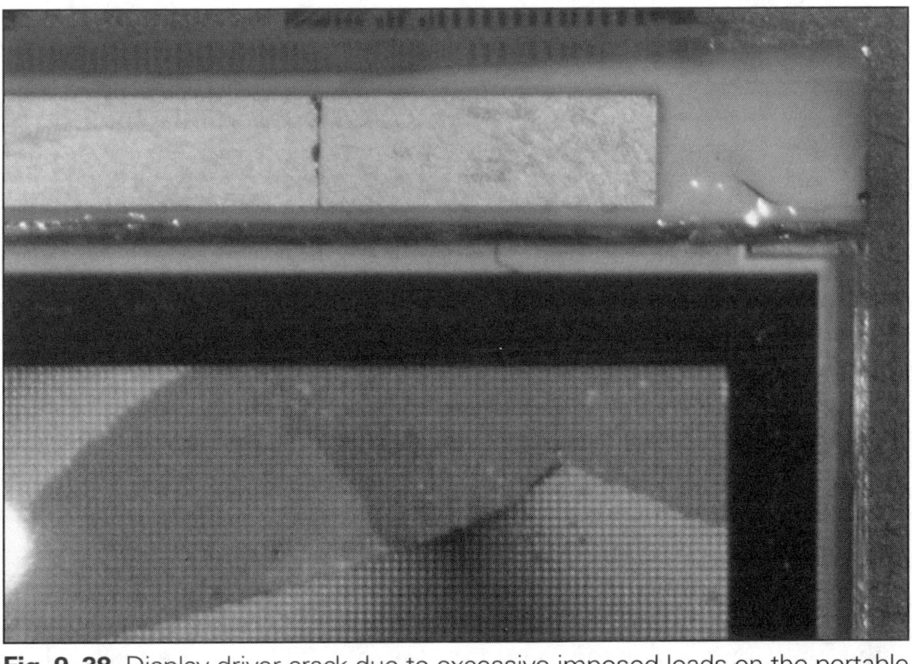

Fig. 9–38. Display driver crack due to excessive imposed loads on the portable electronic device

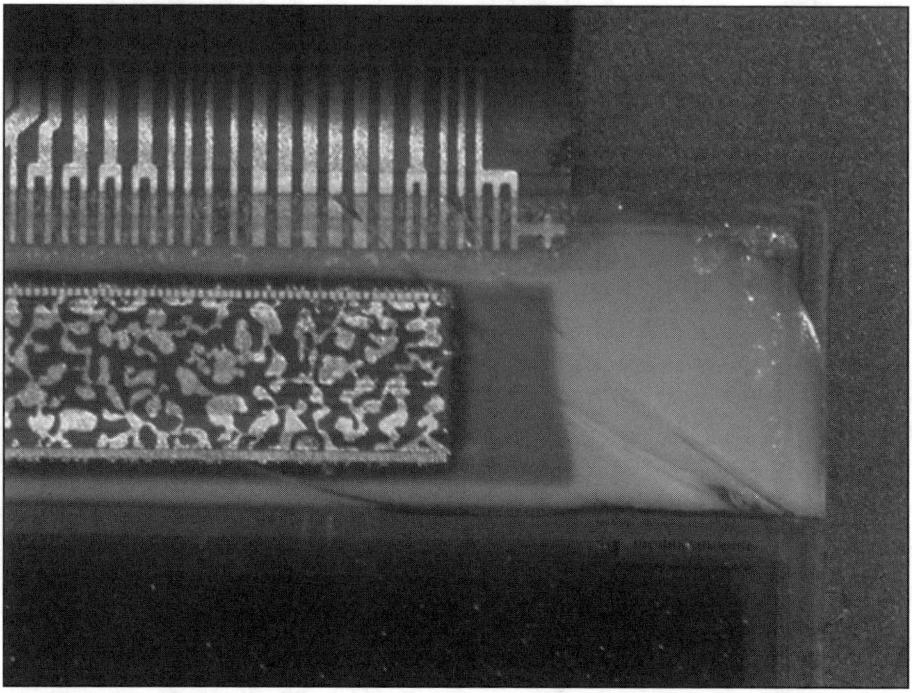

Fig. 9–39. Display glass fracture at the ledge aided by the presence of pre-existing cracks at the glass edge due to the cutting process

Display fractures, on the other hand, are as much a function of the design of the module as the glass-cutting quality. The normal procedure to break the glass laminate is to score it with a sharp diamond tip or stylus and fracture it along the scored line by applying a bending force. However, depending on the process variables such as force exerted during the scoring, residual stress state of the glass, sharpness of the knife, etc., the edge can have ancillary cracks extending a few micrometers from the cut surface. If the size of these cracks exceeds the critical crack length, catastrophic fracture can occur, and example is shown in figure 9–39. In this example, the fatal fracture originated at the ledge between the two glass plates due to the higher stress concentration at the geometric discontinuity and was probably aided by the presence of a pre-existing crack that satisfied a critical threshold for that glass material. While the edge cracks cannot be easily eliminated, a controlled process can help improve the reliability by limiting the size of the cracks due to the cutting process.

One of the common failure symptoms in display-related failures is the presence of missing lines or pixels, and ITO corrosion is a common failure mechanism for this symptom (Wang 2003). The deposited ITO traces are susceptible to corrosion by halogen species and the traces corrode readily in the field, as shown in figure 9–40, if not protected appropriately. ITO films suffer anodic corrosion where the Cl^- and OH^- ions draw electrons from the In–O surface bonds, and the corrosion rate is especially high in acidic environments with pH <7 (Folcher et al. 1997).

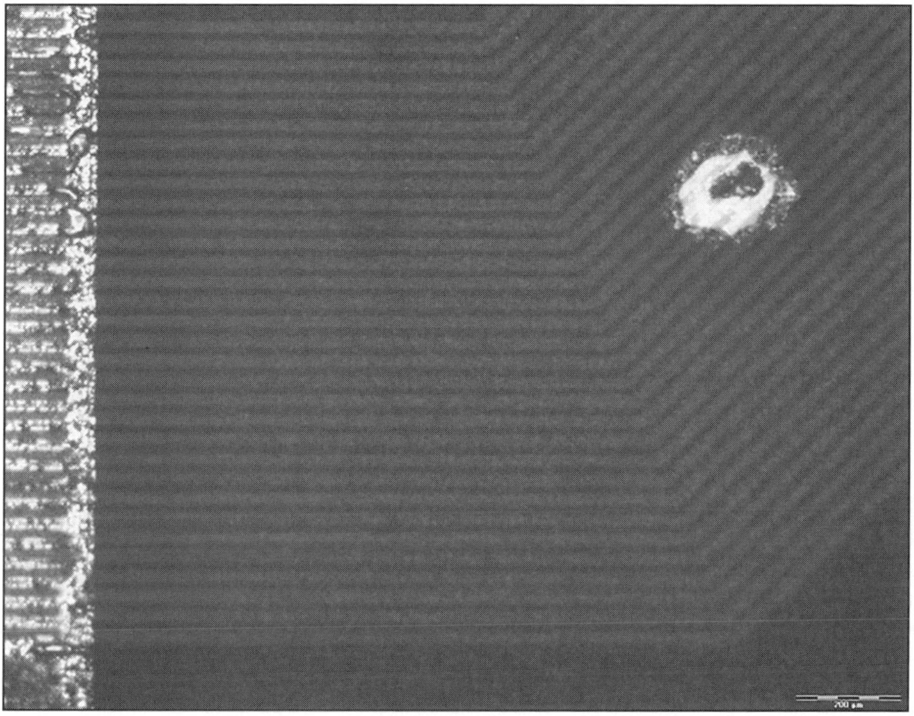

Fig. 9–40. Optical micrograph of ITO corrosion in a display

Reliability Test Practices

Accelerated thermal cycling test practices are influenced not only by the design life and the operating environment but also by the nature of the PWB assembly. In a majority of cases, portable electronic hardware by its very nature has to be small, lightweight, and possess high I/O density. This implies the use of surface laminar circuitry or other HDI PWB technologies. Also inherent is the use of small, low-profile packages.

It has been reported that the industry standard temperature cycle profile, where the upper and lower temperature dwells are invariant, leads to an underestimation of fatigue life (Dishong et al. 2002). It is well known that inelastic strain accumulation is generally proportional to fatigue life. It has been suggested that temperature fluctuations during upper dwell times can reduce elastic strain accumulation, and, as such, using minicycles during dwell times will reduce the maximum inelastic strain. The magnitude of such inelastic strain reduction depends on the number of minicycles and their temperature ranges. Thus, selective superposition of a judicious number of minicycles during the high temperature dwell may enable a more realistic fatigue life prediction.

In addition, portable electronic hardware involving radio communication features can have components such as power amplifiers and RF devices that may run hotter during operation in addition to the thermal exposure imposed by the environment. The thermal effects in such cases can cause excessive growth of interfacial intermetallics, which may be deleterious to the interconnection integrity. Power cycling tests may be much more appropriate in such cases.

Thermal and mechanical stress exposures in portable electronic hardware are rather frequent and sometimes concurrent in contrast to desktop machines, and the effect on product performance can be significant. For example, the interfacial intermetallic growth, which by itself may not affect the solder joint integrity due to the compliance of the alloy, can progressively degrade the mechanical reliability. Thus, separate thermal and mechanical reliability assessments may not reflect the true product performance, as the synergistic effect is not taken into account. The effect of thermal aging on the mechanical reliability can be significant and should be considered in all reliability assessments.

Prevention of failures

The rate of change in consumer electronics technology has been accelerating at an ever-increasing pace, and there is a premium on getting product reliability right the first time. Failures in the field upset today's consumers who demand high quality at a low price. Warranty costs can eat into already slim profit margins in addition to eroding the brand value. Thus reliability not only directly affects the profit margin but can have a financial impact far exceeding the warranty and liability costs. Profitability drives the goal to build a reliable product without any associated delay in time to market or cost of manufacture.

The various package configurations, materials, accelerated test methods, reliability statistics, failure modes, and failure mechanisms have been described in the earlier chapters. While these concepts help in developing an understanding of why and how portable consumer electronic products fail, they do not directly address the issue of preventing failures from occurring in the first place. The goal of this section is to briefly highlight various tools and techniques useful in migrating "reliability"-related activity upstream in the design process. The schematic in figure 9–41 illustrates the "cost of reliability improvement activity" in the product development process and the "cost of failures" as function of the product development cycle. Clearly, it is much more cost effective to incorporate and improve the reliability early in the product development lifecycle rather than after field-related failures. Some of the methodologies used to achieve this goal are the following:

1. Concurrent engineering process
2. Design for
    a. Manufacturing
    b. Assembly
    c. Qualification
    d. Test
    e. Reliability
    f. Environment (green)
3. Continuous quality improvement.

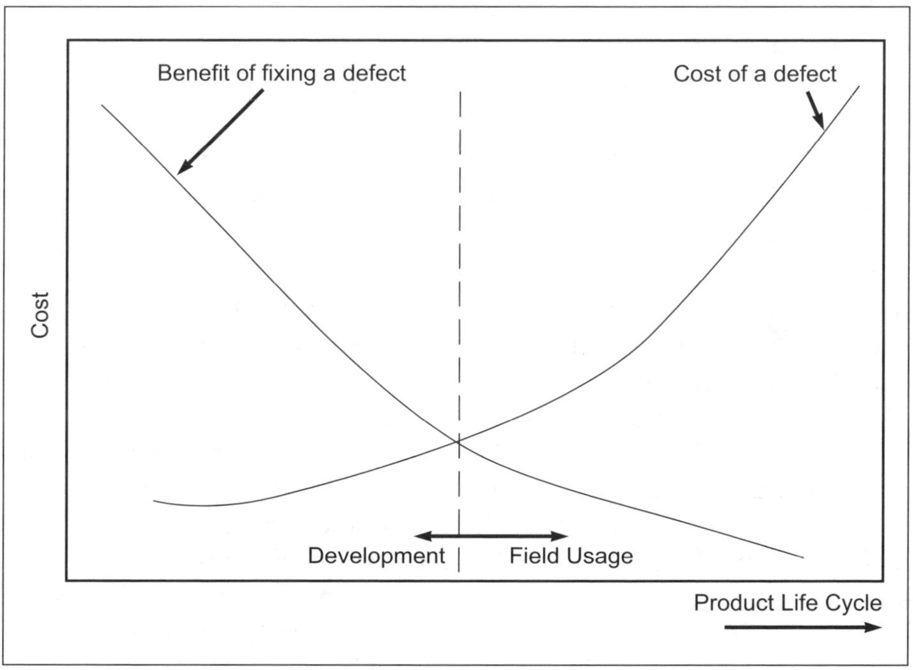

Fig. 9–41. The cost of finding a defect is much smaller and the benefit of fixing a defect is much larger during the development phase than in the field return portion of the product life cycle.

Concurrent engineering process

In the traditional engineering process, various developmental and production areas work on the development in a sequential process. For example, the product design and specifications are given by the marketing and design engineering teams followed by prototype build and test activity where the product assurance and development teams build and test the prototype before transferring it to the manufacturing and quality engineering teams. In such a sequential approach, it is not easy to change upstream decisions and time and effort can be expended in addressing issues that were caused by early design and material decisions. It has been known for some time that 70%–80% of all product defects can be directly related to design issues (Santina 1996). It is also estimated that 75% of the manufacturing cost of a product is determined by its design, 60% of which is derived from design decisions made early in the development process (Holden 1993). Hence, quality and reliability cannot be tested or inspected into a product but needs to be designed or built in. Concurrent engineering methodology (or simultaaneous engineering) provides a framework to "build" reliability into a product and the team consists of marketing, materials, design, manufacturing, test, and reliability engineers that stay with the product all the way from concept to high-volume production with different disciplines taking the lead in different stages of the product development lifecycle.

Design for X (X = manufacturing, assembly, etc.)

Design for Manufacturability. This methodology forces the engineering and other disciplines leading the earlier stages of the product development lifecycle to consider the ease with which the product can be manufactured in high volume. For example, consider the case of discretes such as capacitors, resistors, etc. Recognizing that uniform orientation and placement of these components minimizes rotation during pick and place operation, increases throughput, eases inspection of the solder joint, and simplifies repair, the PWB layout team can make a big impact on the manufacturability of the product. Another example is the discrete land pattern geometry. The quality of the soldering is determined to a large extent by the relationship between the termination and pad sizes and the length of the component. An excessively large pad may result in excessive solder paste, while too small a pad will lead to insufficient solder to make a reliable joint (Horsley 1988).

Design for assembly (DfA). This methodology was developed to help design engineers select designs that are easily assembled. DfA is applicable to portable electronic products that almost always require some operator assembly steps during the manufacture. Standardization, smaller bill of materials (BOMs), and part count directly influence the costs of errors that can be made in assembly. For example, instead of using three or four different closely related screw sizes, a single screw size would reduce cost and the risks of using a wrong screw during assembly. Also, press-fit or snap-fit fasteners can reduce the use of fasteners, making assembly simpler and faster.

Design for test. The trend towards increased functionality packed into continuously shrinking PWB real estate puts pressure on manufacturing yields and the ability to test these components in the assembled state. Thus, test requirements for the product need to be considered in the early stages of the design, especially for components controlling the reliability of the critical subsystems. The advantages of using techniques such as boundary scan method is that the system-level test can be performed on the product for diagnosing a field failure at the customer site and test consistency can be achieved throughout the product design cycle (Adams et al. 1995).

Design for qualification. In the traditional approach, prototypes are tested according to a defined qualification plan to make sure that product reliability meets requirements. Typically, these tests are pass/fail tests that are time-censored and the failures are analyzed for root cause. The design is improved based on the failure analysis and the redesigned product is tested again in an iterative approach to achieve desired reliability. However, in the context of concurrent engineering, the test methodology is modified to understand the physics of failure, and attribute data (pass/fail) are rejected in favor of continuous failure data. The emphasis is on identification and prevention of problems, and in this context conventional "failure analysis" morphs into an exercise to determine the physics of failure.

Design for reliability (DfR). It is an emerging discipline for designing reliability into products. This methodology encompasses several tools and practices. For example, in reliability modeling, the top-level reliability requirements are allocated to subsystems by design engineers and reliability engineers working together using block diagrams and fault trees to provide a graphical means of evaluating the relationships between different parts of the system to failure. Techniques such as failure mode and effects analysis (FMEA), root cause analysis (RCA), accelerated testing (AT), Weibull analysis, etc., are employed in the toolbox for DfR.

Design for green (DfG). Design for green is a new activity initiated by the thrust towards sustainable product development with a view to minimize the environmental cost and meet regulatory approvals. In some areas of the world, notably in Europe and Japan, the OEMs are sometimes asked to pay for the recycling or reuse of the products they sell, and in most cases the OEMs have to pay for these costs along with the product manufacture. In such cases, it makes good economic sense to design the product such that recycling or reuse is facilitated. For example, the use of metals and alloys, plastics, and fasteners that can be recycled with minimum energy ensures that the impact on the environment is taken into account during the product development lifecycle. Another example where DfG makes economic and environmental sense is the design thrust towards low power requirements.

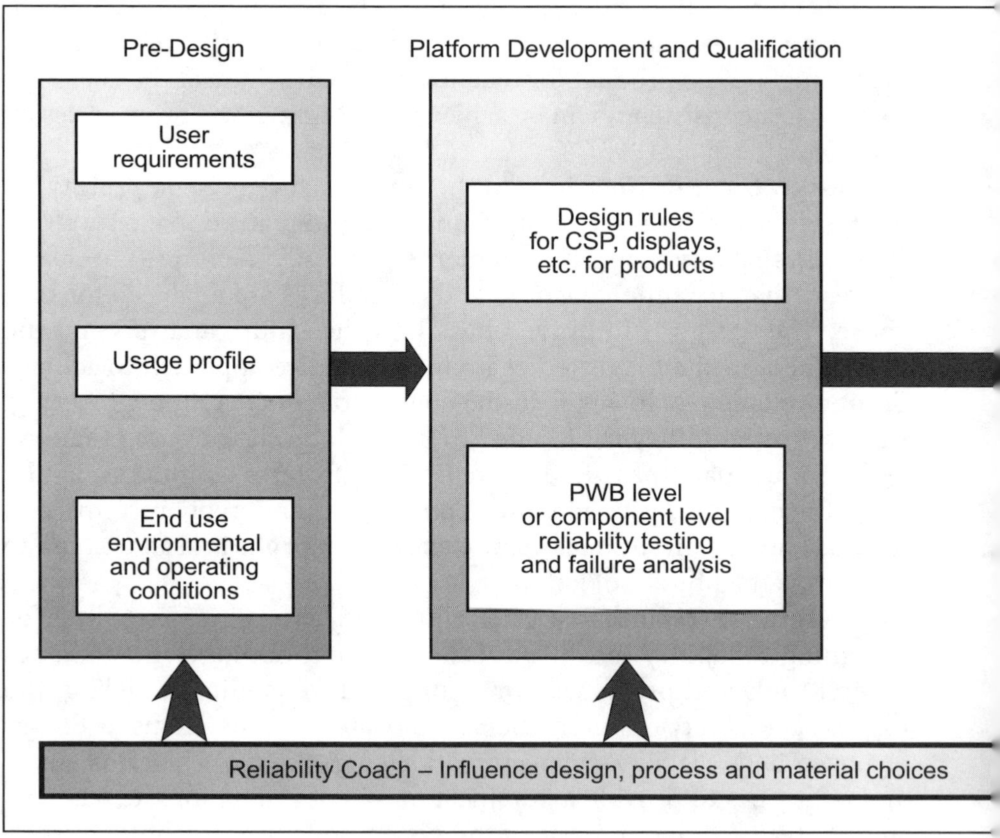

Fig. 9–42. Schematic of a process to prevent failures in portable electronic products

Framework for product development to prevent failures

We outline below a framework by which electronic products can be developed to prevent or minimize failures in the field using the concepts described earlier in this and the previous chapter. The intent is to illustrate the principal elements (as shown schematically in a flow diagram in figure 9–42) at various stages in the development process rather than provide a process and recipe for building failure free products. We will illustrate the steps in this framework for the particular case of drop impact stresses.

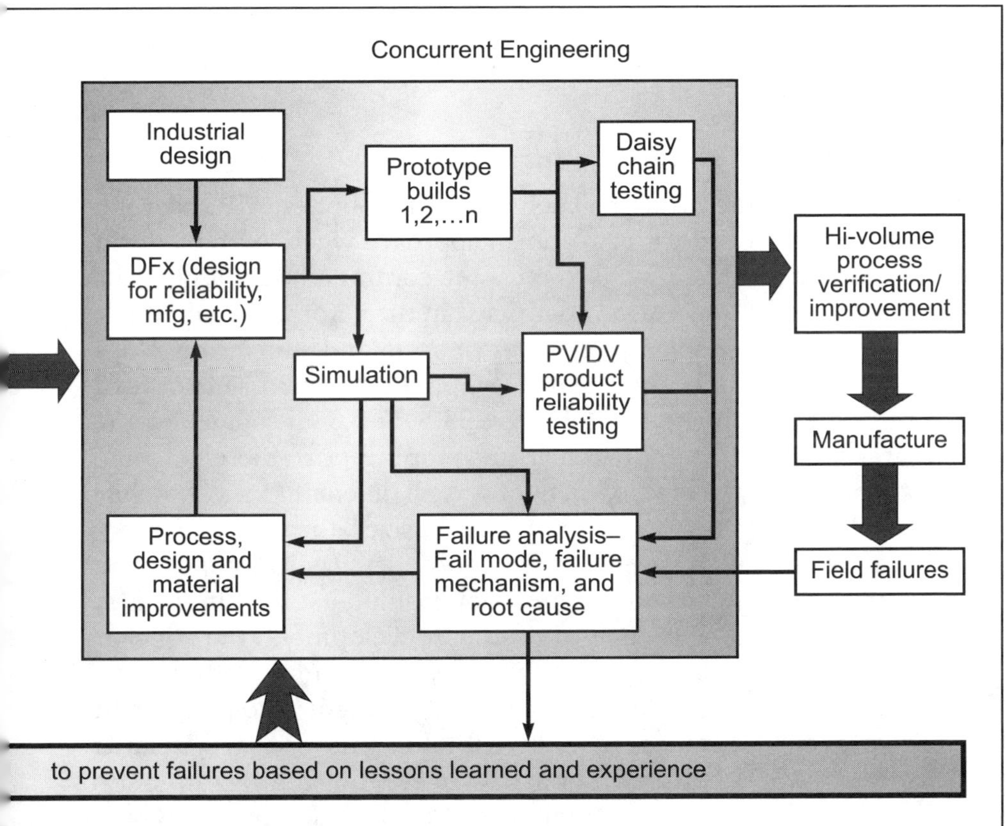

Understanding user requirements and use conditions

A necessary first step in the product development is to get a clear understanding of user requirements with regard to cost, functionality, reliability, etc. Simultaneously, it is essential that the product designers understand the usage profile for their product and the end-use environmental and operating conditions. For example, as explained in chapter 8, the usage profile of mobile phones was collected by employing various transducers embedded in real products and given to representative users for normal usage. This knowledge is essential in defining the bounds of what constitutes "normal" usage, and without this information, defining acceleration factors and pass/fail criteria becomes a fruitless activity. From the study performed by Vakevainen et al. (2001) we know that mobile phones are subject to approximately one drop per week, which translates to about 50 drops per year from heights ranging from 0.5 to 1 μ.

The platform and module approach

Portable electronic products such as mobile phones, MP3 players, etc., are often developed with a platform approach where each platform or generation of products is based on some common underlying technologies. This approach is also widely used in the automotive and other markets because it enables economies of scale and shorter development times. For example, the BMW's E90 platform gives rise to several different models such as the 325i, 328i, 330i, and 335i sedans for the U.S. market. A similar approach in developing portable electronic products can enable the development of several different kinds of mobile phones or MP3 players with different colors, sizes, form factors, functionalities (WLAN, high speed data, Bluetooth, etc.), visual appeal, and user base (voice only, ruggedized, outdoor, fashion user, business user, etc.). In this platform approach to product development, the concept of modules can yield benefits. For example, a particular platform can offer several different display modules with different resolutions and power requirements, modem modules with different wireless standards such as Bluetooth, WLAN (802.11 a, b, g, n, etc.), or different modules of the processor with different amounts of on-board memory (8 MB, 32 MB, etc.) or camera modules with different resolutions (1 Mpixel, 5 Mpixels) etc.

The platform development team would develop these platforms with a 3–4-year time horizon with little detailed knowledge of how the product would even look like. Often, these platform development activities are at the cutting edge of technology, and the reliability requirements are specified at the board or component level rather than at the product level. In such cases, test specimens use standards such as the JEDEC standard for drop test. Figure 9– 43 shows an example of a CSP mounted on a JEDEC drop test board that can be subjected to board-level drop tests from heights of 0.5–1 m. The CSPs are typically daisy-chained at the first- and second-level interconnection but each CSP is on a monitoring net separate from the other CSPs on the board, so it is possible to know exactly when each CSP interconnection has failed. The board typically also has two probe points for each side to deduce which side or corner has an electrical open. The drilled vias at either end of the board are typically used for connecting to the data acquisition system. Board-level daisy-chain testing for assessing first- and second-level interconnection reliability is also evaluated under thermal cycling conditions. The PWB design for thermal cycling applications is very similar to the drop testing case with one exception. In thermal cycling testing, it is necessary to be able to separate out components that fail so that the damage state at failure can be evaluated. Figure 9–44 shows an example of PWB for daisy chain thermal cycling board, and the cutouts around each CSP can be seen. While these cutouts are convenient to separate out failed CSPs and prevent further accumulation of damage in the solder microstructure, they are not used in drop testing situations because the presence of the cutout alters the dynamic behavior of the PWB during drop.

Fig. 9–43. JEDEC drop test board with layout for testing 15 daisy-chained CSPs on single board

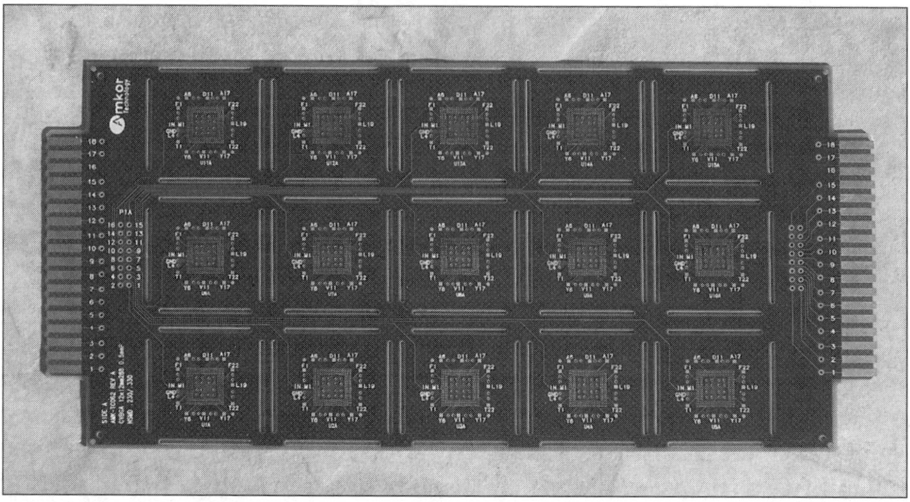

Fig. 9–44. A typical PWB design for daisy chain thermal cycling showing the cutouts around each CSP

The end result of PWB level daisy chain testing typically is design rule generation. For instance, one design rule for a CSP might be that the product design should be such that the PWB strain near the corner ball should not exceed 2,500 μstrain when the product is dropped from a height of 1 m onto concrete. As long as this design rule is not violated for this package, the product developers can expect acceptable reliability in the field under drop conditions.

The advantage of PWB level daisy-chain testing is that the interconnection reliability can be evaluated to the exclusion of the functionality of the component and interaction with other elements of the portable product. Conversely, the disadvantage is that the interaction with the other components and the housing is not easily captured in PWB level daisy-chain testing. This can be a severe shortcoming because screw torque, number of screws, order of screw tightening, etc., can affect the strains imposed on the components on the PWB. For this purpose, both simulation and product level daisy-chain and functional testing are carried out at the product level, which will be discussed later in this chapter.

Concurrent engineering for product development

The first step in the product development concurrent engineering process is defining the product, features, and functionality. The industrial design is typically finalized in conjunction with the product marketing team without excessive involvement of the reliability engineers. The task of the engineering team is to take this industrial design and turn it into reality. Different DfX approaches such as design for reliability, design for manufacturability, design for green, etc., are employed in the design engineering phase of the product development. Typically, the product design maturity is achieved via an iterative design-build-test-analyze approach. The prototype builds are structured such that they are sequentially more representative of the final product. The disadvantage of this approach is that prototypes tend to be very expensive to build and reliability data from the earlier builds is typically very sparse because the product functionality in the earlier builds is minimal and largely unrepresentative of the final product. Consequently, the reliability data that is most representative of the final product is available only at the end of the concurrent engineering process, when making product modifications to improve reliability can be prohibitively expensive both in terms of cost and time to market.

In this context, simulation is increasingly becoming a mainstay of the concurrent engineering process. Indeed, with the advent of faster computers and recent advances in stress analysis by the finite-element method, product reliability simulation is becoming increasingly accurate in identifying areas where the reliability of the design is inadequate in a relative sense if not in absolute terms. The advantage of simulation is that alternative designs can also be evaluated without requiring expensive hardware builds. Further, since the entire portable product is included in the model, the effect on any single component can be studied in isolation and geometry effects such as mechanical interference can be studied in addition to stress and strain. Figure 9–45 shows the finite element solid model of a portable electronic product, in which details of the keypad, display, hinge, and plastics are captured along with the internal components such as CSPs and PWB. The level of detail of the internal components captured in the model can be seen in figure 9–46(a) and (b), which shows the front and back, respectively, of the PWB along with all the major components. Details that are not typically captured in the simulation include the vias, microvias, solder joint microstructure and shape, intermetallic phases, and defects in the solder such as voids.

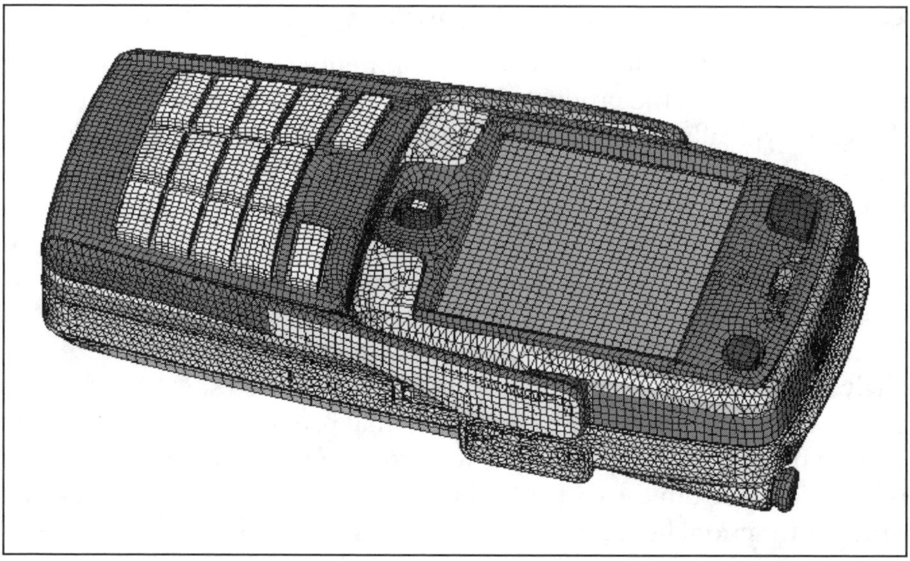

Fig. 9–45. Solid model of a mobile phone used in the product level simulation of reliability in drop and thermal cycling exposure (courtesy of J. Wu)

The simulation can cover any orientation of drop, and the level of stresses and strains in the solid model can be estimated. For instance, for the particular case of drop impact on the back of the product shown in figure 9–47(a) the strains on the PWB contours shown in figure 9–47(b) are highest at the edges of a CSP as shown by the ellipse in figure 9–47(c). In addition, the stress levels can be estimated as shown in figure 9–47(d). As can be seen from the stress values shown in figure 9–47(d), accuracy in the absolute value of the stress is difficult to achieve, but the value of the exercise is in estimating the location of the problem areas. While the prediction of the solder joint stresses is difficult from entire phone simulation, experimental verification of PWB strain predictions is relatively straightforward. In this particular case, strain gages were affixed in the location in figure 9–48(a). The mobile phone was dropped from the same height as was used in the simulation and in the same orientation as shown in figure 9–47(a). The maximum principal strain on the PWB is shown in the plot in figure 9–48(b), and it can be seen that the strain peak is comparable to the value predicted by phone-level drop simulation. In subsequent product-level daisy-chain drop testing, one might encounter solder joint failure as shown in figure 9–49 in the vicinity of the high PWB strain location identified earlier in the simulation and in PWB strain measurements. There is still room for discrepancy between the predicted high strain location and measured failure solder joint failure location because the structure of the package, stiffness of the interposer, surface finish of the PWB and component pads, reflow profiles, and IMC morphology all control final failure. Thus, different components exposed to the similarly high PWB strains can behave differently in terms of solder joint fracture.

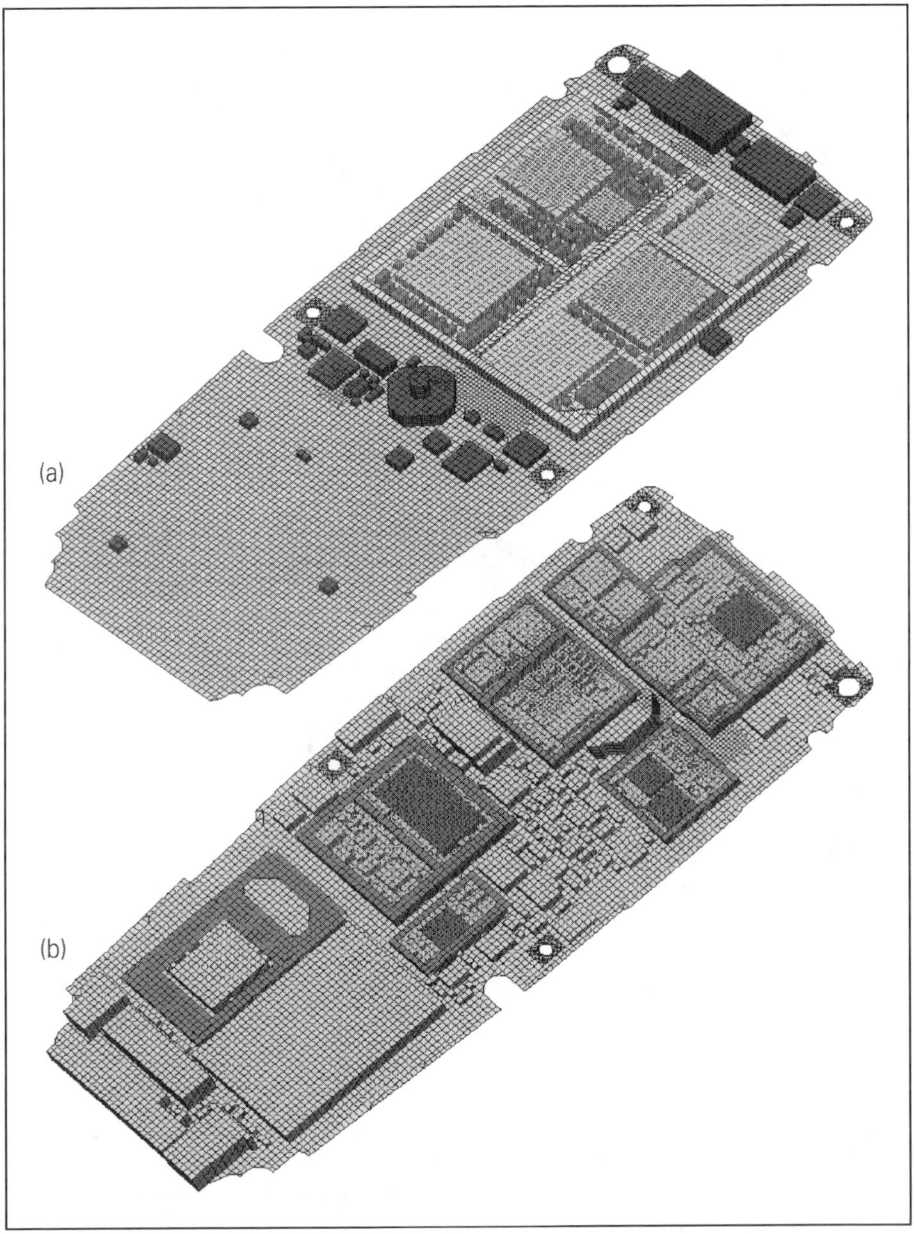

Fig. 9–46. Level of detail captured in the solid model of the simulation showing the (a) front of the board, and (b) back of the board (courtesy of J. Wu)

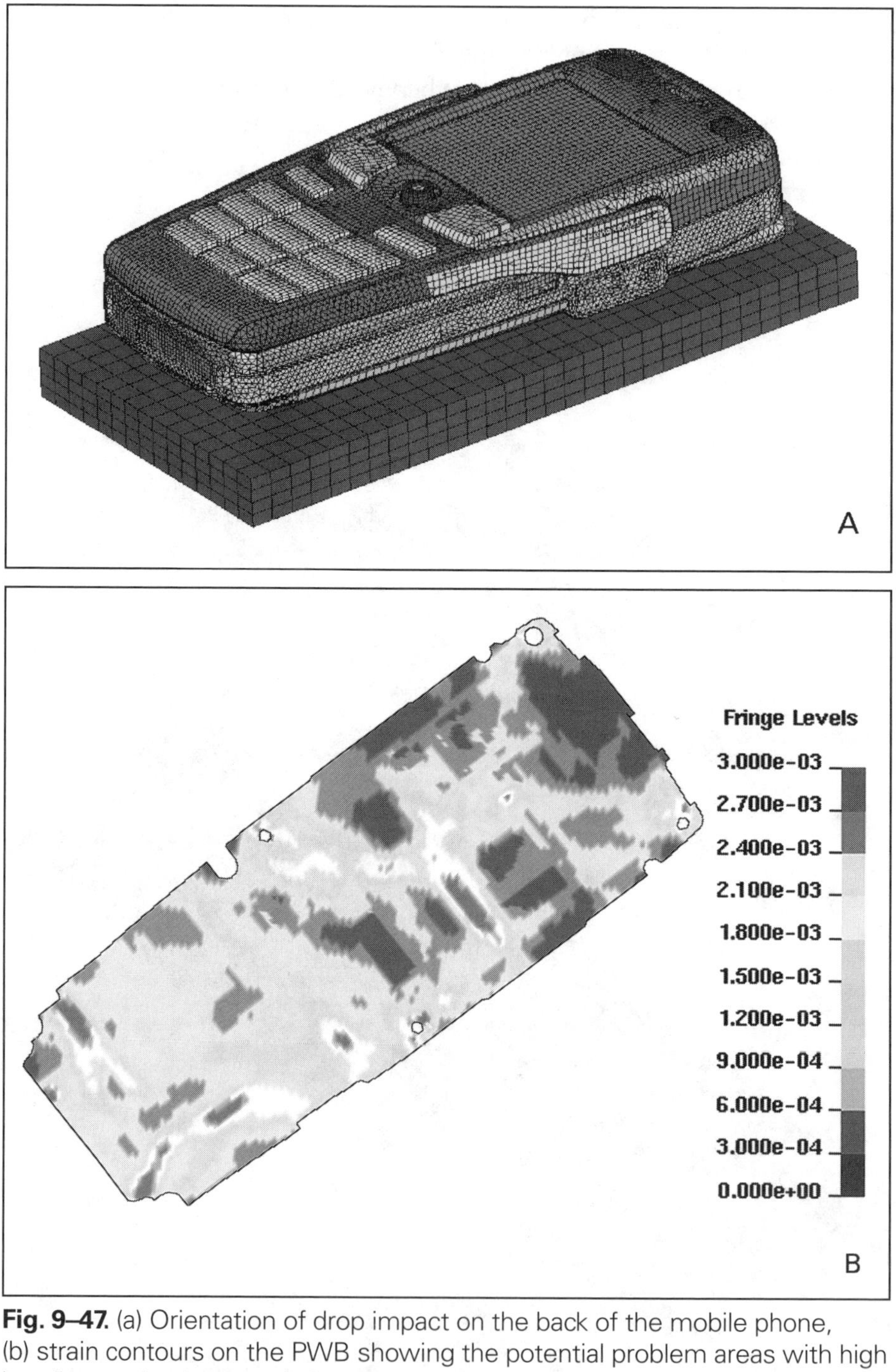

Fig. 9–47. (a) Orientation of drop impact on the back of the mobile phone, (b) strain contours on the PWB showing the potential problem areas with high bending strains (courtesy of J. Wu)

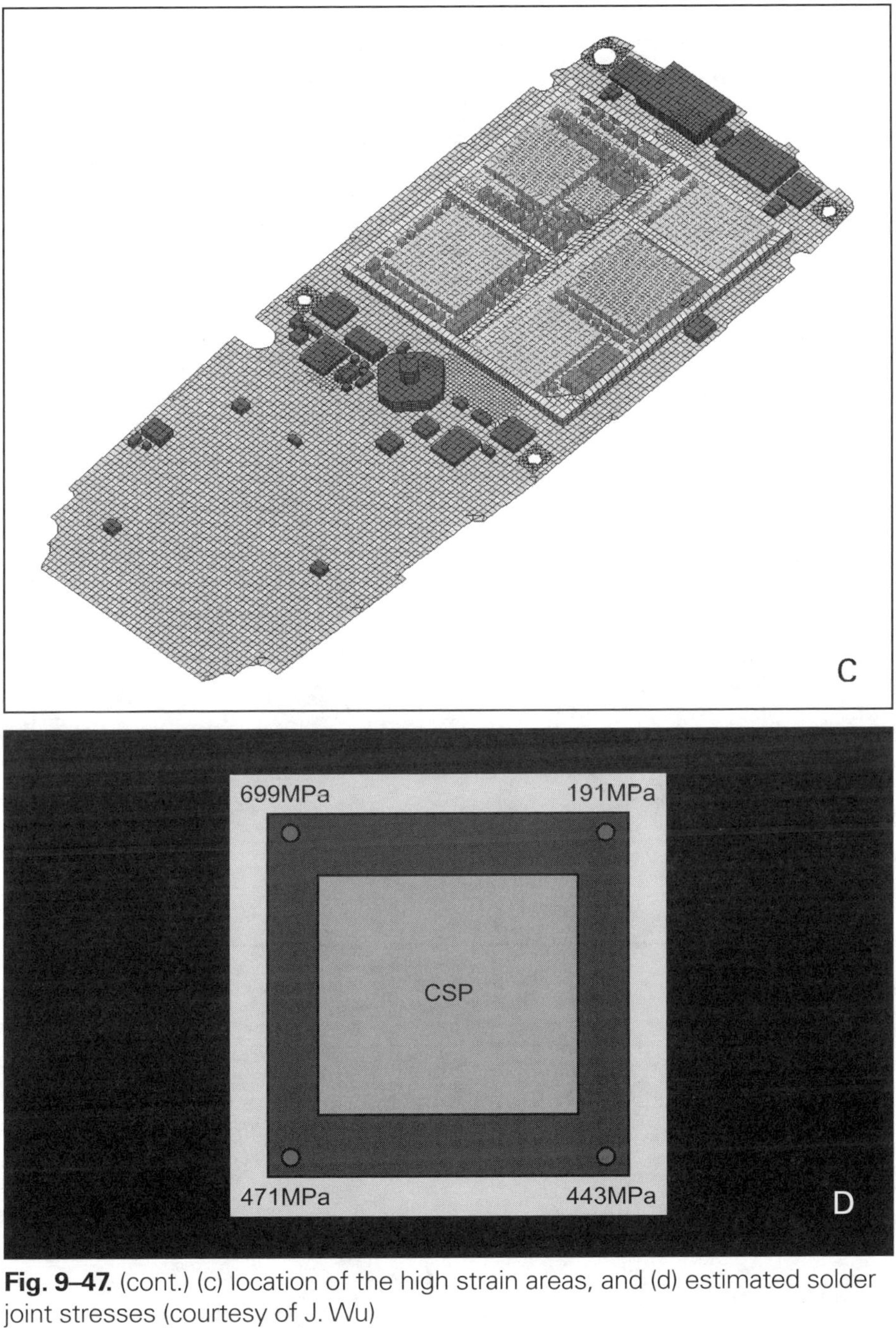

Fig. 9–47. (cont.) (c) location of the high strain areas, and (d) estimated solder joint stresses (courtesy of J. Wu)

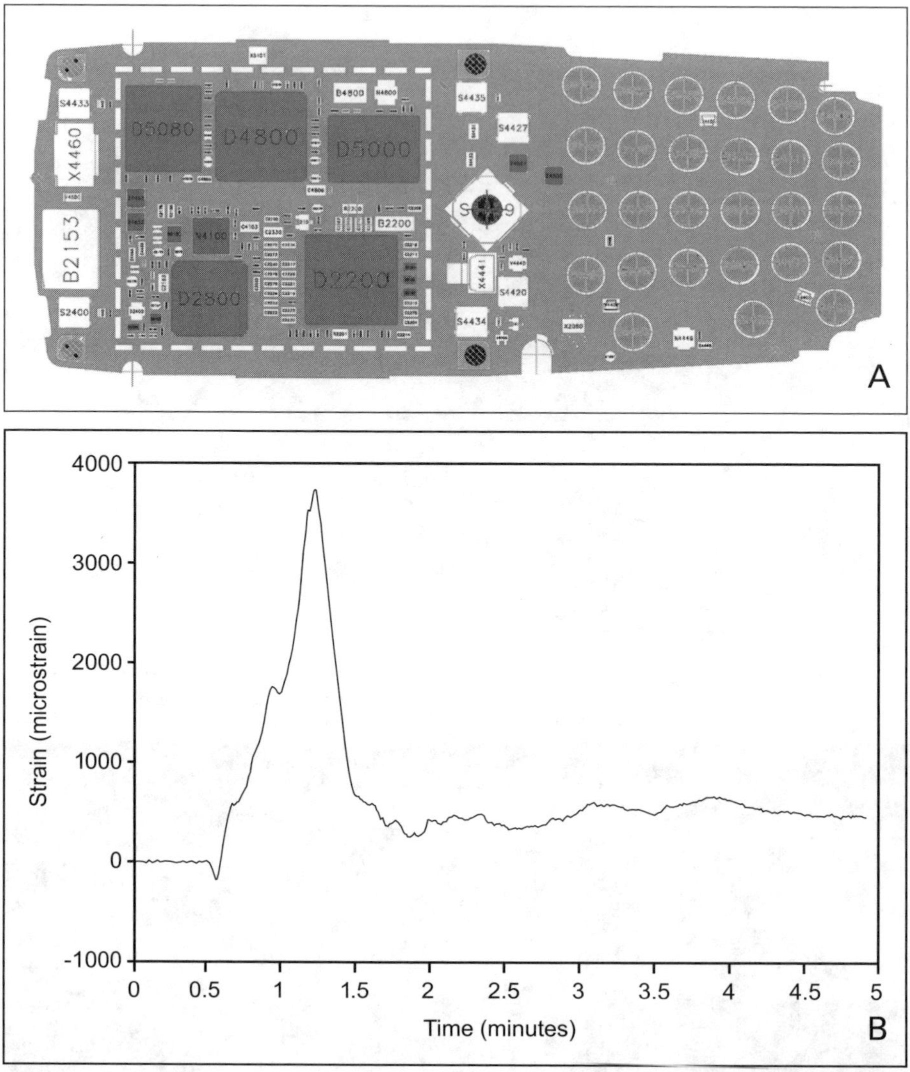

Fig. 9–48. (a) Location of strain gage, and (b) strain measured on the PWB as a function of time during drop impact showing the peak strain (3700 ue) close to that predicted from the simulation contour in fig. 9–47(b) (courtesy of J. Wu)

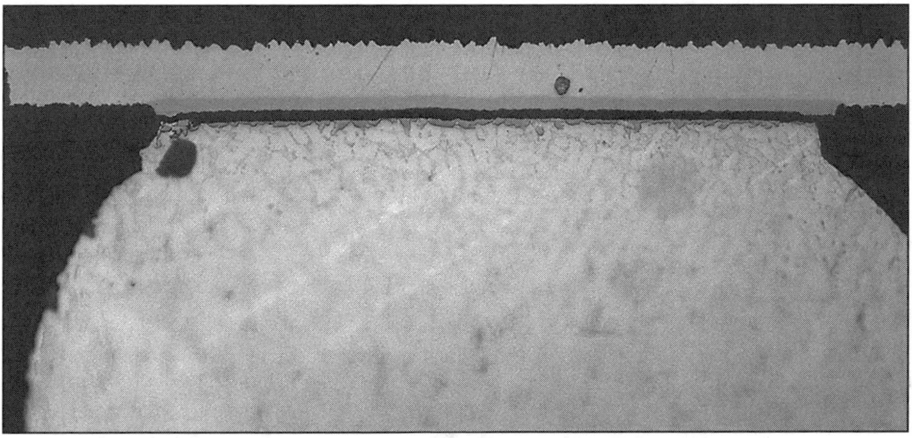

Fig. 9–49. Fractured solder joint in a daisy chain product level drop test that confirmed the CSP failure predictions in high-risk areas of the PWB

Although, simulation is taking on the role of virtual experimentation to reduce time and cost of hardware reliability testing, with the current maturity of simulation tools based on finite-element analysis or closed-form analytical methods, it is still prudent to incorporate a fair amount of experimental validation. The schematic work flow diagram in figure 9–50 illustrates the close cooperation needed between the simulation and experimental design validation teams. First, the critical components are identified based on historical data and board-level drop test results. Simulation is used to estimate the magnitude and location of maximum strains on the PWB and the drop orientations under which this occurs. However, it is not a trivial task to predict the drop life for different components and solder joints because the failure criteria are not well understood yet. Therefore, experimental drop testing is carried out on a system-level drop tester to measure the strain on the PWB at high-risk locations and also document the drop life and failure mechanisms. This information will need to be fed back to the simulation team to calibrate the model and estimate an appropriate failure criterion. Subsequently, designed experiments or parametric simulation are performed to rank the different failure modes in terms of risk of failure in the field. Finally, these design improvements are implemented into the prototype design for all the high-risk failure modes and hardware design validated by a second round of product-level reliability testing.

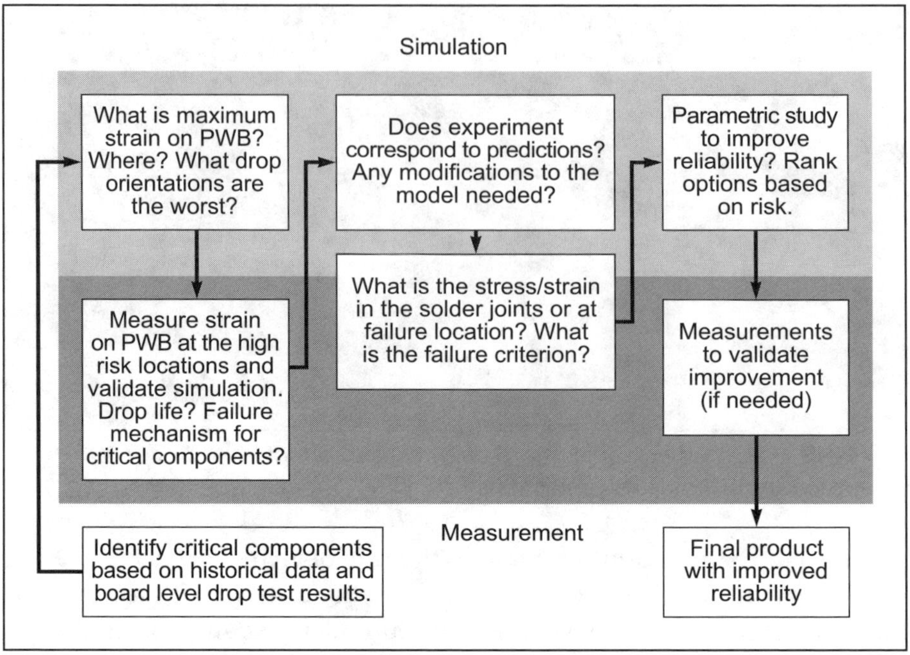

Fig. 9–50. Schematic showing the close cooperation needed between simulation and hardware validation teams

Several reliability tests are also conducted on functional products where the functionality is verified in addition to interconnection reliability. These qualifications, also called PV/DV tests (product/design validation), are often structured to yield pass/fail results rather than a failure-free life in the interests of time. Complete failure analysis of the samples subjected to reliability testing is an integral part of the concurrent engineering product design process.

Reliability assessment and improvement activities are often most effective when knowledge of the product's weakness and key failure modes are shared across the simulation, reliability test, design, and failure analysis teams. Failure analysis activities typically extend past the concurrent engineering phase and well into the manufacturing and product engineering phase. In the manufacturing of the product, several process defects can creep into the product to negatively impact the field reliability. Therefore, it is important that the failure analysis activity cover the high-volume manufacturing defects as well.

Finally, the product is delivered to the customer, and the product engineering phase of the product lifecycle begins. One important activity from a reliability perspective is the failure analysis of the field failures, especially the early fails. It is extremely cost-effective to perform failure analysis on customer returns to understand the following:

- Dominant failure modes: Defect driven, random fails, or wearout failures are useful to understand the design margins and recommend modifications.
- Dominant failure mechanisms: This information is directly useful in developing the appropriate reliability tests and in calculating accurate acceleration factors.
- Improvement actions: This consists of design, process, or material improvement actions that must be implemented to reduce field failure rates and increase customer satisfaction.
- Evaluation of the efficacy of existing reliability tests: Comparison on the failure mechanisms elicited in the field versus the reliability tests performed in the product development process will provide a direct measure of the efficacy of the reliability tests.

Product engineering for sustaining products

One way to improve product reliability is by collecting data about the product under study, understanding the defect Pareto, determine the root cause of defects, and implement improvements to address these root causes. Key elements of this approach are data collection, data analysis, root cause analysis, and engineering or process change management for defect elimination. It is a simple but effective technique for improving product reliability and quality, although the applicability is limited to products that are already suffering reliability issues. However, the lessons learned in continuous improvement are valuable inputs into the concurrent engineering process for avoiding the same mistakes or choices that caused failure in the first place.

Role of reliability coach

Most of the information can be collected into an easily searchable database or converted to an instruction material. However, we have found from experience that such passive approaches are highly ineffective in changing the way designers design products primarily because designers, who are constantly under time constraints, do not find the time to explore such databases and learn from it for day-to-day decisions. An approach that has proved more effective is one that takes people who have years of experience in product design, reliability engineering, reliability evaluation, and failure analysis and inserts them into the design teams. These "reliability coaches" would mentor, coach, and influence the development teams and dispense wisdom gained from years of making and analyzing failures at opportune times. This also fosters continuous learning from the field performance of the products and makes it easier to incorporate lessons learned from geographically far-flung development centers with minimal time lag. A schematic showing the different elements discussed above is shown in figure 9–50.

Summary

As portable consumer electronic hardware becomes more complex with multitudes of functions and increased data handling capacity, further miniaturization and higher levels of integration at all levels of packaging will be a natural trend. The reliability demands will be higher to ensure customer satisfaction and product acceptance. The implications for reliability, failure, and root-cause analysis will be significant. More functions will be integrated into the device. The silicon device thickness will be in the range of 40–50 μm. Stacked devices and folded and stacked packages will be more prevalent with a combination of multiple levels of wire bonding and/or flip chip interconnection. Another emerging trend in packaging is the three-dimensional integration at the wafer level. New materials that will have better mechanical properties and moisture resistance will be developed. More functions will be embedded into the PWB and these may include active, passive, and optical devices, attendant with new embedded interconnection schemes. The PWB technology itself will witness revolutionary changes with thinner and improved materials capable of 10–25 μm vias, 10–20 μm lines and spaces, and structures involving several layers of stacked vias. Consequently, hitherto unknown failure mechanisms are likely to be encountered. As the feature sizes diminish, the distinction between first- and second-level packaging becomes nebulous. Failure analysis, even at the PWB assemblies, will be a formidable challenge.

With shorter product development cycles and faster to market business environment, the need for more automated analytical tools with minimal operator intervention for rapid and repeatable root-cause analysis will increase. Innovative product development practices will be needed to shorten the test durations to accommodate faster development schedules while also preventing failures.

References

Adams, R., P. Spitz, and W. Vahey. 1995. The impact of boundary scan on surface mount technology in design, manufacturing and system test. *Soldering and Surface Mount Technology* 19: 5–11.

Bailon, M. F., A. B. Tarun, M. Nery, and J. Munoz. 2003. Localization of electrical failures from the backside of the die: Failure analysis case studies using infrared emission microscope. Paper presented at the Proc. 10th Int. Symp. Physical and Failure Analysis of Integrated Circuits (IPFA 2003): 193–197.

Barton, D. L., and P. Tangyungong. 2005. Thermal defect detection techniques. *EDFAS Failure Analysis Desk Reference*, ASM Press, Materials Park, OH: 378–397.

Beall, J. 1992. Plasma etching, in *EDFAS Failure Analysis Desk Reference*. eds. T. W. Lee and S. Pabbisetty, Metals Park, OH: ASM.

Biunno, N. 1999. A root cause failure mechanism for solder joint integrity of nickel/ immersion gold surface finishes. *IPC Printed Circuits Expo*, Long Beach, CA: 1–9.

Bradley, E., and K. Banerji. 1996. Effect of PCB finish on the reliability and wettability of ball grid array packages. *IEEE Trans. Components, Packaging, and Manufacturing Technology, Part B* 19 (2): 320–330.

Canumalla, S., and B. P. Schackmuth. 2000. Metrology of thin layers in IC packages using an acoustic microprobe: Bondline thickness. *Int. J. Microcircuits and Electronic Packaging* 23 (1): 146–151.

Canumalla, S. 1999. Resolution of broadband transducers in acoustic microscopy of encapsulated IC packages: Transducer selection., *IEEE Trans. Components, Packaging and Manufacturing Technology* 22 (4): 582 –592.

Canumalla, S. 2005. Thin layer acoustic image interpretation and metrology for microelectronics using a broadband model. *IEEE Trans. Components and Packaging Technologies* 28: 353–365.

Canumalla, S., and L. W. Kessler. 1997. Towards a nondestructive procedure for characterization of molding compounds. Paper presented at the IEEE 35th Int. Reliability Physics Symp. (IRPS), Denver, CO: 149–155.

Canumalla, S., G. A. Gordon, and R. N. Pangborn. 1997. In situ measurement of the Young's modulus of an embedded inclusion by acoustic microscopy. *Trans. ASME J. Engineering Materials and Technology* 119 (2): 143–147.

Canumalla, S., M. G. Oravecz, and L. W. Kessler. 1998. Resolution and loss in acoustic microscopy of encapsulated IC packages. Paper presented at the IEEE 48th Electronic Components and Technology Conf. (ECTC).

Canumalla, S., S. Shetty, and N. Hannan. 2002. Effect of corner-underfill voids on the chip scale package (CSP) performance under mechanical loading. Paper presented at the 28th Int. Symp. Society for Testing and Failure Analysis, Phoenix, AZ: 361–370.

Dermarderosian, A. 1978. The electrochemical migration of metals. Paper presented at the Proc. Int. Soc. Hybrid Microelectronics: 34.

DeVries, J. 2004. Failure mechanism of anisotropic conductive adhesive interconnections in flip chip ICs on flexible substrates. *IEEE Trans. CPT* 27: 161–169.

Dishong, T., C. Basaran, N. Cartwright, Y. Zhao, and H. Liu. 2002. Impact of temperature cycle profile on fatigue life of solder joints. *IEEE Trans. Components, Packaging, and Manufacturing Technology, Part B: Advanced Packaging* 25 (3): 433–438.

Dumoulin, P., J.-P. Seurin, and P. Marce, 1982. Metal migration outside the package during accelerated life tests. *IEEE Trans. Components, Packaging, and Manufacturing Technology* 5 (4): 479–486.

Dunford, S., S. Canumalla, and P. Viswanadham. 2004. Intermetallic morphology and damage evolution under thermomechanical fatigue of lead-free solder interconnections." Paper presented at the IEEE Electronic Components and Technology Conf.: 726–736.

Folcher, G., H. Cachet, M. Froment, and J. Bruneaux. 1997. Anodic corrosion of indium tin oxide films induced by the electrochemical oxidation of chlorides. *Thin Solid Films* 1: 242–248.

Gordon, G. A., S. Canumalla, and B. R.Tittmann. 1993. Ultrasonic C-scan imaging for material characterization. *Ultrasonics* 31 (5): 373–380.

Hageman, R., J. A. Nichols, P. Viswanadham, P. Woodriff. 1982. Comparative interference study for atomic absorption lead determination using constant temperature vs. a pulsed type atomizer. *Analytical Chemistry* 51 (9): 1406–1412.

Harsanyi, G. 1995. Electrochemical processes resulting in migrated short failures in microcircuits. *IEEE Trans. Components, Packaging and Manufacturing Technology* –A 18 (3): 602–610.

Harsanyi, G., and G. Inzelt. 2001. Comparing migratory resistive short formation abilities of conductor systems applied in advanced interconnection systems. *Microelectronics Reliability* 41: 229–237.

Hartfield, C., and T. M. Moore. 2005. Acoustic microscopy for semiconductor package inspection. *ASM EDFAS Failure Analysis Desk Reference*: 268–288.

Holden, H. T. 1993. DFM in PWB Fab. in *SMTA International*: 877–881.

Horsley, S. J. 1988. Designing for surface mount assembly. *Printed Circuit Design*: 8–15.

Inoue, Y., and T. Fukuchi. 1999. Plastic lends flexibility to LCDs while serving needs for thin sturdy devices. *Display Devices* 19: 24–25.

Hageman, R., J. A. Nichols, P. Viswanadham, P. Woodriff, 1979. Comparative interference study for atomic absorption lead determination using constant temperature vs. a pulsed type atomizer. *Analytical Chemistry* 51: 1406–1412.

Kessler, L.W., D. E. Yuhas. 1979. Acoustic microscopy—1979, *Proc. IEEE* 67 (4): 526–536.

Kohman, G. T., H. W. Hermance, and G. H. Downes, 1955. "Silver migration in electrical insulation. *Bell Syst. Technol. J.* 34: 11–15.

Krumbein, S. J. 1988. Metallic electromigration phenomena. *IEEE Trans. Components, Hybrids, and Manufacturing Technology* 11 (1): 5–15.

Lifshin, E., D. C. Joy, J. I. Goldstein, D. E. Newbury, C. E. Lyman, P. L. Kchlin, J. R. Michael. 2002. *Scanning Electron Microscopy and Microanalysis,* Plenum.

Liu, H., C. Basaran, A. N. Cartwright, and W. Casey. 2004. Application of Moiré interferometry to determine strain fields and debonding of solder joints in BGA packages. *IEEE Trans. CPMT* 27 (1): 217–223.

Moore, T. D., D. Vanderstraeten, and P. Forssell. 2002. Three-dimensional X-ray laminography as a tool for detection and characterization of BGA package defects. *IEEE Trans. Components and Packaging Technologies* 25 (2): 224–229.

Rizvi, M. J., C. Bailey, and H. Lu. 2008. Failure mechanisms of ACF joints under isothermal ageing. *Microelectronics Journal* 37: 1101–1107.

Roh, Y., W. Park, and H. Cho. 2003. Correcting image distortion in the X-ray digital tomosynthesis system for PCB solder joint inspection." *Image and Vision Computing* pp. 1063–1075.

Rusanen, O., and L. Samuels. 2003. *Metallographic polishing by mechanical methods,* ASM Press, Materials Park, OH.

Santina, P. 1996. DFM meets ABC. *Circuits Assembly* 7 (9): 32–34.

Schwindenhammer, P., P. Poirier, and P. Descamps. 2006. Microelectronics failure analysis using laser ablation of composite materials in system in package. Paper presented at the IEEE ECTC: 752–759.

Selfridge, A. 1985. Approximate material properties of istotropic materials. *IEEE Trans. Sonics and Ultrasonics* SU-32: 381–394.

Semmens, J. E., and S. Canumalla. 1999. Applications of highly focused, high frequency transducers for evaluation of near-surface flaws and thin packages: smart cards, flip chip, flex circuits, and thick films. Paper presented at the InterPack 99 and PanPacific Conference on Electronic Packaging 99.

Shell, M. K., and S. Golwalkar 1991. Application of infrared microscopy for bond pad detection. Paper presented at the Proc. IEEE IRPS: 152–159.

Siewert, T. A., and W. A. Mark. 1994. The contributions of out-of-plane materials to a scanned-beam laminography image. *Materials Evaluation*: 1194–1198.

Vandervoort, G. F. 2004. *Metallurgical Techniques in Failure Analysis*, ASM Press, Materials Park, OH.

Viswanadham, P., and P. Singh. 1997. *Failure Modes and Mechanisms in Electronics Packages*. New York: Chapman and Hall.

Viswanadham, P., D. R.Smick, J. J. Pisney, and W. Dilsworth. 1982. Comparison of ion choromatography and titrimetry for the determination of sulfur in fuel oils. *Analytical Chemistry* 54: 2431–2433.

Wang, Z. P. 2003. Challenges in the reliability study of chip-on-glass (COG) technology for mobile display applications. Paper presented at the IEEE EPTC: 595–598.

Yasuda, A., H. Yamaguchi, Y. Tanabe, N. Owada, and S. Hirasawa. 1991. Direct measurement of localized Joule heating in silicon devices by means of newly developed high resolution infrared microscopy. Paper presented at the Proc.IEEE IRPS: 245–249.

10

Future Trends in Portable Electronic Products

Introduction

As has been described in earlier chapters, portable electronics is going to be increasingly pervasive and ubiquitous in our lives and activities. Most of the activities that are currently performed using multiple gadgets will be accomplished with fewer and fewer devices. It is already evident that a mobile phone is just not a phone but is also a camera, global positioning system, organizer, financial transactions enabler, multimedia messaging device, cashless buying enabler, and so on. More and more functions will be incorporated into portable electronic appliances through higher levels of integration at the semiconductor level as well as at the first- and second-level packaging and also through miniaturization.

The near-term and long-term trends are discussed in the following section.

Intuitive technology road maps have been linear in their projections. However, historically with every paradigm shift, the technological potential has been exponential. When a paradigm shift occurs in a technology, it takes some initial acceptability and adaptability time, and then the practice and utilization rate is generally exponential. As the technology potential is fully utilized, the growth slows down as shown in figure 10–1.

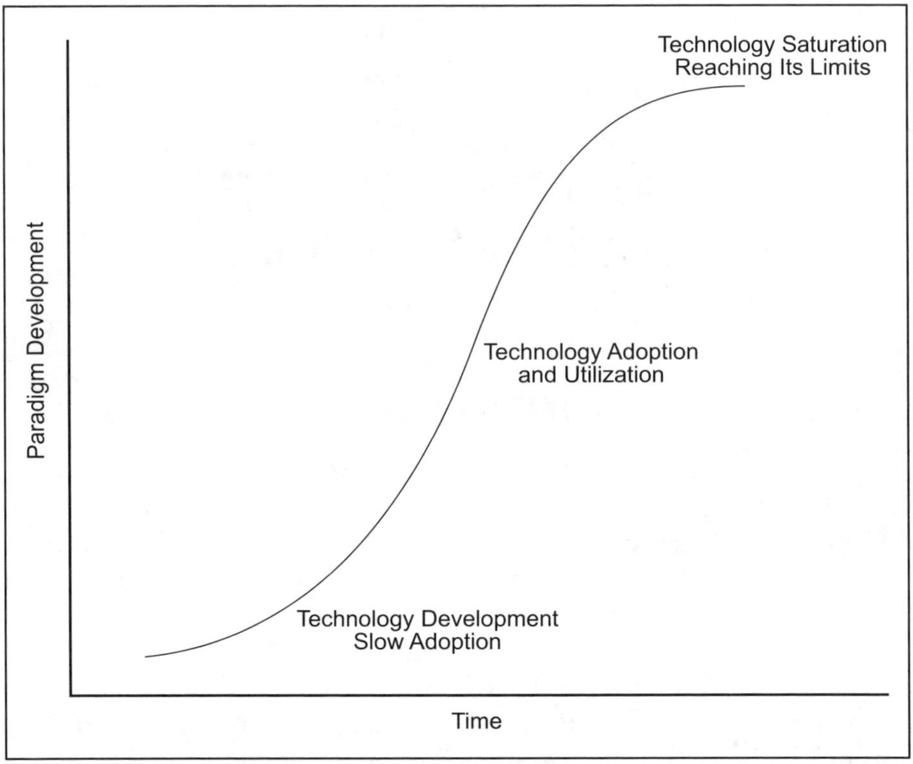

Fig. 10–1. Typical technology evolution and development in a paradigm shift

As a given technology matures and its potential is fully exploited, new product functionality and performance demands may render the technology inadequate. Thus, eventually, the need for a new technology is felt and a new paradigm shift comes into existence. A series of technological paradigms manifest over a period of time. Hence, technology progress as a function of time is represented by a series of S-shaped curves that over lap at the terminations of each, as shown in figure 10–2. Similar trends are seen in the electronics industry, with through-hole giving way to surface mount giving way to area array which is leading in to 3D packaging.

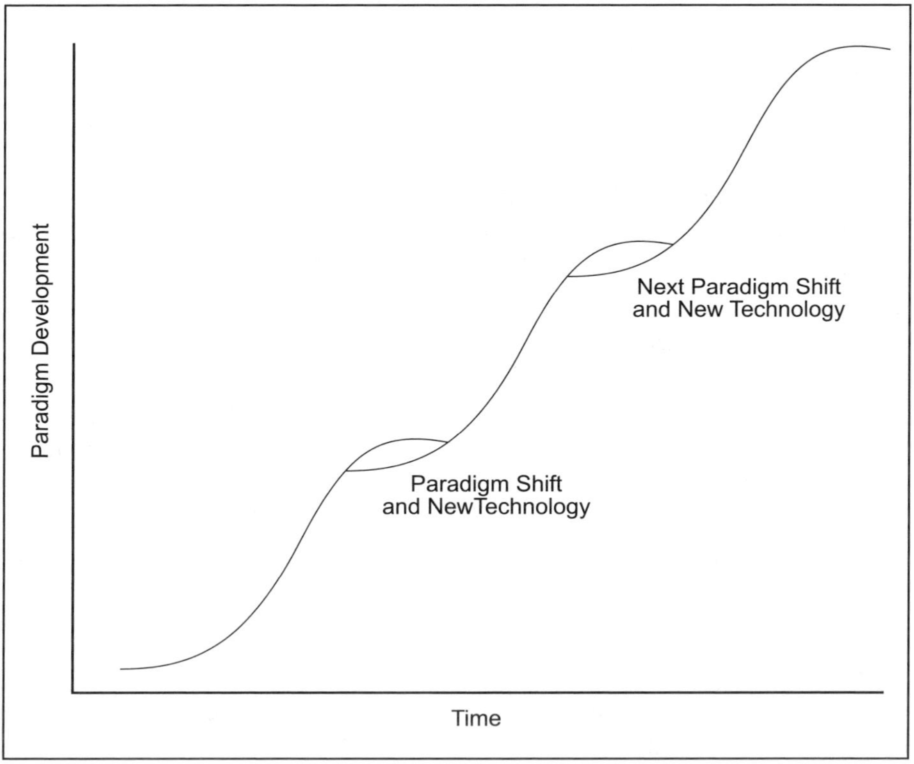

Fig. 10–2. Typical technology developments with paradigm shifts in time

Emerging Trends

Emerging portable electronic products are going to be sensor-intensive in their designs. Given this product and market environment, significant innovations and developments have to occur in several technology areas of electronic packaging of the product. These include semiconductor and first-level packaging, carrier and substrate technologies, display technologies, etc. The drive towards miniaturization and integration in product lends itself to a fresh thinking of the concept of packaging itself, resulting in the elimination of specific packaging modules. An example of this is elimination of data entry key pad in cell phones. In many cases, touch sensor screens are replacing the keyboards, making data entry, scrolling, etc., much more user friendly. Replacement of visual display modules with holographic displays and creation of halos of sound envelopes around the user may be in the horizon. There are several application concepts that drive the technology concepts. Increasing emphasis in the future will be in the convergence of computing and communication. Wireless connectivity will be in greater demand. Innovative display technologies that include flexible as well as interactive features are likely to be in greater demand. Transistor designs are likely to migrate to gigahertz to terahertz to break the speed barriers. It is important to recognize that not all the technologies being explored may come to fruition in all products. Some of them may be specific to certain market segments only, depending on the sophistication, cost, criticality of application, etc. A technology that finds application in a medical or military area may not be cost effective in consumer application. As technologies mature and gain wider applicability, the market volumes are likely to drive the cost down.

First-level packaging

In recent years, the industry has witnessed migration from two-dimensional to three-dimensional packages/architectures to enhance both packaging efficiency as well as packaging density through such concepts as silicon stacking, package stacking (PoP), package in package, etc. While current silicon stacking is limited to three or four chips, the trend will be to continue to increase the number of chips that can be stacked. There are several technical challenges to be surmounted to achieve this goal. Currently, the mainstream silicon wafer thickness is about 100 μm. Wafers of 75 μm thickness are becoming more prevalent. Multiple die stacking may drive the silicon thickness to 25 μm or thinner. Current semiconductor fabrication processes already reached their capability limits. These include such process steps as plating, etching, back grinding and polishing, etc. Testability is another challenge when devices of different functionalities are involved. Electromagnetic interferences and compatibility issues, parasitic effects, etc, can dominate. Emerging stacking concepts such as through-silicon-via (TSV) are likely to alleviate some of the impediments.

Embedded active silicon in rigid substrates, though not widely practiced, has already been demonstrated by Imbera Technologies, General Electric, and others. More recently, embedding a 25-μm-thick chip in a two-layer flex carrier constituting an ultrathin chip package (UTCP) has been demonstrated. The UTCP is then encased in a standard two-layer flexible carrier. The system did not require any high-density carrier for package-to-carrier interconnection. These embedded systems are 60 μm or less in thickness and are flexible and can be used in wearable electronics such as for monitoring heart rate, or muscle activity, etc. (Christina D'Airo 2009). Embedded active devices as a packaging trend will become prevalent especially in wearable electronics.

Eventually, the industry is likely to migrate to a universal, single, compact chip that is multifunctional, and energy efficient for cell mobile communication, once the compatibility and testability issues are worked out.

Three-dimensional integration incorporating radio frequency, DRAM, MPU, sensor, logic, analog circuitry, etc., are likely to become prevalent in the next decade. Engineered memory resistors can offer controllable resistance for switching circuits.

Elimination of first-level packaging, namely packaging the silicon in order to facilitate mounting on to the printed wiring board (PWB), is likely to be practiced more extensively in the future. This significantly reduces materials, processing efforts, product weight, volume, and the associated cost. Embedding the actives as well as passives into the carrier PWB will be the trend. For this to be cost effective and reliable, the processing has to be at a high level of sigma in terms of allowable defects per million opportunities (DPMO). Rework and repair activities will also be significantly reduced as a result.

Printed wiring board technologies

Laminate and PWB technologies will migrate to extremely thin and very high density interconnect technologies. These involve microvias of the order of 25 μm and lines and spaces as fine as 15–25 μm in width to accommodate circuit routing densities on the surface layers. Many of these technologies may still use conventional PWB technologies as the subcomposite layers.

However, more and more portable products will be under considerable market pressure to reduce weight, volume, and cost. Flexible laminates offer not only weight and volume advantages but are also more amenable to three-dimensional circuitries, offering an advantage of conformability. Thus the trend will be more towards flex circuits or rigid flex laminates. Also, embedding passives and actives in the laminates will afford enormous advantages in creating not only portable electronics but also wearable ones. Portable and wearable electronics will find innumerable applications not only in consumer electronics but also in military portable electronics and medical electronics for vital functions monitoring, critical communication with the service providers, doctors, etc. The technology and industry trend in the coming years will be migration towards highly miniaturized, sensor-intensive, intensely integrated, wearable, portable, and wireless electronics.

Thus it can be envisioned that PWB and laminate technology will migrate towards very high density routability, more flexible and conformable laminates, and active and passive integration into the laminate itself, and also have embedded optoelectronic components.

The industry will experience a convergence of first-level and second-level packaging with high-density interconnect and embedded active and passive devices as an emerging trend.

Roll-to-roll printed wiring boards

To be cost effective and faster-to-market, efficient high volume manufacturing techniques are evolving. One of them for the printed wiring board is the roll-to-roll manufacturing of thinner laminates. This is, of course, more feasible with flexible laminate. While in concept this is very attractive, as is the case with all emerging technologies, several practical hurdles have to be overcome to render it. Some recent developments in this area are highlighted in the ensuing paragraphs.

As has been described in earlier sections, the circuit pattern is generated using a photo-imageable resist on the copper clad rigid core blank, exposing with a mask followed by develop, etch, and strip processes. The feature sizes, namely, the lines and spaces are mostly limited to about 150 μm. Thus conventional contact and proximity printers, steppers, etc., used for large area applications at their limits of capability are not likely to meet the emerging needs. Incremental improvements in the practicing technologies may provide finer feature capabilities but at the risk of yield losses. Those techniques that enable producing 25 μm features may not be adequate for 1 to 5 μm level features. One of the recent methods utilizes projection lithography.

In this technique, flexible carrier substrate, for example copper on polyimide, is fed onto a stage where the substrate and a mask containing the pattern are mounted rigidly and held on a stage with a vacuum diffuser. The mask and the substrate are scanned in unison through the object and image fields of the projection lens. The illumination system consists of an XeF or XeCl pulsed excimer laser source that is coupled to the optical elements. The pulsed laser beam passes through the defined mask, then through the projection lens system and impinges on the organic polymer. The impinging UV laser radiation breaks down the polymer in that area into simpler and smaller molecules. These molecular fragments are kinetically scavenged and thus create the desired pattern.

The take-up and supply rollers enable a measured amount of slack on either side of the patterning region to facilitate precise alignment and exposure. After exposure, the vacuum is released, the web advances towards the take-up roller and the next segment of the roll of materials for photo-ablation. This process repeats until all the material is processed. The laser, scanning stage, alignment system, and the web handling are all carried out with precision and accuracy by a computer based interface system. A conceptual schematic of the system is shown in figure 10–3.

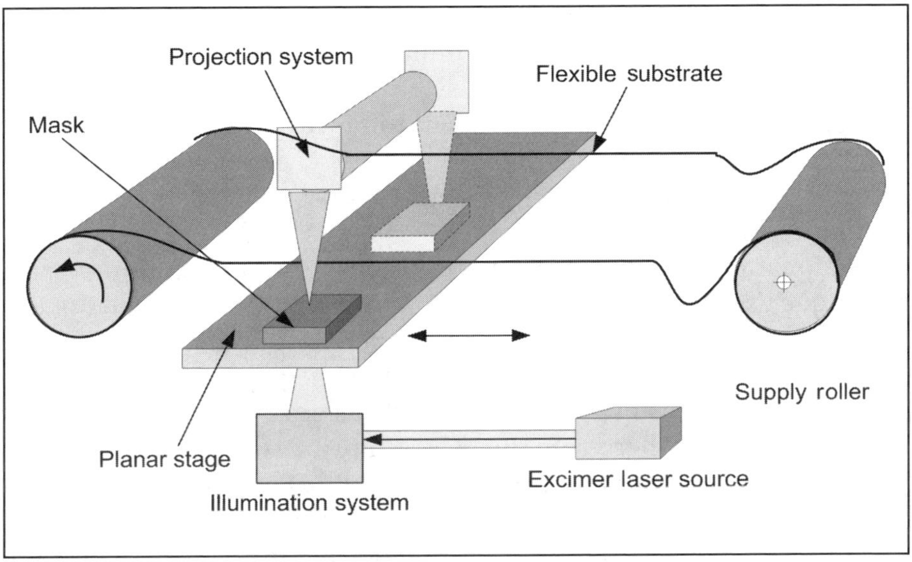

Fig. 10–3. Conceptual schematic of roll-to-roll PWB processing

The system is reported to have a throughput of 6 sq.ft/min (120 panels/hr) with a resolution of ±10 μm. A capability of 30 μm traces was demonstrated in the fabrication of a flip chip substrate by this roll-to-roll process (Jain K et al. 2005). Photoablation can be faster than reactive ion etching and drilling, and is cleaner and more precise.

The needs of near-term portable, wearable, and mobile electronic products are likely to be satisfied by three to four layer flexible laminates with lines and spaces in the range of 15–10 μm. These are similar to the developmental needs of flat panel display technology that uses thin film transistor back planes. Flexible back panels are used to drive the OLED displays, integrated circuits, MEMS, MOEMS, fiber optic communication links, microfluidics, etc., which are all likely to be fabricated on single flex carriers, and roll-to-roll processing can indeed become a process of choice. Capabilities of 10 μm features on 1.5 m × 1.8 m panels and 1 μm features on 0.6 × 0.6 m panels was also indicated.

Roll-to-roll processing provides attractive features like greater throughputs, lower contamination levels, higher yields, and minimized load-unload operations. However, it should be recognized that several manufacturing challenges need to be overcome. Metal clad polymer films can have residual surface tension stresses between the polymer and metal. As some of the metal is etched away during the circuitization, some of the residual stress gets released resulting in dimensional changes in the form of distortion and warpage. The distortions can be symmetric or asymmetric. The extent of such dimensional instability or changes is dependent on the material set as well as the process and can be as much as 2,000 ppm. Dynamic scale compensation with anamorphic x-y scale compensation is sometimes employed. Static charge build up, kinking, dimples, and scratches can be yield detractors and need to be carefully monitored and controlled.

Nanopatterning of polymer films is another roll-to roll processing technique used in the area of high-resolution plastic display and organic transistor fabrication laminates. In one example, a 95 μm thermoplastic cellulose acetate film substrate is passed between two rollers at a web speed of 0.2 to 5 m/minute. Flexible nickel shims wrapped around the top and bottom rollers constitute the imprint stamps. The imprints stamps consist of lines and dot structures of 150 nm to 10 μm for that specific application or product. The top roller is heated by an electrical heater located inside the roller and the bottom one is heated by a hot air stream. The temperature of the rollers is maintained in the vicinity of the softening temperature of the carrier film. In the case of cellulose acetate, the temperature is maintained in the range of 85–105°C. As the film is squeezed between the rollers at about 8 MPa pressure it softens and the imprints are created in the film. The imprints are created on both sides of the film simultaneously. The imprints are examined by both optical microscopy and by atomic force microscopy techniques (Makela et al. 2007).

Inkjet printing of PWB circuitization

Jetting of inks for printing has been extensively used for printing on paper for many years. Jetting of the fluids and suspensions is generally done by one of three techniques: thermal, piezoelectric, or continuous. In the thermal process application, a current pulse results in a mini explosion creating bubbles. These bubbles propel the ink onto the paper. As the bubble collapses any surplus ink is sucked back from the surface.

In the piezoelectric process, the piezoelectric crystal causes the stream of liquid to break into droplets at regular intervals. By translating the print head as it jets the fluid, desired patterns can be generated. Application of ink-jetting technique to create conductive patterns on laminates is a relatively recent phenomenon. The following are several advantages:

- The technique eliminates conventional masking, expose, develop, etch, and strip process steps
- Lower energy consumption
- The process is fully additive
- The process is non-contact in nature
- Reduced number of process steps
- Minimum or no material wastage
- Suited to printing on flexible laminates
- Production flexibility
- Design and routing changes can be made even at the last minute and hence greater flexibility
- Real time production of digital design
- Three dimensional printing
- Lower cost

Inks can be aqueous or organic-solvent-based and can be acidic or basic in nature. Organic esters, ketones, hydrocarbons, and even such compounds as N-methyl-2 pyrrolidine have been attempted. For successful ink-jetting several material properties and process parameter variables have to be carefully considered, evaluated, and optimized.

The conductive particles in the ink are generally silver nano-powders. There is considerable experience in the industry with silver-based conductive inks. Oxidation of silver does not cause serious impediments, as silver oxide is also conductive. Carbon is also some times used. Also, silver inks provide acceptable electrical conductivities for electronic circuits. Silver trace resistance of 3 $\mu\Omega$.cm compared to 1.6 $\mu\Omega$.cm for bulk silver were reported.

The following are some of the aspects that need special attention:

- Compatibility of the ink and print head
- Interaction of the substrate and the ink
- Metal or particle loading
- Drop volume

- Temperature control
- Surface energy
- Viscosity of the ink
- Evaporation rate of the ink
- Boiling temperature of the ink
- Drive voltage of the print head

The higher the viscosity of the fluid, the higher the drive voltage requirement is of the print head. On the other hand, low viscosity can cause splashing of the material. Evaporation rate of the material should be high enough to dry the print fast and at the same time slow enough as not to cause print head clogging. Surface energy affects the contact angle and hence the wetting of the fluid to the surface. The equilibrium contact angle is the resultant of interfacial forces at the three phase contact line formed by solid, liquid, and vapor. For a given volume of the fluid, the drop height depends on the area of spread or wetting. It was reported that application of electric field between the droplet and the surface, electrostatic forces acting on the ions in the liquid tend to reduce the contact angle, thus enhancing the wettability of the surface (Esinenco D et al. 2006)

An example of acceptable inkjet material properties and process parameters are shown in table 10–1.

Table 10–1. Example of Acceptable Inkjet Material Properties and Process Parameters

Parameter or property	Value
Metal content	55 wt percent
Density	1.66 g/cc
Drop volume	14 pL
Driving frequency	1–10 kHz
Viscosity	5–15 mPa.sec @ 20°C
Drive voltage	20 volts
Drop velocity	6 m/s

Drop volumes are generally in the range of picoliters and vary from a couple of picoliters to about 50 picoliters. With one picoliter print head, 20 μm wide features can be printed. Drop velocity depends on the drive voltage, and hence a calibration of drive voltage versus drop velocity is essential to obtain drop jetting consistency and uniformity

(Nishi S. 2007). For stable jetting of nanoparticle dispersed inks, it is important to ensure prevention of particle coalescence, aggregation, and/or precipitation.

Real time high accuracy three-dimensional imaging and shape monitoring techniques are required to monitor roll-to-roll manufacturing of flexible electronics. It is important to monitor the processing qualities of deposition, patterning, and packaging of flexible electronic assemblies. Also, 3D imaging systems are useful in evaluating mechanical and thermal deformations of electronics and associated reliability implications of flex assemblies. A real time, non-contact high accuracy, low cost, broad range, full field 3D shape measurement technique based on fringe projection profilometry was recently described (Wang Z and S. B. Park 2008).

Several ink jetting processes and roll-to-roll PWB fabrication and display technologies have been evolving in recent years to meet industry demands. Only select representative methods are described to elucidate the general principles. Current developmental efforts are aimed at verifying proof of concepts, prototype fabrication, and process development. Several manufacturing aspects to enable cost-effective manufacturing are still in their infancy and are maturing slowly. Reliability of these technologies under a variety of operating loads such as thermal, cyclic bending and repetitive stretching, mechanical shock, mechanical vibration, humidity and corrosive gas environments, is not fully evaluated. As many of the portable and mobile electronic products have shorter design lives, they can be expected to meet the performance criteria when manufactured with optimized designs, robust manufacturing techniques, and properly chosen materials sets.

Design convergence

Historically, there are four design groups operating in electronic packaging: the chip designer, the package designer, the industrial designer, and the PWB designer. Traditionally, independent road maps are generated by each group indicating where the respective technologies are headed as a function of time. Many a time they are not in unison. Each is developed without much knowledge of each other's requirements. PWB design is guided by the interconnection requirements and has little to do with the semiconductor design requirements. In recent years, with the development of high speed-circuits, high I/O devices, and RF modules, there is an increasing need for the codesign efforts. A perfectly designed chip may not perform if the package design is incompatible.

A similar thing can happen in the case of PWB. A perfectly designed package may not function on a given board. As the silicon devices become increasingly complex, multifunctional, and high I/O, there is a need for high density-carriers and substrates such as HDI and sequential build-up multilayer microvia PWBs. The entire interconnect system, namely, chip-to-package-to-board has to be taken into account in the design process to be successful. Pin allocations are very important. Thus a design convergence is essential for modern-day electronic products.

As a given chip or package is used in a variety of products and the end product PWB design is not obvious at either the chip or package design stage, it becomes imperative to utilize concepts of virtual prototyping. One utilizes the design that is known in combination with design data as yet unknown to generate a mockup of the PWB and then utilize that to optimize the PWB–package or PWB–chip, (in case of direct chip attach) interconnects. The importance of this concept is easily evident when one considers the effort needed to change either the chip design or package design after the packages are manufactured. The costs and efforts associated are huge and immense. Thus, it is imperative that the different design groups begin interactions at a very early stage in the product design phase. Silicon, package, and PWB design convergence is of paramount importance and will be the trend of the future product design efforts. As the consumer and portable electronic products drive technology with continued evolution of high-density laminates, the boundaries between the different design groups, of necessity, become nebulous. Manufacture of high-performance multitechnology portable electronics systems driven by consumer market require design convergence (Viklund 2008). Also, advances in semiconductor manufacturing technology are likely to migrate from the 32-nm node to 11- or 8-nm nodes in the next 10 years.

Display technology

In display technology, organic light emitting diode (OLED) based displays are most likely to dominate the future, at least in the short run, while holographic displays may take over in the long haul. OLED displays do not require backlighting like liquid crystal displays. They provide bright, clear images with image contrast of 10,000:1. They have low switching rates and dissipate very low power. In addition, they have microsecond response rates.

They are relatively light in weight and are extremely thin, usually of the order of 0.2–0.3 mm. It is expected that they will be available both in passive and active matrix formats. In the active matrix format (AMOLED), the cathode, organic, and anode layers are stacked above a low-temperature polysilicon substrate layer which contains thin-film transistor circuitry. The pixels in AMOLED can be turned on and off at least three times faster than the speed of traditional motion picture film. A passive matrix is relatively simple in design and construction and is less expensive. The adaption of OLED displays is anticipated to take place in the portable and mobile electronic products since they require smaller size displays which can be produced with better yield than the larger units (Allan 2008).

A typical OLED consists of (1) an emissive layer, (2) a conductive layer, (3) a substrate and anode, and (4) a cathode. A schematic of a two-layer OLED is shown in figure 10–4. The emissive and conductive layers are made of organic molecules that conduct electricity. These have conductivities in a wide range, from that of conductors to insulators. Thus they are also considered as organic semiconductors. One of the first OLED organic polymers was poly p-phenylene vinylene.

When a voltage is applied across the electrodes of an OLED, electron flow is initiated from the cathode to the anode. Thus, cathode provides electrons to the emissive layer and the anode withdraws the electrons from the conductive layer. The anode provides electron holes to the conductive layer. The emissive layer becomes negatively charged, while the conductive layer becomes rich in positively charged holes. Electrostatic forces bring the electrons and holes towards each other and result in recombination, and this occurs close to the emissive layer. It is because, in organic semiconductors, holes are more mobile than the electrons. This recombination results in a drop in the electron energy levels accompanied by an emission of radiation whose frequency is in the visible region, and hence the name emissive region.

Indium–tin oxide (ITO) is a commonly used anode material and is transparent to visible radiation with a low work function to facilitate injection of holes into the adjacent polymer layer. Aluminum or calcium is the common cathode material, which also has low work function to inject electrons into the polymer layer.

Multilayer OLEDs can have several layers to improve the device efficiency. As well as conductive properties, layers may be chosen to aid charge injection at electrodes by providing a more gradual electronic profile or block a charge from reaching the opposite electrode and being wasted.

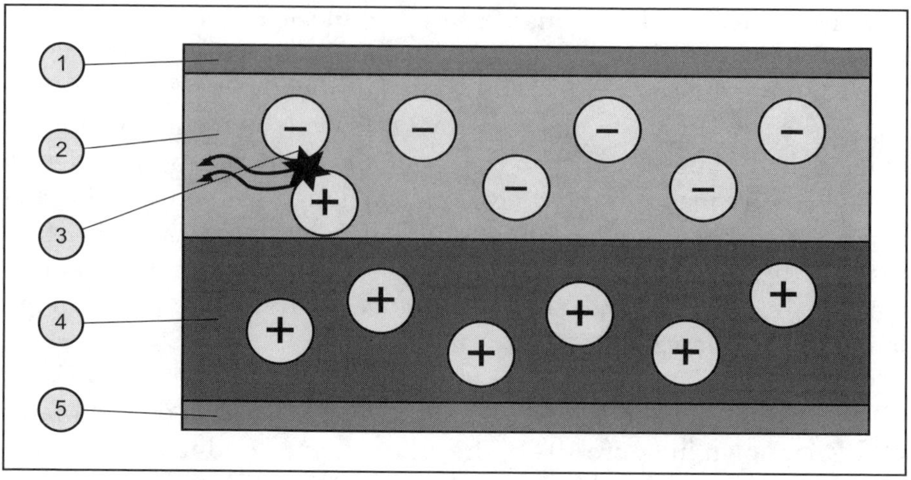

Fig. 10–4. Schematic of a two-layer OLED: 1. cathode (–), 2. emissive layer, 3. emission of radiation, 4. conductive layer, 5. anode (+)

Flexible displays are another emerging technological trend. Several flexible displays have recently been entering the market place. Especially in the portable electronics arena, stylus, pen, or finger-touch input appears to be a preferred user interface compared to key pad strokes. Flexible displays are lighter in weight and offer advantages of curved and odd-shaped viewing surfaces. They can also be shatterproof and are likely to be more rugged than their glass counterparts in terms of impact resistance.

Flexible displays may involve arranging thin-film transistors on a flexible metal foil in conjunction with a color filter on a flexible layer and employing electronic inks, thus constituting the display unit.

Glass-free active matrix display on a flexible substrate with touch screen capability was recently demonstrated. The display utilizes thin-film transistor (TFT) technology, polyethylene naphthalate (PEN) flexible films, and VizPlex ink laminate technology as the basis technologies.

Flexible touch-screen display technologies are likely to proliferate in portable and mobile consumer electronics owing to their light weight, ruggedness, eventual low cost, ease of use, and elimination of key pads, etc. Also, they use less power, as energy is consumed only when touched (Dirjish 2009).

Electronic packaging is likely to witness notable advances in the areas of molecular electronics, bio- and medical electronics, sensor-intensive automotive electronics, portable military electronic hardware, display technologies, telecommunications, etc., in the coming years (Kinkaid 2007).

There is a significant market pressure for the development of flexible displays, and the applications include electronic papers, billboards, medical labels, smart cards, and a host of other applications.

Energy sources

Batteries that power the portable electronics constitute the heaviest component of the appliance. Emerging trend will be towards development of lighter and more efficient batteries with longer life, as well as other alternatives. Super or ultra-capacitors, a subset of capacitors, offer an alternative energy source. A standard capacitor sandwiches a dielectric substrate between two metal electrode plates. This dielectric, depending on the application, can be composed of oxides of aluminum, titanium, tantalum, etc., or an organic polymer such as polyethylene, polypropylene, etc., depending on the capacitance and voltage specifications. This single-layer topology has capacitance that is related to the size of the capacitor.

This problem is eliminated by employing double layers where a second layer of dielectric that acts in parallel is added in the same package. Electric double-layer capacitors (EDLC) employ such materials as carbon aerogels, carbon nanotubes, select conductive polymers, etc., that exhibit higher storage capabilities. These materials are extremely thin in configuration and offer an extremely large surface area capable of storing large amounts of energy. A schematic is shown in figure 10–5 (Dirjish 2008).

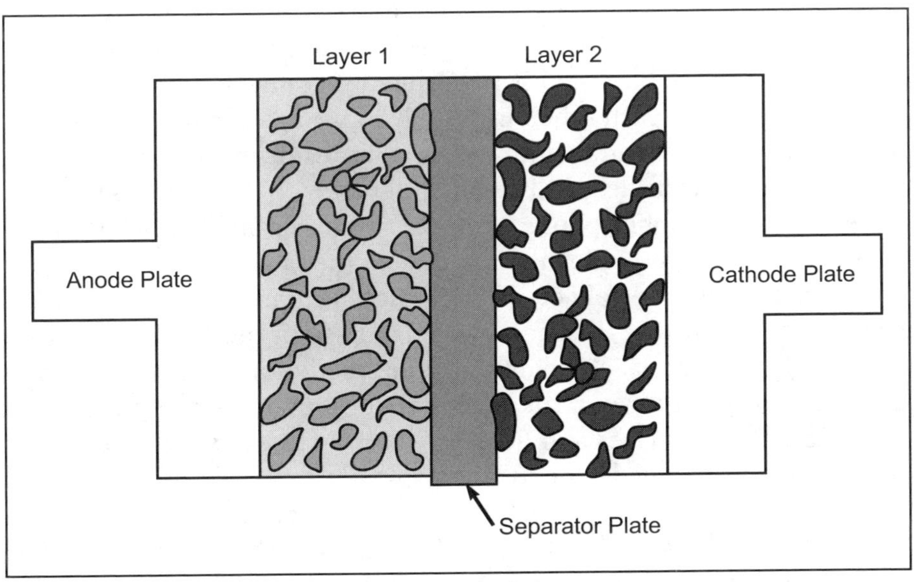

Fig. 10–5. Schematic of a super or ultra-capacitor

These *super caps* can have several advantages. Super capacitors with values of 5 F or more are finding applications as energy sources replacing batteries in many portable and hand-held products. While current batteries can only withstand about 1,000 charge/discharge cycles during their life, super caps can offer a 1,000,000 or more charge/discharge cycles. This will substantially reduce the number of battery replacements needed, and their disposal into landfills, thus providing an environmentally friendly alternative. Also, super capacitors have low charging cycles, as much as 98% efficiency, and a higher endurance to electrical short circuits (Bell 2009).

A concept that will be explored with intensive efforts is that of self-powered electronics especially in the area of mobile electronics. Harnessing solar energy to charge the device's energy source, conversion of user-generated kinetic and thermal energy into electrical energy, and use of piezoelectric devices will be explored and can be expected to be the energy harvesting trends.

A significant trend that is proliferating in all portable and mobile electronics systems is micro-electro-mechanical systems (MEMS) based sensors. A variety of sensors are already incorporated into these devices. The trend will be to combine more than one type of sensor into a single package in order to miniaturize the product, reduce weight and volume, and also aid cost-effective manufacturing.

Recently, mobile phones that harness the low levels of electromagnetic radiation from the surroundings have been announced. Although, the current capability for power conversion and storage is only adequate for stand-by mode, the future looks bright for wireless charging devices.

Another inexorable trend is the aspect of connectedness of devices that need to talk to each other. Especially in the age of cloud computing, where the user's data and computing power reside in the cloud servers, users of electronic devices will seek to interact with their data seamlessly. For example, the mobile phone, personal computer, and television will be able to display and provide interaction with the same set of photographs or videos that may be collected with a digital camcorder, media phone, or webcam but reside on a server or database in the "cloud". Another example for seamless activity is social interaction networks, which will be synchronized between the personal computer, mobile phone, and television. Both services and devices will have to evolve to cater to the social networking needs of people.

Innovative methods and techniques are needed to extend the miniaturization and integration trends. Attendant with this technological trend will be the increase in the number of users, and hence very high unit-volumes of these products will need to be manufactured at affordable cost and also acceptable reliability. For example, almost a billion cell phones are sold every year in the world. That is just one type of portable electronic appliance. Taking into account laptops, camcorders, digital cameras, digital electronic toys, and a host of other portable electronic appliances, the volumes are extremely huge. Newer models with advanced features are going to replace earlier versions more frequently than ever before. The older ones enter either recycling units, garbage dumps, or landfills. Thus, disposal or recycling of electronic waste in an environmentally acceptable manner can become a formidable technical challenge.

Miniaturization and integration

Electronic packaging has gone through a series of evolutionary steps involving innovative concepts in design, materials, and processes. The original concept of system on a PWB with individual components such as actives, passives, connectors, and other elements still continues to a great extent with most electronic systems where the constraints of space, volume, and weight are not severe. However, the proliferation of consumer electronics and the emergence of portable electronics and personal information processing systems components have shifted the focus to achieving higher and higher packaging densities and packaging efficiencies. This thrust has resulted in migrating from single-sided coarse-pitch insertion mounting to double-sided fine-pitch surface-mount technology through a paradigm shift. The constraints associated with fine-pitch and ultrafine-pitch leaded packages and the conventional PWB technology, sometimes referred to as *perimeter paralysis*, have given way to still another paradigm shift, namely, area-array packaging with finer lines, finer spacing, and microvias with high-density interconnect technology. An outcome of this paradigm shift is the proliferation of ball grid array (BGA) packages with numerous variations in design, materials, and construction. Concurrent with incorporating more and more transistors on a single semiconductor device, in accordance with Moore's law, integration at the package level has continued. Some of the first efforts in the early 1970s and 1980s included the development of multichip modules, where the individual ICs were assembled in a horizontal array on a multilayer substrate such as co-fired low- or high-temperature ceramic. Further developments led to the incorporation of organic dielectrics with sputtered or electrodeposited conductors into the system. While multichip modules (MCMs) served the high-end business systems and military systems well, their penetration into consumer electronics has neither been pronounced nor significant due to several constraints including cost.

As the insatiable desire to increase the packaging efficiency and packaging density continued, new forms of packaging emerged. Pluggable devices for portable memory auxiliary devices such as Ethernet and font cards have evolved promoted by the Personal Computer Memory Industry Card Association (PCMCIA). Thus, there evolved a product called PCMCIA cards, later called PC cards. These utilized low standoff, fine-pitch leaded packages in their design. Thin quad flat packs as well as type I and II thin small outline packages (TSOP) emerged with relatively high packaging efficiencies. However, owing to larger chip-to-package ratio and very small amount of plastic molding compound encasing the silicon, these packages have a lower thermal coefficient of expansion (CTE), hence a greater CTE mismatch with the organic PWB, resulting in poor board-level reliability. Evolution of different types of chip-scale packages with built-in stress relieving interposers followed to meet the ongoing demand for greater packaging efficiency.

As the aerial real estate has been fully utilized in maximizing the packaging density, three-dimensional packaging involving thinner silicon devices was explored, resulting in silicon stacking, popularly known as *system in a package* (SiP). Initial efforts comprised stacking memory devices and slowly migrating to heterogeneous IC technologies. These include combinations of memory, logic, RF, analog, digital signal processors, etc. Inter-chip path lengths are considerably reduced, thereby offering significant enhancements in performance. These are immensely important in portable electronic appliances such as PDAs, MP3 players, cell phones, mobile games, laptops, and palmtops, where multiplicity of functionalities and immensity of performance are in great demand.

As discussed in an earlier chapter, individual ICs are stacked and interconnected by a plurality of techniques such as side metallization, wire-bonding, flip chip, and combination of wire-bond and flip chip. These techniques are generally termed *non-through silicon via technologies* (non-TSV). Other three-dimensional non-TSV architectures, of course, include package on package (PoP) and package in a package (PiP). While the system in package architecture has been in vogue for quite some time, mechanical, material, electrical, and thermal challenges still persist with the introduction of any new system and need to be addressed taking into account the fabrication, assembly, characterization, and manufacturing aspects. Choice of materials must meet the thermal, electrical, mechanical, and thermomechanical considerations. Some of the electrical parameters of significance are dielectric constant, loss factors, temperature coefficient of resistance and capacitance, insulation resistance, electrical conductivity, etc. Interconnection and assembly reliability depend strongly on the CTE, modulus, etc, of the materials set. These include the silicon, the spacers, the die attach adhesives, molding compounds, substrates, etc.

Silicon thickness and overhang on the spacer can impose the limitations on wire bondability. While 2 mm overhang may be acceptable for a 150 μm or thicker die in wire bonding, a 75 μm-thick die may require zero overhang. Also, due to the piezoresistive effects of silicon, assembly-induced stresses can affect functionality and transistor performance depending on the die thickness.

The interconnection materials' bulk properties, like those of Sn/Pb, Sn/Ag/Cu, their intermetallic phases, and their microstructure, play an important role in the ultimate performance of the designed system. The performance not only includes electrical but also thermal cycling, shock, mechanical drop, vibration, etc. The nature of the interconnection failures depends not only on the CTE mismatches but also on the stiffness of the assemblies. While ductile fractures may be the norm in thermal fatigue, brittle fractures characterize drop loading failures.

From the signal integrity point of view, higher speed means constraints on package size and the number of substrate layers. Each signal line needs a nearby power or ground path for reference and also for cross-talk shielding. Package design should take into account higher IC power delivery. Voltage fluctuations during switching strongly depend on parasitic inductances. Higher density of wire bonds implies greater coupling. Isolation of digital and RF is important to reduce electromagnetic interference and compatibility. Wireless signal speeds are anticipated to reach 10 GHz in order to meet future transmission needs.

As one combines heterogeneous functionalities into a single package, potential heat transfer and thermal management aspects have to be carefully addressed. Natural and forced air convection can be utilized with an appropriate heat sink design. Heat can also be dissipated, albeit to a small extent, through the solder balls/joints and the carrier. Heat dissipating modules need to be located in the proximity of heat transfer paths. Both package-level and system-level thermal management approaches need to be considered.

Stacking of up to six or eight chips has been reported. Stacking the chips and at the same time meeting the total package thickness requirements imply the use of thinner chips. Currently, wafers are thinned to a thickness of 50–70 μm. Chips as thin as 20 μm are projected for 2015. Wafer thinning by back grinding can yield a thickness of 100 μm. This is accomplished by coarse grinding followed by fine grinding. Wafers are polished by chemical and mechanical polishing. Special pads and NH_4OH solutions are used concurrently to perform mechanical and chemical polishing. Wet etching with solutions of HNO_3 and HF mixtures and/or dry etching with fluorine and oxygen plasma are also employed.

Dicing of the wafer into individual ICs can introduce several defects such as chipping of the front and back surfaces, delamination of mechanically brittle interlayer dielectrics such as the low-k dielectrics, microcracks, etc. Newer methods such as water-jet-assisted or guided laser dicing are also being explored.

Most of the stacked devices are predominantly wire-bonded with a few involving a combination of flip and wire-bonding interconnections. Pyramidal, staggered, and stacks with spacers in between have been popular configurations. Wire-bonding devices generally are prone to parasitic and inductance effects, impacting their electrical performance. Ultrathin ICs with dice as thin as 3–10 μm and a stack of 100 chips are anticipated in the future.

Despite numerous advantages and significant advances in the die-stacking and wire-bonding capabilities, there are still some cost-constrained impediments. Some of these include known good die (KGD), testability prior to die stacking, lower yields, single source issues, inability of upgradeability, difficulty of integrating dissimilar devices, and infrastructure issues.

Package on package and package in a package

These packaging concepts have been described in some detail in an earlier chapter. They employ either wire-bonded packages, or flip-chip packages, or a combination thereof. PoP or PiP alleviates some, if not all, of the impediments mentioned earlier and has several advantages. The chief among them are the following:

- A simpler process
- Faster time to market
- Testability of individual packages
- Product upgradeability
- Design simplicity
- Memory screening

The disadvantages may include inability to meet the package thickness standards, bigger package size, lack of infrastructure, and non-expandability. Use of package stacking and package within a package is expected to continue as a near-term trend in heterogeneous device integration.

As has been indicated earlier, wire-bonded structures are generally inferior in performance to flip-chip bonded structures. However, the number of devices that can be stacked is limited. A recent development in chip stacking efforts is the through silicon via (TSV) technology. This subject has also been briefly described in the chapter on packaging. Some additional aspects are addressed here, as this could become a trend of significance in the near future.

TSV envisages stacking and interconnecting dies, chip carriers, and/or wafers. The inter-chip interconnections are accomplished by filling the vias with a bondable conductor material. The method affords the shortest length and hence highest performance. The interconnection can even be further increased by area-array configuration of the I/Os. In addition, heterogeneous function integrations are also accomplished.

Via drilling through the silicon is accomplished by a variety of techniques. These include wet etching, deep reactive ion etching (DRIE), cryogenic DRIE, etc. In cryogenic DRIE, the wafer is cooled to about −110°C prior to etching. This prevents ion migration and reduces side etching. Subsequent to via formation, the vias are coated with an insulating film to isolate the silicon and the conductors. This film is generally formed using plasma-enhanced chemical vapor deposition (PECVD) of tetraethoxy silane type of oxides. After forming the insulating layer, the vias are metallized and filled with Cu, Mo, or W using one or more processes like CVD, plating, etc., depending on the aspect ratio of the vias. Wafer-to-wafer bonding is accomplished either by SiO_2, metal–metal bonding, or polymer adhesive bonding. The polymers that are commonly used are benzocyclobutene, polyimides, parylene, etc.

There are two major process schemes in TSV, namely via first and via last, and in each of these there are variations within, depending on the company. In the via first scheme, one path way is drill, fill, bond, and thin the wafer; the second one is drill, fill, thin, and then bond the wafer. In the via-last scheme, one variation consists of bond the wafers, thin them, drill, and then fill the vias. In the second approach, thin the wafers, bond them, drill, and then fill. In the third variation, thin the wafer, drill the vias, fill the vias, and then bond the wafers.

The materials, processes, and the infrastructure are still in the development stage and TSV technology, when fully developed, can become the technology of choice for miniaturization and heterogeneous function integration for the system in a package.

System-on-chip (SoC)

The ultimate integration and miniaturization involves packaging all functionalities on to a single semiconductor device, namely, a system on chip (SoC). Naturally, given the limitations of the near-term semiconductor fabrication techniques, the chip is, by design, going to be larger than the currently available ones. It is a technology that envisages integration of several functionalities such as memory, logic, analog, digital, and radio frequency all on one device structure. Generally, the efforts are focused on a specific application domain or target. It is technologically an extremely complex endeavor depending on the application. Design complexity and challenges increase with every new generation application. Multitudes of aspects related to device micro-architecture and system-level aspects have to be addressed concurrently. Several aspects are to be taken into consideration. These involve performance, power dissipation, testability, electromagnetic interference, soft error rates, etc. Memory subsystem and software subsystem are to be mutually and concurrently taken into account in the application development. An SoC design may require as many as 10–20 million gates to be integrated. Higher speeds of operation and shrinking metal layers can cause severe IR drops. These IR drops on power and ground distribution can cause timing failures. A 10% drop in voltage in a 180-nm design, it was indicated, can increase gate delays by 8%.

In terms of packaging, the system flip-chip interconnections provide significantly lower IR drop than wire-bonded interconnections and hence are to be preferred. Compromises have to be made in regard to the ball pitch that can be best accommodated by the available substrate or board technologies at an affordable cost. Simultaneous thermal and mechanical modeling is needed to ensure performance and reliability. Since the SoC concept involves embedded processors, hardware and software co-design, verification, and simulation become critical and are to be addressed in order to avoid incompatibilities at a later stage in the product development. While hardware facilitates performance, software affords features and flexibility to the user.

Modern portable electronic appliances continue to demand greater and greater levels of integration. Many portable electronic devices such as cell phones and the like have Bluetooth enabled products, wireless headsets, WLAN, etc., and integration of mixed signals, digital, analog, RF, etc., requires on-chip integration. RF circuits generally take up more than 40% of a cell phone PWB real estate. With higher levels of integration, this area can increase even more. When high-speed digital switching induces noise or spiky signals, they get injected into the common substrate. The noise can affect sensitive analog circuitry on the same chip. The end result can be silicon failures, yield loss, etc., for the mixed-signal RF integrated systems. In addition, cycle time, cost, defect rates, etc., are to be taken into consideration.

Multiple CPU architectures become pervasive in order to facilitate flexibility and manage and distribute power. Considerable engineering effort to support the development of complex SoCs is still needed. Connectivity among the various component parts such as hardware ID, memory blocks, mixed signal, custom blocks, and external I/O cells requires planning and execution. Interconnect wiring depends on wiring capability as well as efficiency. There is a relation between power dependency and the associated tradeoffs. Further, a tight control of power dissipation is needed. Metal conductor resistance increases with scaling and the delay contribution from the interconnect continuously increases, and hence timing estimation can become critical.

Owing to the considerable complexities involved, current offerings in SoC are limited or confined to relatively simpler systems. In the interim, system on package (SoP) will continue to meet the needs of the industry (Tummala and Swaminathan 2008).

Nanomaterials

The word "nano" in Greek means "dwarf." As a prefix in metrics, nano is a billionth (10^{-9}) of a metric. Particles with dimensions of the order of 100 nm or less are regarded as nanoparticles. A technology that harnesses these materials and their unique and exceptional physicochemical properties is generically termed *nanotechnology*. It is an enabling technology with a wide range of potential applications, and portable electronics is no exception. The applications span a broad spectrum from scratch-resistant coatings to semiconductors and optoelectronics. Some pertinent and potential applications relevant to portable electronics are highlighted in the ensuing paragraphs.

Electronic packaging industry is going to be greatly benefitted by the emergence of nanotechnology with single- and multiwall carbon nanotubes, nanocomposites, and other nanomaterials with unique physical and chemical properties. Use of nanomaterials is being explored vigorously. Semiconducting nanowires are being explored for nanocomputing, and nanomaterials are also being experimented with in order to reduce the size of Flash memory devices. Nanowire transistors have been indicated to be amenable to high-performance electronics.

Nanomaterials are also being explored for battery technology, as these structures enormously increase the surface area for energy storage.

Owing to their unique properties, they can be tailored to be used as filler materials for no-flow underfills. Dielectric layers with improved reliability can be constituted with nanomaterials. Nanoparticle size interconnection alloys, owing to their high surface energy, can be sintered at much lower temperature to effect interconnections.

Another important area that is actively explored is their use as jetting materials for producing thin-film conductive patterns (Kim and Moon 2005). In one case, an aqueous dispersion of silver nanoparticles is ink-jetted to form conductive patterns after sintering at as low a temperature as 300°C (Kamishnay et al.). UV-curable, solventless compositions with binders, acrylic monomers, photoinitiators, and nanoparticle fillers are emerging. Ink-jettable solder mask materials with an average particle size of 300 nm are also reported (Dietz 2008). Nanoparticle coatings for scratch resistance on optical surfaces are already being practiced in the industry.

While the current portable electronic devices utilize tens of gigabytes of memory, emerging ones with complex new functionalities may require several hundred gigabytes to terabytes. Use of structurally linked nano-sized dots consisting of discrete balls made of several hundred nickel atoms and capable of discrete 0 and 1 states enables the fabrication of postage-stamp-size devices that can store up to 5 Tbytes of information.

Nanomaterials find potential applications in consumer electronic appliances in the following areas.

- RF shielding with nanocomposite materials
- Nanoscale coatings for displays
- Nanostructured chip cooling systems for thermal management
- Nanostructured battery electrodes

Ambient sensor networks based on nanotechnology enable faster data transfer rates. OLEDs fabricated of 100-nm-thick layers of electrically conducting organic molecules are increasingly coming into use, with an emphasis on improving reliability and performance.

Tiny bundles of highly conductive carbon nanotubes are being explored as vertical interconnects in logic gates. Carbon nanotubes behave as ballistic conductors with current densities as high as 1000 A/cm^2 compared to about 100 A/cm^2 currently available and show potential application in the fabrication of superior semiconductor devices.

Design and fabrication of nano-emissive field emission displays using large arrays of carbon nanotubes are being actively explored. Carbon nanotubes (CNT) are known to be highly efficient electron emitters. When the carbon nanotube emitters are positioned behind the phosphor dots of a display unit, they emit electrons by a process known as *field emission*. These have a very high redundancy factor. Field emission displays are very thin, about 3.3 mm, and are expected to consume far less energy than conventional LCDs. These displays are anticipated to have several preferred features like (a) low power consumption, (b) vivid colors, (c) wider viewing angle, (d) fast response time, (e) low voltage, and (f) low cost. Several technical issues such as consistency of the carbon nanotubes, high-volume manufacturing, longevity of the phosphors, and reliability are yet to be solved (www.nanoforum.org/educationtree/electronics-computers.htm).

Wearable electronics

Wearable electronics has many connotations and some authors have defined wearable electronics as any apparel into which electronic functions are unobtrusively incorporated. They are sometimes called *smart textiles*. Lifestyle evolution and transformation requiring instant communication ability and access to information at an affordable price is the impetus to the development and deployment of wearable electronics. Wearable electronics have to be very flexible, light, washable, comfortable, robust, and reliable. By far the most pervasive wearable electronic device is the Bluetooth head set. A jacket with a built-in ear gear and microphone into the collar, a body area network with wiring in the garment and integrated with a GSM phone, and an MP3 player with a unified remote control is an example of a wearable electronic product. Such devices have been invading the portable consumer market in recent years.

Wearable electronic devices, in general, incorporate many of the traditional technologies and include the following:

- ICs
- Interface
- Communication
- Energy/power
- Data management

Current limited offerings in portable electronics incorporate the conventional silicon semiconductor technologies in plastic packages. It is anticipated that these may soon reach their limits of applicability and adaptability, and non-silicon-based technologies may be needed. Molecular electronics may be needed and are being explored.

An interface is an appropriate medium for transacting information among many devices and the wearer, and also between the wearer and the outside world. Interface technologies include multiplicity of sensors such as environmental, physiological functions, etc., global positioning systems (GPS), antennae, receivers, etc.

Communication is essentially transfer of information. It can be short-range, such as among the different devices embedded, or long-range between different users via the Internet or a network protocol.

Batteries and power supplies are generally the heaviest components in an electronic appliance. But, power supplies for wearable electronics need to be lighter, more efficient, rechargeable, amenable to be integrated into the apparel, and at the same time robust enough to endure wear and care. Current energy sources to power these devices include lithium metal hydride and lithium ion battery packs with relatively short usage life before recharge, and these are not likely to meet the future needs. Powering of wearable electronics is discussed later in this chapter.

Information storage and processing constitutes data management, and magnetic, optical, or solid-state storage technologies are employed. It is important to recognize that the need for higher storage capacity is expected to increase exponentially with every new generation of devices. Migration from gigabyte to terabyte memory appears imminent, and newer, higher capacity storage devices involving nanotechnologies is anticipated.

Wearable electronics are of different kinds. Some are block-based where the requisite devices are interconnected and are attached to the clothing. These devices can be detached at will. Then there are those in which the electronics is embedded into the apparel with microelectronics, fiber-based technologies, etc.

There are multitudes of applications for wearable electronics, and these are growing. They can monitor and communicate physiological conditions of the wearer. They can enable direct, on-demand physician's advice and treatment, thus rendering effective and efficient health services. A wearble cardiovascular defibrillator with chest harness and a hip pack can provide emergency medical aid to patients prone to heart attacks (Martin et al. 2000). Wearables with artificial muscles can be deployed to enable limbs and arms to become more mobile for physically-challenged individuals. Cochlear implants have been under development to help the hearing impaired to regain lost functionality of damaged cells by transmitting signals to underlying nerve structures. Integrated biosensor layers can monitor body temperature, blood pressure, heartbeat, etc.

Portable military hardware is a promising market place for wearable electronics. Suitable apparel can be designed so that it can provide guidance through inhospitable terrain, provide the position of enemy or allied forces, etc.

Wearable electronics utilize smart textiles where optical fiber wiring is incorporated and integrated into the fabric itself. These textiles can be knitted, woven, or nonwoven. Knitted textile structures are either warp- or weft-knitted and optical fibers or electrical wiring are to be integrated in a straight line interlacing with loops, taking care to see that severe mechanical deformation does not occur. With bending curvatures larger than critical values, optical fibers are integrated in a serpentine pattern into the textiles, creating warp-knitted and segmented yarn structures. Fiber materials have a wide range, varying from silica to all polymer. Optical fibers are cylindrical dielectric waveguides consisting of a concentric core, clad, and jacket layers. The core and clad are transparent, while the jacket provides mechanical support and hence can be nontransparent or opaque. Critical bend parameters need to be established for the different types of fibers/wires (El Sharif 1997, 1989; Winterhalter et al. 2005; Tao 2004).

Several wearable electronic devices with smart textiles are already in the market place and are making headway in penetrating the portable electronics field, although not in great numbers.

The success of wearable electronics is dependent on advances mainly in (a) the development of textile technology platforms, (b) new semiconductor technology, (c) higher levels of miniaturization and integration, and (d) sensor technologies, all of which will enable high-volume manufacturing of highly reliable multifunction gadgets.

Compliant mechanics

Cost-competitive pressures continue to prevail in the portable electronics industry due to consumer demand for cheaper, faster, and better products. Also, owing to the huge product volumes involved, even tiny savings in materials or process simplification can amount to huge cost savings and hence profitability. An area that is gaining considerable attention is that of compliant mechanics. The principle involves designing a mechanical device that traditionally required many constituent parts with fewer parts. Movable joints made of a plurality of members are replaced by a smaller number of flexible members to accomplish the same movement. An example is a vice grip. It normally requires seven parts assembled and connected together with bolts, nuts, screws, or rivets. The same device has been possible to be designed with a single piece with a few compliant members.

Many times, parts are injection-molded as single pieces. Another example of compliant mechanics is the high-performance bicycle brake assembly. Two pin joints and return springs are integrated into a single flexible strip of titanium or stainless steel. Traditionally, a four-bar linkage achieves the desired motion and a return spring is incorporated to disengage the brake when let go off the handle.

The following are some of the benefits that accrue from this approach.

- Cost reduction
- Reduction in parts count
- Reduced assembly time
- Simplified manufacturing process
- Reduced weight
- Reduced maintenance

Energy is stored in the form of strain energy in the flexible members, similar to that in a spring. This energy is released or transformed in a different manner at a later time. When an archer draws the bow string with the arrow, the string is deflected and the energy is stored as strain energy until released. When the string is released, the strain energy is translated into kinetic energy of the arrow. The mechanical component design and fabrication of portable electronic product is considered fertile ground for exploration of compliant mechanics that can lead to innovation of new designs to accomplish parts reduction, assembly efficiencies, cost reduction, and product versatility.

Another incarnation of compliancy is the concept of shape-changing or deformable electronics. One embodiment of this concept is the “Gummi” or bendable computer which has no mechanical parts (Schwesig et al. 2004). These concepts are also known as “flexible electronics” and are not limited to digital newspapers or roll-up displays (Rogers and Bao 2002). One challenge of deformable electronics is the user interaction, because traditional pen, mouse, and keyboard user input becomes a real challenge as these products become smaller and more nonplanar.

One enabler of flexible electronics that presents exciting challenges is the area of interconnection. Recent work by Lacour et al. (2005) demonstrates the approach of distributing rigid subcircuit islands over a polymer surface and then fabricating active devices on these islands. These islands are interconnected with stretchable metallization made of a thin gold film patterned on the elastomer base. They show that the circuit remains functional even at strain levels of more than 10% although the resistance does increase as a function of the applied tensile strain.

Sensor technology

MEMS have been in use for the last four decades in a variety of electronic applications to enable switching, sensing impact or pressure, and detecting light, and as couplers for movable micro-mirrors, accelerometers, gyroscopes, etc. As portable and personal electronic appliances tend to be sensor-intensive in their embedded features, incorporation of miniaturized MEMS devices is a natural trend. Some examples are the incorporation of three-dimensional piezoelectric sensors to protect the hard disk drives in laptops in the event of mechanical drop and image stabilizers in digital cameras. However, future functionality and reliability enhancements into portable electronic appliances require extensive use of sensors. MEMS devices show great promise. Simple mini-structures, actuators, and sensors are seeing wider usage.

The following is a short list of the type of sensors that are likely find application.

- Pressure
- Inertial accelerometers
- Chemical
- Magnetic
- Radiation
- Optical

As portable electronic devices are by design personal information processing devices, MEMS devices find multitudes of applications for sensing a host of parameters, such as DNA, blood pressure, humidity, temperature, and toxic gases, to name a few.

Micro-needle and micro-fluidic drug delivery system MEMS are finding applications in personal healthcare systems.

Power technologies

A whole range of technologies have been developed at ever-increasing rates to make the current portable electronics revolution possible, with the exception of power (and battery technology). For example, from 1990 to 2003, while disk storage density has increased by a factor of approximately 4,000, the energy density of batteries has only increased by a factor of three (Starner and Paradiso 2004). An examination of key mobile computing enablers such as the disk capacity, CPU speed, memory (RAM), wireless data rates, and battery energy density reveals the lagging power density of the batteries, as shown schematically in figure 10–6.

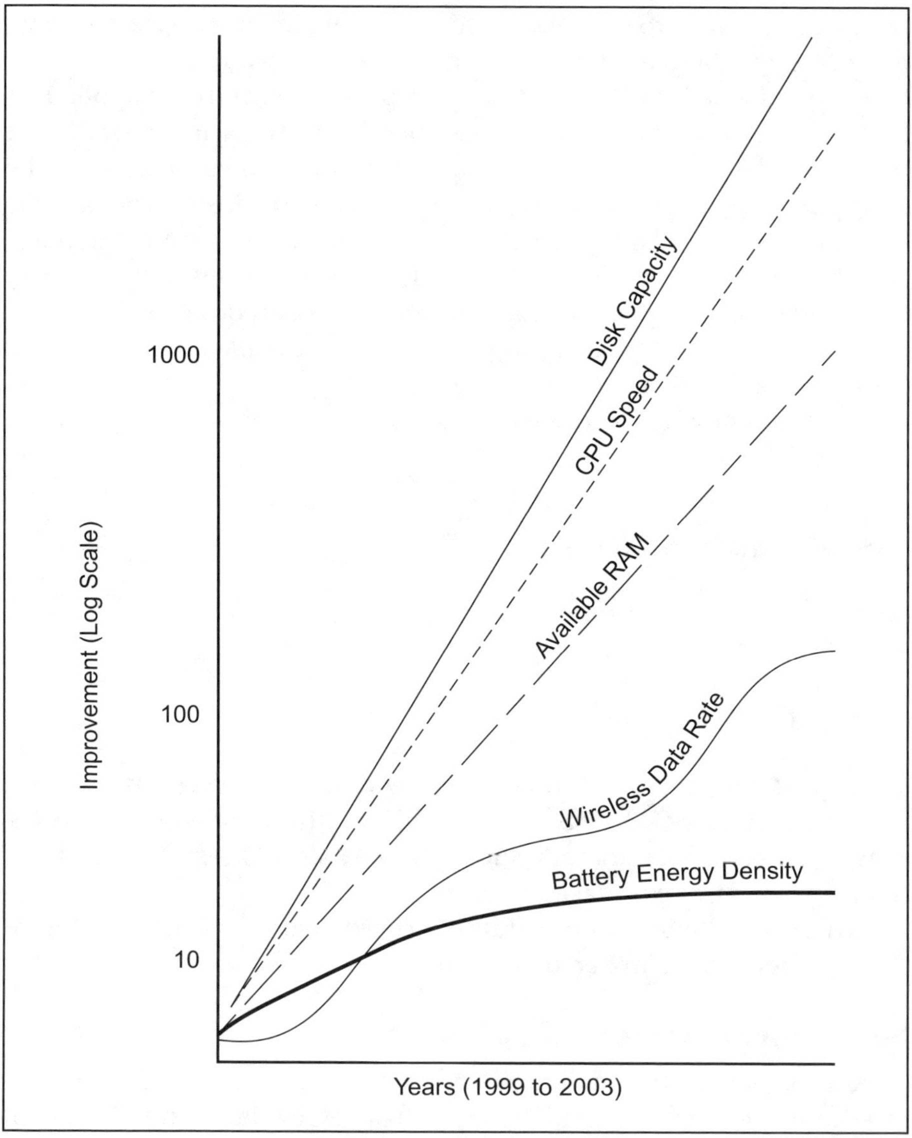

Fig. 10–6. Change in key enablers for mobile computing from 1990 to 2003

Some of the enabling technologies and user requirements, such as CPU speed, display size, application complexity, use time, place, etc., further ratchet up the power requirements, and we may be approaching the scenario where the most limiting factor in the design and use of portable electronics will be the power requirements. Even today, for the lightest portable electronic products, the battery is one of the heaviest components, and can represent up to 40% of the weight of a phone, for example.

Since, the development of new products is typically constrained by the maximum allowable size, weight, and cost, the product designers should consider the budget for the battery prior to the industrial design and functionality mapping. When numerous peripherals are used and they require their own power source, supplying power takes on added complexity. An extreme case is that of wearable electronics, where the battery also needs to be unobtrusive. Ideally, the battery charging should be achieved just by performing everyday actions without requiring plugging into a wall socket. An inductive charging system could wirelessly charge the electronics by the act of the body moving or by the act of placing the jacket on a hanger. Potential power generation solutions being investigated include the use of piezoelectric materials in the shoe inserts and heels (Kymissis et al. 1998), rotary generators in the shoe heel (Hayashida 2000), and hydraulic pumps (McLeish and Marsh 1971). Another approach is to scavenge power from the energy stored in a human body using *in vivo* fuel cells (Keanelly 2000; Ritter 2001) that oxidize blood glucose to provide power in the range of 1 mW. It is conceivable that small amounts of power can also be obtained from the heat given off by the human body, and there are several examples of personal electronic devices such as watches running off of human body heat, such as the Seiko Thermic wristwatch. Since people tend to move around, inertial microsystems have long been used for power generation, with commercial examples most prevalent in the realm of wristwatches and flashlights that can be charged by shaking the product. Windup cellphone chargers have also become widely available as of this writing, and the phone user can charge the phone battery enough to talk for a few minutes by turning the crank for less than a minute (Anonymous 2001).

An incomplete summary of different alternatives to battery power for portable electronics is provided in table 10–2. While these numbers are sufficient for short usage times of portable products, continuous operation of multifunctional portable electronics remains a challenge.

Table 10–2. An incomplete list of alternatives for obtaining power from the users of portable electronic products

Source of Power	Power
Scavenging energy stored in human body using in vivo fuel cells	1 mW
Thermoelectric generator coupled to human body heat	0.2 to 1 W
Inertial or vibrational generation of product	10mW to 1.5W
Crank driven battery chargers	1 to 2 W
Piezoelectric shoe and heel inserts	150 mW
Rotary generator in shoe	250 mW

A related problem is the dissipation of heat. Apart from the problems related to microprocessor scaling, there is the problem of dissipation of heat from portable electronic devices, which are increasingly smaller, flexible, closer to the body, and requiring ever-higher processing power to perform the multitude of functions. For example, thermal management in high-end laptops that are also ultra-portable adds considerable expense in the design and manufacture of these devices. A rough rule of thumb until recently was that processors exceeding the 40-W range result in a cost penalty of approximately a dollar for every additional watt (Tiwari et al. 1998) in computing applications. The heat dissipation problem is further exacerbated in portable electronics because active cooling methods such as fans, refrigeration, etc., carry a very severe cost and battery power requirement penalty. There is also a desire to keep portable electronic products cooler than mobile computing products because of their proximity to exposed skin, which is sensitive to heat.

Because of the portable nature of the electronic products, however, some alternatives have become more attractive. For example, with a wearable electronic product, the enhanced air flow caused by user movement can better cool an arm-mounted device. The use of thermal reservoirs to store the heat generated during the use of portable electronic products is also an option for applications where the peak usage is limited to a few minutes. Phase change materials, which undergo an endothermic change in phase from solid to liquid or liquid to gas, are other attractive options to capture the heat during peak usage and divest heat when the device is not in peak usage. An ideal material would have its first phase change temperature at approximately body temperature, and its second phase-change temperature at approximately 40°C would ensure temperature plateaus at the standard user comfort temperature and another at the maximum allowable operating surface temperature. The consideration of thermal dissipation is best inserted not only at the hardware design stage but also at the software and application development stages. For instance, the software could be written in such a way that performance scaling and task scheduling is done with battery power and thermal dissipation in mind.

Recyclable electronics

Billions of portable electronic devices are manufactured every year and are brought into use. In spite of the best manufacturing practices, if there is a small percentage of them, even at 99.5% yield, that are defective, the number of defective units will be 5 million per every billion produced. That is indeed a huge number to discard into landfills. As newer models with advanced features are introduced, older products become obsolete and are generally discarded. In addition, consider the field returns, upgrades, user-damaged products, etc. As a result, enormous quantities of electronic products and devices end up in landfills. The term for all the electronic appliances that are discarded and ending up in landfills is appropriately *electronic waste* or *e-waste*. Every year, millions of tons of e-waste is generated. In US alone, 2–3 million tons are involved. These contain precious metals such as gold, platinum, and palladium; copper, ferrous alloys such as stainless steels and alloy-42; nonferrous alloys such as phosphor bronze, solders, brass, aluminum alloys, and lead-free alloys; and a host of organic polymers such as plastics, epoxies, polyimides, polycarbonates, ABS, organic PWBs containing bis-phenol A-type bi-functional and tetra-functional and other epoxies, adhesives, paints, etc. Often fully functional devices also end up in landfills owing to product obsolescence. Some of these materials are either toxic by themselves or their degradation products. If the e-waste is not properly handled or processed, it can lead to an ecological nightmare. Current estimates are that only 15%–17% of the e-waste is recycled. In order to address the issue, some countries are introducing legislation. Some others are establishing a recycling charge on products being sold to fund waste collection and recycling. Thus increasing awareness is brought to bear on this prodigious problem.

The recycling process consists of several important steps. Initially, the received e-waste products are tested for functionality. If some of them are in good condition and were discarded only due to obsolescence, they are resold to appropriate agencies where they can be put to use again. The nonfunctional appliances are examined for repairability. Repairable ones are brought into working condition and are sold to appropriate agencies for re-entry into the market. The nonrepairable units are then dismantled, the individual constituent components are tested for functionality, and the good parts are segregated and used for further production. The nonfunctional parts are then separated into various categories such as metals, glass, plastics, etc. Some of the material is subjected to precious metal recovery means, and others are treated with appropriate safe disposal processes. Figure 10–7 is a schematic of a possible recycle/reclamation scenario. Thus, the amount of e-waste that finally ends up in landfills is significantly reduced.

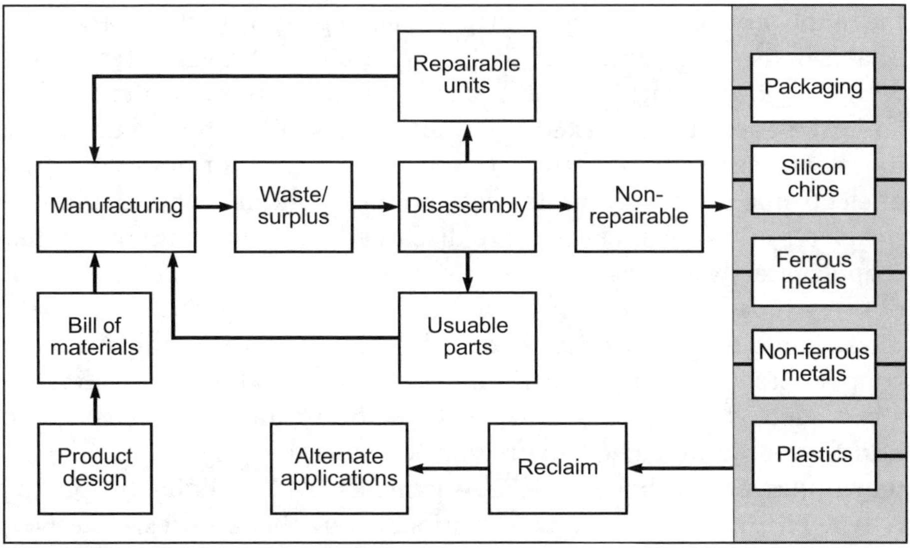

Fig. 10–7. A recycle/reclamation scenario

An important aspect is the cost associated with the various steps of the recycling process. One of the main operations is the product disassembly. This is an area where the original product design can be of immense help. If the product design is carried out with a focus on recycling and reclamation, it will facilitate lower ultimate cost of the product as well as a safer environment. The following are some of the measures that design groups should consider.

- Choice of nonhazardous materials.
    - Nontoxic PWB materials
    - Nontoxic metals and alloys
    - Minimizing soldered batteries

- Choice of fasteners and connectors
    - Easily removable
    - Minimum number fasteners
    - Pluggable instead of hard soldered
- Reusable plastics
- Optimal placement locations of components
- Disassembly with minimum number of tools

Future trends in electronic packaging, especially in portable electronic products is the increasing emphasis on recycling, reclamation, and a greener environment in all stages of product cycle from cradle to the grave. Another aspect, namely, design for recycling, is added to the DfX process and incorporates the disassembly, reuse, and recycle concepts. The product manufacturer and the consumer both live on the same planet, and both groups will be demonstrating increased vested interest in the environment they live in.

Biodegradable electronics

In view of the environmental concerns related to the e-waste disposal, electronics industry has begun exploration, investigation, and evaluation of alternative electronic materials for a safer environment. Biodegradable materials are being considered to replace some of the organics used. Biodegradation is the degradation or braking down of organic substances by enzymes produced by living organisms. When the degradation occurs in the presence of oxygen, it is termed aerobic, and in the absence of oxygen it is termed anaerobic degradation. Some materials like detergents degrade in an aerobic environment and may not in an anaerobic environment. Conversion of organic materials into minerals is called biomineralization. Also, microorganisms secrete biosurfactants to enhance degradation. Thus, all organic materials degrade over a period of time. The time for complete degradation depends on the nature and chemical structure of the organic substance and also how compact they are. Table 10–3 shows the times of degradation for some common products, arranged in increasing order of time for degradation. It can be seen that there is a range even within a class of compounds. The degradation times can be dependent on the nature of the soil, its acidity or alkalinity, porosity, the water content, etc.

Table 10–3. Time of degradation for some common products

Category	Material	Time for Degradation
Organics	Fruit peels (banana, orange, etc.)	A few days to a month
	Sugar cane pulp	1–2 months
	Cotton rags	1–5 months
	Biodegradable plastic cups or bags	2.5 months
	Woolen socks	1–5 years
	Milk cartons	5 years
	Plastic bags	10–20 years
	Leather	25–40 years
	Plastic bottles	20–450 years
	Styrofoam	> 1 million years
Inorganics	Aluminum cans	200–500 years
	Tin cans	50–100 years
	Tinned steel	550 years
	Glass	1–2 million years

Source: http://behealthyandrelax.com/2007/how-long-does-it-take-to-degrade

Some standards are already being put in place for acceptability of degradation levels. One U.S. standard specifies 60% degradation in six months, while a European standard indicates a more stringent 90% degradation in three months.

A vegetable-based plastic based on polylactic acid (PLA) has been demonstrated for the fabrication of a Sony Walkman product. Some of the materials that are being explored are corn-based plastics. They are being tried as the case materials for laptops, printers, monitors, cell phones, covers, and the like. PLA-based plastics are filled with magnesium silicate, an insoluble powder, to provide mechanical strength and prevent shrinkage. Corn-based plastics degrade giving rise to carbon dioxide and water over a period of time.

Chemical mechanical polishing is a technique that is used for providing the final back finish in wafer processing and is carried out with polishing pads. The pads are made of nonbiodegradable materials like polymer foams, nonporous polymer sheets, and sintered thermoplastic particles. Once the polishing is accomplished, these pads are discarded, which enter solid waste streams and landfills. Newer pad materials consisting of glycolic acid, lactic acid, and butylene succinate that are biodegradable are coming into use. PLA-based materials are tailor-made to provide a range of molecular weights, hardness, crystallinity, and thermal properties.

The biodegradable flexible MEMS system utilizing electroactive polymers is an area of intense activity. Ionic polymer composites, gel polymers, electro-rheological fluids, electroactive papers, etc., are being explored. Electroactive paper is based on cellulose and has the ability to transport ions on the imposition of an electric field. The paper is coated with a thin layer of gold on each side. As a voltage is imposed, sodium ions in the paper migrate towards the positive electrode carrying the water molecules along. The cellulose fibers in the paper also exhibit piezoelectric properties and undergo shape change due to the applied voltage. The ionic migration, water molecule transport, and the piezoelectric effect result in a physical deformation of the paper. The piezoelectric properties have been enhanced by incorporation of carbon nanotubes. The in-plane contraction or strain is a function of the applied voltage. This deformation phenomenon of the cellulose-based electroacive paper is considered beneficial in the fabrication of biodegradable sensors. Conductive polymers such as pyrroles, polyaniline, carbon nanotubes, etc., combined with emeraldine-based polyanilines are used as coatings on electroactive papers to enhance their properties (Yun 2006; Kim 2006).

Cross-linked polymers selected from plant materials such as wood resins, crop oils, lignin, and tannins, as well as polysaccharides resins and combinations thereof are being explored as technology materials, as many of these are biodegradable (U.S. Patent 6339116)

In corn-based products, corn is first milled and converted to destructured starch and, with suitable chemical modification, converted to amorphous amylase and amylopectin. Polymeric complexing agents from vegetable oil intermediates are then combined to prepare layered starch structures that can constitute base materials for corn-based dielectrics. Cellulose, starch, and glucose from corn, wheat, potatoes etc., and proteins and fatty acids from vegetable oils constitute some basic biodegradable raw materials. Polyhydroxy alcanoates, PLA, etc. are increasingly coming into use.

Electrostatic discharge (ESD) is a concern in the fabrication of electronic devices and several ESD prevention protocols are in place. One such is the use of ESD protection bags. These are generally nickel-coated low density polyethylene bags and are not biodegradable. Newer biodegradable static dissipating polyester plastic bags are coming into use. These bags are impregnated with a carboxylate amine salt vapor corrosion inhibitor which continually releases a corrosion inhibiting vapor that protects metals.

Hemicellulose is a material obtainable from corn, straw, wood, and the like. A biopolymer xylan is extracted to prepare xylophone, which also acts as an oxygen barrier. These hemicellulose-based materials are being explored as biodegradable packaging materials.

Other biodegradable plastics under consideration are polybutylene succinate (PBS), polyvinyl alcohol, and starch, and the polyester poly(3-hydroxybutyrate-co-3-hydroxyvalerate) (PHBV). Filler materials like talc and carbon are added to PBS to provide better mechanical bending strength, as well as conductivity. Filed PBS polymer is used as shipping container trays for large-scale integrated devices. PBS has been shown to degrade in about six months in water at a specific temperature and pH (Takafumi 1977).

Suggested Reading

1. Gil, B., B. C. Cavenett, R. L. Aulombard, R. Triboulet, G. Leising, and F. Stelzer. 1977. *Semiconductors and organic materials for optoelectronic applications.* (European Materials Research Society Symposia Proceedings), New York: Elsevier Science.
2. Peng, Y. Q., et al. 2008. *Charge carrier transport in organic semiconductor thin film devices.* Nova Science Publishers, Hauppauge, NY.
3. Chuang, S. L. 2009. *Physics of photonic devices.* Wiley Series in Pure and Applied Optics. Hoboken, NJ: John Wiley and Sons.
4. Yersin, H. ed. 2007. *Highly efficient OLEDs with phosphorescent materials.* New York: Wiley-VCH.
5. Lyshevski, S. E. ed. 2007. *Nano and molecular electronics handbook.* Nano- and Microscience, Engineering, Technology, and Medicines Series. 1st ed. CRC Press, Cleveland, OH.
6. Hanson, G. W. 2007. *Fundamentals of Nanoelectronics.* U.S. ed. New York: Prentice Hall.

References

Allan, R. 2008. OLEDs put on quite a display. *Electronic Design* 565 (21).

Anonymous. 2001. Turn the handle and talk. *The Economist* December 8: 10.

Bell, A. 2009. Capacitor powered electronics is at the core of green design. *Electronic Design* 57 (8): 43–44.

D'Airo, Christina. 2009. UTCP package enables wearble electronics. *Electronic Products* 51(11): 13–14.

Dirjish, M. 2009. University of researches roll out first flexible display with touch screen. *Electronic Design* 57 (8): 47.

Dirjish, M. 2008. Ultracapacxitors branch out into wider markets. *Electronic Design* 56 (24): 35–39.

Dietz, K. 2008. Fine lines in high yields (Part CLVII): Nanoparticles in electronics. *CircuiTree* 21 (10): 20–21.

El Sharif, M. A. 1989. On fiber sensor and modulator. IEEE Trans. Instruments and Measurements, Vol. 38: 595–598.

El Sharif, M. A. 1997. Fiber optic sensors for soldier's smart uniform. *3rd ARO Workshop on Smart Structures*. Blacksburg, VA: Virginia polytechnic and State University: 27–29.

Esinenco D., I. Codreanu, and R. Rebigan. Sept. 2006. Design of inkjet print head based on electrowetting for printable electronic applications, Proc. International Semiconductor Conference, vol. 2: 27–29.

Hayashida, J. 2000. Unobtrusive integration of magnetic generator systems into common footwear.

Jain. K., M. Zemel, M. Klosner, and S. Raghunandan. 2005. Flexible electronics and displays: high resolution roll-to-roll projection lithography and photoablation processing technologies for high throughput production, Proc. IEEE Vol. 93, No. 8: 1500–1510 (2005).

Kamishnay, A., et al. Ink jet printing of metallic particles and microemulsions. *Macrmolecular Rapid Communications* 26 (4).

Keanelly, C. 2000. Power from the people breaks the hold of batteries and plugs. *The New York Times*, August 3.

Kim, H., and J. Moon. 2005. Highly conductive inkjet printed nano silver particles for printable electronics. *Electrochemical and Solid State Letters* 8 (11).

Kim, J. 2006. *Possibility of cellulose electroactive EAPap Actuator*. Incheon, South Korea: Inha University.

Kinkaid, K. 2007. Flexible displays open new windows of opportunities. *Laser Focus World* 17.

Kymissis, J., C. Kendall, J. Paradiso, and N. Gershenfeld. 1998. Parasitic power harvesting in shoes. Paper presented at the IEEE Int. Symp. Wearable Computers: 132–139.

Lacour, S. P., J. Jones, S. Wagner, T. Li, and Z. Suo. 2005. Stretchable interconnects for elastic electronic surfaces. *Proc. IEEE* 93 (8): 1459–1467.

Makela T, V. Lambertini, T. Haatainen, P. Majander, and J. Ahopetto. Nov 5–8, 2007. Continuous two-sided roll-to-roll nanopatterning of a polymer, microprocesser and nano-technology. Digest of papers: 426-427.

Martin, T., E. Jovanov, and D. Raskovic. 2000. Issues in wearable computing for medical monitoring applications—A case study of a wearable ECG monitoring device. Paper presented at the Int. Symp. Wearable Computers, ISWC 2000, Atlanta, GA.

McLeish and Marsh, J. 1971. *Human Locomotor Engineering*: 126–132.

Nishi S. 2007. Direct metal patterning by ink jet technology, Proc. of IEEE Polytronic Conference: 297–300.

Ritter, S. 2001. Biofuel cells get smaller. *Chemical and Engineering News* 79 (36): 10.

Rogers, J., and Z. Bao. 2002. Printed plastic electronics and paperlike displays. *J. Polymer Science, Part A: Polymer Chemistry* 40: 3327–3334.

Schwesig, C., I. Poupyrev, and E. Mori, 2004. Gummi: A bendable computer. Proc. Sig CHI Conf. on Human Factors in Computing Systems, Vienna: 263–270.

Starner J. and J. Paradiso. 2004. Low power electronics design. C. Piquet, CRC Press, Cleveland, OH.

Takafumi, H. 1997. *Biodegradable materials for LSI shipping materials*. Fujitsu Laboratories.

Tao, X. M. 2004. *Wearable Electronics and Photonics*. U. K.:Woodhead Publishing

Tiwari, V. et al. (1998). Reducing power in high performance microprocessors, *Proc. Design Automation Conf.* 732-737.

Tummala, R. R., and M. Swaminathan. 2008. Introduction to System on Package (SOP) *Miniaturization of the Entire System.* New York: McGraw Hill.

Viklund, P. 2008. PCB and package convergence. CircuiTree 21 (9): 42–44.

Wang Z and Park S. B. Three-dimensional real time and high accuracy In-line monitoring system for roll-to-roll manufacturing, IEEE.

Winterhalter, C., J. Teverovsky, P. Wilson, J. Shade, B. Farrell, W. Horowitz, and E. Tierney. 2005. Development of electronic textiles to transport data and power in future U.S. military protecive clothing systems. *J. ASTM International* 2 (7): 1–9.

Yun, S. 2006. *Micropatterning on cellulose EAPap for biodgradable MEMS.* Incheon, South Korea: Creative Research Center for EAPap Acuator, Inha University.

A
Appendix

Standards and Specifications

Standards and specifications are essential in building quality products in a consistent manner in any industry. There are several standards and guidelines documents that are generated by the IPC and JEDEC organizations for the benefit of the electronics industry and are available at their respective Web sites. The following is a short list of selected documents pertaining to consumer and portable electronics. The reader should refer to the IPC and JEDEC Web sites for a complete list of available literature, specifications, and guidelines.

Number	Title of document
IPC 2221A	Generic Standard on Printed Boards
IPC 2223B	Sectional Design Standard for Flexible Printed Boards
IPC 4202	Flexible Base Dielectrics for Use in Flexible Printed Circuits
IPC 4553A	Specification for Immersion Silver for Plating Printed Circuit Boards
IPC 4554	Specification for Immersion Tin for Plating Printed Circuit Boards
IPC 6013B	Quality and Performance Specifications for Flex Printed Boards
IPC A6006	Acceptability of Printed Wiring Boards
IPC /JPCA 6202	Performance Guide Manual for Single and Double Sided Flex Printed Wiring Boards
IPC 4563	Resin Coated Copper foil for Printed Board Guidelines
IPC 6016	Qualification and Performance Specifications for High Density Interconnect Layers or Boards
IPC 4104	Specification for High Density Interconnect Micro-via Materials
IPC 4101C	Specification for Base Materials for Rigid and Multilayer Printed Boards
IPC 4821	Specifications for Embedded Passive Devices Capacitor Materials for Rigid and Multilayer Printed Boards
IPC 6010 Series	Family of Board Performance Documents
JSTD 003 B	Solderability Test for Printed Boards
A6006	Acceptability of Printed Boards

9691 A	User Guide for IPC-TM-650 2.6.25 CAF Resistance Test (Electrochemical Migration Testing)
IPC 5702	Guidelines for OEMS in Determining Acceptable Levels of Cleanliness of Unpopulated Printed Boards
IPC 9201	Surface Insulation Resistance Handbook
IPC 6012B	Qualification and Performance Specification of Rigid Printed Boards
TR 432	Solderability Evaluation of Printed Boards with Coatings for Long-term Storage
TR 464	Accelerated Aging for Solderability Evaluations
IPC 610D	Acceptability of Electronic Assemblies
IPC 9701	Performance Test Methods and Qualification Requirements for Surface Mount Solder Attachments
J STD 012	Implementation of FC, CSP Technology
J STD 013	Implementation of Ball Grid Array and Other High-Density Technology
J STD 020-D	Moisture Sensitivity Classification for Nonhermetic Solid State Surface-mount Devices
JESD 22- A100 C	Cycled Temperature Humidity Bias Life Test
JESD D22-A101C	Steady State Temperature and Humidity Bias Test
JESD D22-A102C	Accelerated Moisture Resistance Unbiased Autoclave
JESD D22-A103C	High Temperature Storage Life Test
JESD D22-A104D	Thermal Cycling
JESD D22-A105C	Power and Temperature Cycling
JESD D22-A106B	Thermal Shock
JESD D22-A110C	Highly Accelerated Stress Test (HAST)
JESD D22-A118	Accelerated Moisture Resistance—Unbiased HAST
JESD D22-A121A	Measuring Whisker Growth of Sn and Sn Alloy Surface Finishes
JESD D22-B104C	Mechanical Shock
JESD D22-B111	Board Level Drop Test Method for Handheld Electronic Products
JESD D22-B112A	Package Warpage Measurement of Surface Mount Integrated Circuits at Elevated Temperature
JESD D22-A102C	Accelerated Moisture Resistance—Unbiased Autoclave
JSTD -020-C	Moisture/Reflow Sensitivity Classification for Non-hermetic Solid State Surface Mount Devices

B

Appendix

Acronyms and Abbreviations

Acronym/ Abbreviation	Description/Explanation
ACF	anisotropic conductive adhesive film
AES	auger electron spectroscopy
AF	acceleration factor
AFR	average failure rate
AHP	analytical hierarchy process
ALT	accelerated life test
ALIVH	any layer inner via hole
AMOLED	active matrix OLED
ASIC	application specific integrated circuit
AT	accelerated testing
BGA	ball grid array
BCB	benzocyclobutene
BCC	bumped chip carrier
BLM	ball limiting metallurgy
BLP	bottom leaded package
BLR	board level reliability
BT	bismaleimide triazine
BOM	bill of materials
BTA	benzotriazole
C4	controlled collapse chip connection
C4NP	controlled collapse chip connection new process
CAD	computer aided design
CAF	conductive anodic filament
Cdf	cumulative distribution function
CI/CB	confidence interval/confidence bound
CM	contract manufacturer
CNT	carbon nanotubes
COB	chip on board

CPU	central processor unit
CSP	chip scale package
CTE	coefficient of thermal expansion
CVD	chemical vapor deposition
DAC	digital to analog conversion
DC	direct current
DES	develop, etch, and strip
DFG	design for green (environment)
DFQ	design for quality
DFM	design for manufacture
DFR	design for reliability
DFT	design for test
DIG	direct immersion gold
Dicy	dicyandiamide
DNP	distance from neutral point
DPMO	defects for million opportunities
DRAM	dynamic random access memory
DRIE	deep reactive ion etching
DSC	differential scanning calorimetry
DSP	digital signal processor
DV	design validation
DVD	digital video display
ECM	electrochemical migration
EDLC	electric double layer capacitor
EDX	energy dispersive X-ray
EGME	ethylene glycol monomethyl ether
ESCA	electron spectroscopy for chemical analysis
ESD	electrostatic discharge
EMC	epoxy molding compound or electromagnetic interference
ENIG	electroless nickel-immersion gold
ENIPIG	electroless nickel palladium immersion gold
EPROM	erasable programmable read only memory
ETS	European test symposium
FEOL	front end of line
FIB	focused ion beam
FLGA	fine pitch LGA
FMEA	failure mode and effects analysis
GB	grain boundary
GPS	global positioning system
GSM	Global System for Mobile Communications
IC	integrated circuit
IEC	International Electrotechnical Commission
IPC	Association Connecting Electronic Industries
IR	infrared

IREM	IR emission microscopy
ITO	indium tin oxide
HASL	hot air solder level
FC	flip chip
FR	fire retardant
HDI	high density interconnect
HIC	humidity indicator card
IMC	intermetallic compound
JACS	just about chip size
JEDEC	joint electron device engineering council
JESD	JEDEC standard
JMP	statistical analysis software by SAS institute
KGD	known good die
LCC	leadless chip carrier
LCD	liquid crystal display
LGA	land grid array
LIMS	laser induced mass spectrometry
LOC	lead on chip
LTCC	low temperature co-fired ceramic
MCM	multi-chip module
MEMS	micro-electro-mechanical systems
MOEMS	micro-opto electro mechanical systems
MHz	megahertz
MLE	maximum likelihood life estimate
MP3	MPEG-1 audio layer-3
MTTF	mean time to failure
NDSP	near die size package
NED	near eye display
NSMD	non-solder mask defined
OEM	original equipment manufacturer
OLED	organic light emitting diode
OSP	organic solder preservative
ODA	oxidianiline
PBGA	plastic ball grid array
PC	personal computer
PCB	printed circuit board
PBISS	package-to-board interconnection shear strength
PDA	personal digital assistant
PCMCIA	Personal Computer Memory Industry Association
PEM	plastic encapsulated circuit
PEN	polyethylene napthalate
PEP	portable electronic product
Pdf	probability distribution function
PFA	physical failure analysis

PECVD	plasma enhanced CVD
PI	polyimide
PID	photo-imageable dielectric
PIH	pin in hole
PIP	package in a package
PLA	polylactic acid
PLCC	plastic leaded chip carrier
PMDA	pyro mellitic dianhydride
POP	package on package
PTFE	poly-tetrafluoro-ethylene
PTH	plated through hole
PV	product validation
PWB	printed wiring board
QBAM	quantitative B-mode acoustic microscopy
QFN	quad flat non- lead
QFP	quad flat pack
RAM	random access memory
RCC	resin coated copper
R2R	roll-to-roll
RF	radio frequency
RGB	recrystallized grain boundary
RIE	reactive ion etching
RNT	reactive nano-technology
ROHS	reduction of hazardous substances
ROM	read only memory
SAC	Sn-Ag-Cu (tin-silver-copper)
SEM	scanning electron microscopy
SIMS	secondary ion mass spectrometry
SIR	surface insulation resistance
SMI	stretchable moldable interconnect
SRAM	static random access memory
SLC	surface laminar circuitry
SMD	solder mask defined
SLR	single lens reflection
SMT	surface mount technology
SOC	system on chip
SOIC	small outline integrated circuit
SON	small outline no lead
SRS	shock response spectrum/spectroscopy
SSOP	shrink small outline package
STELLA	stretchable electronics for large area applications
T/C	thermo-compression
TAB	tape automated bonding
T/S	thermo-sonic

TAL	time above liquidus
TFBGA	thin fine pitch BGA
TFT	thin film transistor
TGA	thermo-gravimetric analysis
TMA	thermo-mechanical analyzer
TMBDA	tetra methyl butane diamine
TQFP	thin quad flat pack
TSOP	thin small outline package
TSV	through silicon via
TZOP	thin zero outline package
UEM	universal energy management
US	ultrasonic
USON	ultra thin small outline no-lead
UTCSP	ultra thin CSP
UV	ultraviolet
VLSI	very large scale integration
VSOP	very small outline package
VSSOP	very small shrink outline package
WLP	wafer level package
WEEE	waste in electrical electronic equipment
WLAN	wireless local area network
XPS	X-ray photoelectron spectroscopy

C

Appendix

Selected Source Books in Electronic Packaging

The books listed below, while not exhaustive, provide background information on several pertinent aspects of electronic packaging. They cover such aspects as general microelectronic packaging and assembly technologies; specific topics, such as soldering, lead free soldering, general reliability, and solder joint reliability; specific packaging concepts, such as fine pitch surface mount technology, area array technology, and chip scale packaging; advanced topics, such as system on package; and some general references to portable electronics, flexible electronics, etc.

1. *Fundamentals of Microsystems Packaging,* by R. R. Tummala, McGraw Hill Books, New York, 2001.
2. *Microelectronics Packaging Handbook,* edited by R. R. Tummala, E. Rymaszewski, and A. Klopfenstein, Parts I, II, and III, Chapman and Hall Publishing, New York, 1997.
3. *Electronic Packaging and Interconnection Handbook,* edited by C. A. Harper, McGraw Hill Books Inc. New York, 1994.
4. *Surface Mount Technology,* by R. P. Prasad, Chapman and Hall, New York 1997
5. *Ball Grid Array Technology,* edited by J.H. Lau, McGraw Hill Inc, New York, 1997
6. *Area Array Interconnection Handbook,* edited by Karl J, Puttlitz and Paul A. Totta, Kluwer Academic Publishers, Norwell, MA, 2001.
7. *Chip Scale Package,* by J. H. Lau and S. W. Ricky Lee, McGraw Hill Co, 1999
8. *Tape Automated Bonding,* edited by John Lau, Van Nostrand Reinhold, New York, 1992.
9. *Handbook of Fine Pitch Surface Mount Technology,* edited by J. H. Lau, Van Nostrand Reinhold, New York, 1994.
10. *Printed Circuits Handbook,* by Clyde F. Coombs Jr, McGraw Hill Books Company Inc, New York, 1998.
11. *Microvias,* by J. H. Lau, and S. W. Ricky Lee, McGraw Hill, New York, 2001.
12. *Micro-and Opto-Electronic Materials and Structures,* edited by E. Suhir, Y. C. Lee, C. P. Wong, Volumes 1 and 2, Springer Science and Media Inc. New York, 2007.

13. *Failure Modes and Mechanisms in Electronic Packages*, by Puligandla Viswanadham and Pratap Singh, Chapman and Hall Publishing, New York, 1998.
14. *Handbook of Lead-Free Solder Technology for Microelectronic Assemblies*, edited by K. J. Puttlitz, and K.A. Stalter, Marcel-Dekker Inc, New York, 2004.
15. *Plastic Encapsulated Microelectronics*, edited by M. G. Pecht, L.T. Nguyen, and E. B. Hakim, John Wiley and Sons, Wiley InterScience, New York, 1995.
16. *Introduction to System on Package*, by R. R.Tummala and M. Swaminathan, McGrawHill Companies Inc, New York, 2008.
17. *Soldering in Electronics*, by R. J. KleinWassink, Electrochemical Publications, Isle of Man, British Isles, 1989.
18. *Solder Joint Reliability*, edited by J. H. Lau, Van Nostrand Reinhold, New York, and New York, 1991.
19. *Reliability, Maintainability, and Availability Assessments*, by M.O.Locks, ASQC Quality Press, Milwaukee, WI, 1995.
20. *Solder Joint Reliability of BGA, CSP, Flip Chip, and Fine Pitch SMT Assemblies*, by J. H. Lau, and Yi-Hsin Pao, McGraw Hill Companies Inc, 1997.
21. *Practical Reliability Engineering*, by P. D. T. O'Connor, John Wiley and Sons, New York, 1991.
22. *Portable Electronics—Product Design and Development*, by Bert Haskell, McGraw-Hill Companies Inc, New York, 2004.
23. *Damage Prediction of Portable Electronics*, by Dhananjay Panchgade, VDM Verlag 2009.
24. *Portable Electronics—World Class Designs*, by J. Donavan, Elsevier Publishing, Burlington, MA 2009.
25. *Flexible Electronics: Materials and Applications (Electronic Materials: Science & Technology)*, by W. S. Wong and A. Salleo Springer Science and Business Media Inc. 2009.

Index

A

D

G

H

I

J

K

L

M

N

O

P

Q

R

S

T

U

V

W

X

Y

Z

About the Authors

Dr. Sridhar Canumalla is a Principal Engineer with Microsoft in the Entertainment and Devices division. He is involved in failure analysis and reliability improvement of hardware consumer electronic products. Prior to joining Microsoft, he worked at Texas Instruments as Package Development Reliability Engineer, where he was involved in flip chip and wire bond IC package development projects for consumer electronics applications as reliability engineer. He has also worked at Nokia R&D as Senior Specialist. While at Nokia, he was involved in the development of several mobile phone products. He was responsible for Reliability Testing and Failure Analysis lab management. Sridhar has also served as Staff Scientist at Sonoscan Inc., where he worked on advancing quantitative acoustic microscopy techniques in microelectronic failure analysis.

Sridhar received his PhD in engineering science and mechanics from Pennsylvania State University in 1995 for his work on understanding the fatigue and fracture behavior of metal matrix composites using destructive and nondestructive characterization techniques and mechanics. He was awarded first prize in the Graduate Research Exhibition in 1990, and is a member of Tau Beta Pi, an engineering honor society. He also has a B.Tech. degree in Mechanical Engineering from the Jawaharlal Nehru Technological University, Hyderabad.

Sridhar has more than 40 technical publications in journals and conference proceedings in the areas of IC packaging, reliability engineering and testing, robust design, solders, acoustic microscopy, acoustic emission, materials characterization, fatigue and fracture, and processing of composite materials. He has also published papers in the areas of corrosion and renewable energy, holds one patent and is the coauthor of three book chapters. He is a senior member of IEEE Components, Packaging and Manufacturing Technology Society and a regular reviewer for IEEE Journals. He is active in the Applied Reliability subcommittee of the Electronic Components and Technologies Conference, and is a past chair.

Dr. Puligandla Viswanadham recently retired from Nokia Research Center, Irving, Texas, as Principal Scientist and was involved in second level electronic packaging that included reworkable underfills, chip scale packaging, lead (Pb)-free solders and reliability investigations. Prior to joining Nokia he worked as Senior Member of the Technical Staff at Raytheon TI Systems and Texas Instruments and was involved in the ball grid array and chip scale package technology assembly development and implementation, and at IBM as an advisory scientist in Rochester, MN, Endicott, NY, and Austin, Texas, facilities. At IBM, Viswanadham was involved in the process development and qualification of surface laminar circuitry, assembly and reliability of fine pitch quad flat packs, thin small outline packages, tape automated bonding, and area array technologies. Also, while at IBM Austin he was manager of site Analytical Laboratories from 1989–1990. As a member of the Materials and Process Engineering group at IBM Rochester, Viswanadham was involved in corrosion studies, analytical methods development, plating studies and contamination control. Prior to joining IBM, his research activities included high temperature chemistry and thermodynamics of binary and ternary chalcogenides, atomic absorption, slag-seed equilibria in coal fired magneto-hydrodynamics energy generation, and astrophysics. During

1959–1968 he worked as Senior Research Assistant at the Astrophysical Observatory and India Meteorological Department.

Viswanadham has authored or co-authored more than 130 technical publications in journals, symposium proceedings, and trade magazines. He is coauthor of the book Failure Modes and Mechanisms in Electronic Packages published by Chapman and Hall in 1998. He also authored or coauthored nine book chapters in the areas of microelectronic packaging, fine pitch surface mount technology, ball grid array technology, chip scale packaging, corrosion in microelectronics, reliability of portable electronics, and other topics.

He received the First, Second, and Third plateau IBM invention achievement awards for several of his patents and invention disclosures, and an IBM excellence award. He also received fourth level IBM Technical Author Recognition Award for his publications. From 1974–1978, he was on the faculty of Ohio Dominican College Columbus, OH, as assistant professor, where he taught chemistry and physics

Puligandla Viswanadham has a PhD in chemistry from the University of Toledo, Toledo, OH; an M.Sc. in chemistry from Saugor University, Saugor, India; and a B.Sc. in chemistry from Sri Venkateswara University, India.

He is a member of the Surface Mount Technology Association and served as a member of board of directors, technical committee, and is currently on the journal review committee. He has given several professional development courses and tutorials at SMTA, IPC, and IMAPS technical conferences

His current institutional affiliation is Mechanical and Aerospace Engineering, University of Texas at Arlington as a member of the adjunct staff teaching failure modes and mechanisms in electronic packages and chip scale packaging. He is also currently involved in the preparation of a book on the fundamentals and essentials of electronic packaging.